STUDY GUIDE AND SOLUTIONS MANUAL

LEO J. MALONE
St. Louis University

to accompany

BASIC CONCEPTS OF CHEMISTRY

7th Edition

Cover Photo: ©David Fleetham/Taxi/Getty Images

To order books or for customer service call 1-800-CALL-WILEY (225-5945).

ISBN 0-471-21523-6

Printed in the United States of America

10 9 8 7 6 5 4 3 2 1

Printed and bound by Victor Graphics, Inc.

Table of Contents

Preface to the Student Study Guide

A Study Guide has one main purpose: To help you tie the material together in such a way so that you will not only be successful on exams but actually "know" the material. It is understood that your first thought is probably just concerned with doing well on exams. (Your instructor is more worried about having you learn the material.) To this end, the first topic in this guide is devoted solely to the secrets of how to do well on exams. Please read the section on "How to 'Ace' Chemistry." It may be the most important part of this Study Guide to you.

The chapters in the Study Guide parallel those in the Seventh Edition of *Basic Concepts of Chemistry* by Malone. No new material is presented here. The Study Guide is designed to be most effective if referred to immediately after a *Section* is covered in the text. In the text, the related topics in a Section are grouped under Learning Check A, Learning Check B, etc. In the Study Guide, the same Section is covered under Review Section A, Review Section B, etc. After carefully studying these related topics in the text, refer to the Study Guide for a summary of the Section and for the questions in the Self-Test. After checking the answers to the test, refer back to the text for more exercise problems, then proceed on to the next Section in the text. After completion of the whole chapter, work the Chapter Summary Self-Test in the Study Guide. This Guide can also be used in other fashions such as a review and as a source of sample exams before a test.

The general headings within each chapter of the Study Guide are as follows:

Review Section A, B, or C

The material in each chapter is divided into two or three subsections labeled A, B, and C. Under each review section in the Study Guide, there are four subdivisions.

OUTLINE AND OBJECTIVES

The important concepts in each text section are shown in outline form. Chapter objectives are included opposite the particular topic. Read each topic and ask yourself what it is about. Refer to the objective to see what you need to be able to accomplish.

SUMMARY OF SECTIONS

Read through this section with careful attention to any worked-out examples. If you get stuck, go back to the text for the more thorough explanation. The summary is a somewhat informal and brief description of the material covered. This section is not meant to be a short course in chemistry, or a substitute for the text. It does not stand alone because all points and details are not explained. It is meant to test your ability to read effortlessly and smoothly through chemistry material. If the reading is difficult, you need more work in the text. Some general questions concerning the sections precede the discussion. New terms are shown in bold type for easy reference.

NEW TERMS

As you read through the list of new terms, recite to yourself the meaning and significance of each. If a term is not familiar, refer to the text chapter or the glossary (Appendix F) in the text.

SELF-TEST

This is probably the most important section of the review. You are now ready to answer some multiple choice questions and to work some problems on your own. Space has been left after each problem for your calculations. Most of these problems are very similar to the worked-out examples in the text. Multiple choice and matching can seem confusing but they are excellent tests of your knowledge. It is one thing to write an answer when asked, but it requires more knowledge to pick a correct answer from among incorrect but similar answers. Learning to answer multiple choice questions correctly is a science in itself. The problems should serve as a good measure of how you will do on an exam.

Review Section C or D

CHAPTER SUMMARY SELF-TEST

This test is reserved for your use after all of the chapter has been covered. The intention of this exam is to pull the various review sections together. Problems in this section usually require use of two or more concepts developed in the chapter.

Answers to Self-Tests

After you have completed each self-test, compare your answers to those provided. The answers to problems include solutions and in many cases information on how the problem was worked. If you are still hazy on how a type of problem was worked, refer back to the text.

Solutions to Black Text Problems

This section is a supplement to Appendix G in the text. It contains the worked-out solutions to all of the quantitative problems that appear in black print in the text.

Good luck in your start in chemistry. I hope this Study Guide, as well as the text, will aid you in your efforts. Mostly, I hope you find this subject fascinating.

Foreword: How to "Ace" Chemistry

There is good news and bad news about the study of chemistry. First the good news: Chemistry is not nearly as tough as its reputation leads you to believe (seriously). Now for the bad news: Many of you will probably have to give serious effort to adjusting your study habits and time commitments in order for the good news to be true. If this course helps you to achieve good study discipline, the bad news turns out to be good news also.

Although there are many reasons why some people don't do well in chemistry, there are three reasons that will be mentioned here because they are *correctable.*

(1) Insufficient background
(2) Poor study habits and discipline
(3) Excessive nervousness on exams

Let's take these three points one at a time.

Insufficient Background

If this is your problem, you are not alone. In fact, you are in the right place. The text and Study Guide were written especially for such students. The text assumes that you have no previous experience in chemistry, and that you will need a careful and logical introduction to each new concept. It *is* assumed, however, that you have had (and can apply) one year of high school algebra as well as the ability to deal with percent, proportional relationships, exponential notation, and perhaps logarithms. Many of you will need some review in one or more of these areas. The appendices in the text are provided so that you can identify and repair the mathematical rusty spots. If serious deficiencies in math and algebra exist, it is recommended that the study of chemistry be delayed until you have taken some college level math.

Poor Study Habits

The importance of proper study habits cannot be overemphasized. We've already discussed this idea to an extent in the Prologue of the text, but you are beginning the study of a highly ordered discipline where one fact builds on another. This requires the same highly ordered approach to its study. *There is no other way (unless you're a near genius).* The following suggestions may be of some help.

1. Make a schedule with a regular time set aside for the study of chemistry. Two 1-hour sessions are much more effective than one 2-hour session. The brain tends to become mushy after a while, and it is easy to be distracted. Be sure to schedule one study session as soon after each lecture as feasible. Stick to your schedule.

2. Read the text ahead of your lecture. You may not fully understand the material, but you will at least have a nodding acquaintance with some new terms and concepts. You'll be more receptive when the topics are covered in class.

3. Take clear and concise notes in lecture but don't try to copy everything. You will miss too much if you can't confine notes to sample problems and high points. Write problems worked in class in such a way that you can cover them up later and rework them. (Remember, they are the most likely types of problems that you will see again on a test.) Reading ahead helps cut down on notes. If you know it's in the book, why write it all down again? Use about 2/3 of the paper when you take notes so that you can later make other notes and reminders next to those that you took in lecture.

4. Shortly after class, in your scheduled study time, go over your most recent notes. The sooner you firm in your mind the material from the lecture, the easier it will be. Rework any problems from lecture to see if you can do them on your own. You'll notice that what seemed easy when your instructor did it may not be so easy when you try it alone. You don't want to find this out the night before the test. Reread the appropriate section of the text, working through all sample problems and thinking through all new concepts or definitions. Make a note when you come across material or a problem that you do not understand. Ask your instructor about specific questions or problems that require help.

5. Make good use of this Study Guide and the end-of-chapter exercises in the text. In working problems, think about what you've done. What kind of a problem is this? How did I recognize it and would I recognize it again? What concept does it illustrate? What was given and what was requested? How did I work it? Would I recognize the problem if it were turned around (i.e., the answer was given and the question was what you were asked to find)? If you don't work all of the problems in the text, at least think through and analyze all of them so that there won't be any new twists that may surprise you later.

Nervousness on Exams

Good! Now you have a good organized approach to the study of chemistry. That's the most important step. Will you now do well on all of the exams? In most cases, the answer is yes! But, taking exams is a science in itself. All instructors have heard an unhappy student say something like "I studied hard, I knew the material, but I froze on the exam and couldn't do a thing." This situation happens, but it can be avoided with some effort. Everyone is and should be a little nervous before a test just as an athlete is nervous before a game. However, the nervousness should improve the performance, not hinder it. Again, a few suggestions may help.

1. The most important point relates to the previous discussion on study habits. Regular work on the material beats "cramming" the night before the exam. That time should be reserved for a comprehensive review and good night's sleep. It seems that there is a lot of peer pressure to stay up all night to study for an exam. Everybody does it or so it seems. Unfortunately, many of these students who study into the "wee hours" find themselves in a zombie-like trance during the test. Numbers and definitions are going through their minds in a scrambled nightmare. In this state of mind, it may take the whole 50 minutes just to write your name. Being totally alert and rested is an obvious advantage on any test.

2. Don't dramatize the situation. "If I don't do well on this exam, I won't do well in the course. If I don't do well in the course, I'll flunk out of college. If that happens, I'll never get a decent job, my parents will disown me, and my girl or boyfriend will take a hike. If this happens, I might as well wander off alone into the mountains. Therefore, if I don't do well on this *one* test, I'll have to go to the mountains." Relax! This exam may be important, but you can work yourself into a helpless

frenzy by overthinking the situation. Just do your best; there will be other tests and opportunities. Think of the basketball player who coolly sinks a free throw, with one second left, to win the game in front of 10,000 hostile, screaming fans. She just blocked out the screams and the importance of the free throw and did her best. You can do your best on a chemistry exam, and you won't even have 10,000 hostile people screaming at you.

3. In your preparation and review, think "test." How would you ask the questions if you were making up the test? In fact, why not make up a test and take it within a certain time limit? Better yet, you and a friend could get together and make up tests for each other. Take the test in 50 minutes (or whatever time limit you have on exams), and see how you do. Remember it's one thing to work problems when you have all evening and quite another to do several different types of problems in a relatively short time interval. It is very important to practice doing problems under test conditions, that is at deliberate but careful speed. Review your lecture notes for concepts and problems emphasized for your best clues as to what will be on the exam.

4. A commonly heard complaint goes something like this: "I could have worked that 10-point problem at the end, but you called time just as I turned to it." Solution: Work that easy 10-point problem first. There is no rule that says that you cannot work Problem 2 until you have completely finished Problem 1. Learn to skim through a test picking out the easy problems (or ones that you successfully anticipated) for your first effort. Don't worry about unfamiliar or tough problems yet. If the gist of a problem or question isn't immediately obvious, don't even finish reading it - move on. When you find the easy problem you're looking for, carefully read the question and do the problem. It may be the last problem on the test but that's fine. There are three beneficial results of this procedure:

   (1) You won't get caught as time is called with easy ones undone.
   (2) It makes you feel good to know that you've scored and that you've done some problems correctly. (On the other hand, if you get rattled by focusing on the hard ones first, you may not recognize the easy ones later.)
   (3) You know the extent and emphasis of the test, and you can budget your remaining time accordingly.

   After doing easy ones, go back over the test and read the problems more carefully. Pick the easiest of the remaining problems for the second effort. Finally, save the hardest for last. Then you can focus on them without worrying about what's on the next page. Try to save some time to look for math errors, however. Make sure you read the problems correctly, but don't spend time trying to talk yourself out of correct answers. Chances are that your first approach was correct and that the teacher was not looking for some extremely subtle and obscure interpretation of the problem. In summary, how you take a test can do much to reduce the possibility of a mental "freeze-up." If you continue to have problems taking exams, talk to your instructor. Most college counseling centers have workshops or private counseling on test anxiety.

5. When you get your test back, don't just toss it aside - it holds the key to the final. Go over the test carefully and find out what you are doing right and what you are doing wrong. Learn how to do all the problems that you missed as soon as possible. Try not to keep making the same mistake(s). Before you file the exam away for future reference, try taking it again in the allotted time.

1

Measurements in Chemistry

Review Section A *The Numbers Used in Chemistry*

OUTLINE	OBJECTIVES
1-1 The Numerical Value of a Measurement	
1. Precision and accuracy	*Distinguish between the precision and the accuracy of a measurement.*
2. Significant figures	*Determine the number of significant figures in a measurement.*

1-2 Significant Figures and Mathematical Operations

1. Mathematical operations — *Express the result of a mathematical operation to the proper decimal place (addition and subtraction) or to the proper number of significant figures (multiplication and division).*

2. Rounding off numbers — *Be able to round off measurements to a specified number of significant figures.*

1-3 Expressing Large and Small Numbers: Scientific Notation

1. Scientific notation — *Convert a very large or very small number to scientific notation.*

2. Scientific notation in calculations — *Carry out mathematical operations with numbers expressed in scientific notation.*

SUMMARY OF SECTIONS 1-1 THROUGH 1-3

Questions: *What numbers in a measurement are valid? How many numbers should we write in a measurement? What do we do if we have to add or multiply numbers that do not have the same precision? How do we handle very large or very small numbers in chemistry?*

Chemistry, a physical science, is concerned with quantitative laws. Since quantitative laws are based on **measurements**, it is necessary for us to establish how to appropriately express a measurement and eventually how to carry out mathematical operations on specific measurements. A measurement is composed of a numerical value and a **unit**. In this first section of the review we will concentrate on properly expressing the numerical value of the measurement.

Accuracy reflects how close a measurement is to the true value. Measurements are also expressed to various degrees of **precision** which are reflected by the **significant figures** in the number. Since a zero may or may not be a significant figure, it is necessary to be familiar with the certain rules. For example, the following numbers represent measurements. The zeros that are significant are underlined.

$$13.\underline{0}7 \quad 1\underline{0}.7 \quad 0.3\underline{0} \quad 1\underline{0}.3\underline{0}1 \quad 1\underline{0}30 \quad 2\underline{0}.\underline{0} \quad 0.02\underline{0}$$

You should also know that there are different rules concerning the proper form for an answer for addition and subtraction as opposed to those for multiplication and division. A common mistake is to apply the rules on the number of significant figures to addition and subtraction where, in fact, this is not important. For example, in the addition example that follows, the answer is rounded off to the same number of decimal places as the measurement with the fewest number. In the multiplication example, the answer is rounded off to the same number of significant figures as the multiplier with the fewest number of significant figures.

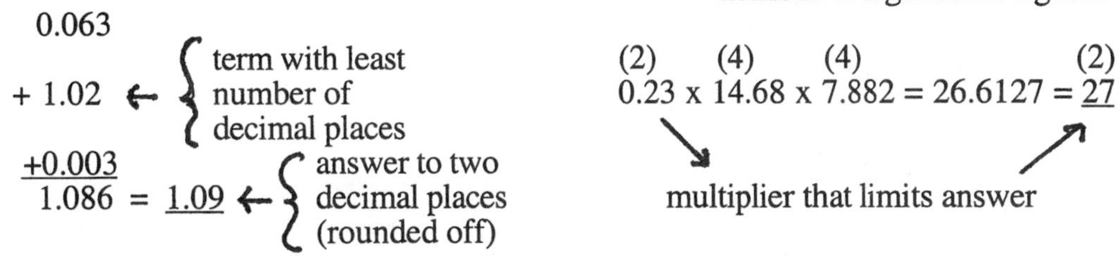

Exponential notation is introduced as a welcome tool to express very small or very large numbers without writing a lot of zeros. In **scientific notation** the number is written with one digit to the left of the decimal point multiplied by 10 raised to a power. The power is known as the **exponent** of 10. In later chapters we will assume that the student is not only familiar with scientific notation, but also with the mathematics of manipulating numbers expressed in this fashion. Refer to Appendix C in the text for additional help in this area if needed.

NEW TERMS

Accuracy
Exponent
Measurement
Precision
Scientific notation
Significant figure
Unit

SELF-TEST

A-1 Multiple Choice (There is only one correct answer for each question.)

___ 1. A container is known to have a volume of one quart. Which of the following measurements of the volume is the most precise?

(a) 1.0 qts (b) 0.98 qts (c) 1.052 qts (d) 0.9 qts

___ 2. How many significant figures are in the following measurement? 0.03010 cm?

(a) 6 (b) 5 (c) 4 (d) 3 (e) 2

___ 3. When 78.5 cm^2 is divided by 2.654 cm, the answer is:

(a) 29.577995 cm
(b) 29.5 cm
(c) 30 cm
(d) 29.6 cm
(e) 29.58 cm

___ 4. When the numbers 0.508, 25.0 and 1.188 are added, the answer is:

(a) 26.7 (b) 26.6 (c) 27 (d) 27.0 (e) 26.696

___ 5. The number 0.0070×10^{-8} expressed in scientific notation is:

(a) 7×10^{-11}
(b) 7×10^{-5}
(c) 7.0×10^{-11}
(d) 7.0×10^{-10}
(e) 7.0×10^{-5}

A-2 Significant Figures

1. How many significant figures are in each of the following numbers?

157 _______ 0.50 _______ 120.0 _______

103 _______ 0.209 _______ 1006 _______

0.02 _______ 1200 _______

2. Assume that the following numbers represent measurements. Express the answer for the calculation to the proper number of significant figures or decimal places.

(a)
$$\begin{array}{r} 0.0375 \\ 0.12 \\ +0.001 \\ \hline 0.1585 \end{array} = \_\_\_\_\_\_$$

(b)
$$\begin{array}{r} 9700 \\ -180 \\ \hline 8520 \end{array} = \_\_\_\_\_\_$$

(c) $\dfrac{14.222 \times 0.3010}{0.200} = 21.40411 = \_\_\_\_\_\_$

(d) $(0.6)^2 = 0.36 = \_\_\_\_\_\_$ (e) $(6.0)^2 = 36 = \_\_\_\_\_\_$

A-3 Scientific Notation

1. Express the following numbers in scientific notation.

(a) $0.0042 = \_\_\_\_\_\_$ (b) $47900 = \_\_\_\_\_\_$

(c) $0.00040 = \_\_\_\_\_\_$ (d) $0.037 \times 10^{-6} = \_\_\_\_\_\_$

(e) $1560 \times 10^{-8} = \_\_\_\_\_\_$

2. Express the following as normal numbers without scientific notation.

(a) $4.83 \times 10^{-3} = \_\_\_\_\_\_$ (b) $3.77 \times 10^{4} = \_\_\_\_\_\_$

(c) $0.091 \times 10^{-2} = \_\_\_\_\_\_$ (d) $915 \times 10^{4} = \_\_\_\_\_\_$

3. Carry out the following calculations. Express the answer to the proper number of significant figures or decimal places.

(a) $(14.6 \times 10^{-3}) + (0.141) + (237 \times 10^{-5})$ _____

(b) $(18.3 \times 10^{-6}) \times (0.00320 \times 10^{-3})$ _____

(c) $\dfrac{392 \times 10^{-22}}{0.022 \times 10^{24}}$ _____

(d) $\dfrac{(0.0631 \times 10^{6}) \times (1.009 \times 10^{8})}{(7.71 \times 10^{-4})}$ _____

Review Section B *The Measurements Used in Chemistry*

OUTLINE / **OBJECTIVES**

1-4 Measurement of Mass, Length, and Volume

1. English and Metric Units — *Name the basic metric units and their symbols.*
2. Prefixes and symbols of the metric units — *Write the relationships of desired units having the prefixes kilo, centi, and milli.*
3. Mass and Weight — *Distinguish between mass and weight.*

1-5 Conversions of Units by the Factor-Label Method

1. Conversion Factors — *Construct conversion factors relating units within one system of measurement or between two systems of measurements.*
2. Conversion Problems — *Make conversions between units using the factor-label method (dimensional analysis). Know how exact relationships affect a calculation.*

1-6 Measurement of Temperature

1. Temperature
2. The Fahrenheit, Celsius, and Kelvin Scales — *Convert among the Celsius, Kelvin, and Fahrenheit scales.*

SUMMARY OF SECTIONS 1-4 THROUGH 1-6

Questions: *What units do we use in chemistry to express dimensions, mass, and volume? Is there a convenient way to convert between units of measurement? How do we measure heat intensity?*

The units of measurement used in chemistry are those of the **metric system**. Metric units also serve as the basis of the SI system which are universally recognized units. U.S citizens can become more comfortable with the metric units by relating the magnitude of units for **length, mass** or **weight**, and **volume** to the more familiar English units. Table 1A, which follows, summarizes the more common relationships within and between the English and metric systems.

Many chemistry problems involve conversion of a measurement in one system of measurement to that of another. The **factor-label (dimensional analysis)** method for working problems of this nature is introduced in this chapter. It is a powerful problem solving tool when used consistently. The method can be illustrated by conversions within and between English and metric units. Basically, this method involves the inclusion of units in all operations as well as the numerical value of the measurement. In this fashion, the units lead us to the proper mathematical operation and thus to the correct answer. For a simple one-step conversion, the given measurement is multiplied by a **conversion factor** that transforms it to the desired measurement. Conversion factors are usually expressed as **unit factors** which relate to *one* of a unit.

Table 1A

RELATIONSHIPS BETWEEN UNITS

	English	Metric	English-Metric
Length	1 ft = 12 in. 3 ft = 1 yd 1760 yd = 1 mile	1 m = 10^3 mm 1 m = 10^2 cm 1 m = 10^{-3} km	2.54 cm = 1 in. (exact)
Volume	2 pt = 1 qt 4 qt = 1 gal 42 gal = 1 barrel	1 L = 10^3 mL 1 L = 10^{-3} kL 1 mL = 1 cm^3	1.057 qt = 1 L
Mass	16 oz = 1 lb 2000 lb = 1 ton	1 g = 10^3 mg 1 g = 10^{-3} kg	453.6 g = 1 lb

[What's given] x (Conversion Factor) = [What's requested]

The conversion factor is constructed from an equality of equivalency that relates the given and requested measurements. An example problem follows.

Example B-1 Conversion of Pounds to Grams

Convert 0.78 lb to grams.

PROCEDURE

A conversion factor is needed that relates lb to g. From Table 1A, this is 453.6 g = 1 lb. The conversion factor expresses the relationship in factor or fractional form.

$$\frac{453.6\ g}{lb} \quad \text{or} \quad \frac{1\ lb}{453.6\ g}$$

Choose the proper conversion factor that cancels the given unit (lb) and leaves the requested unit in the numerator (g).

SOLUTION

(Given) (Conversion Factor) (Requested)

$$0.78\ \cancel{lb} \times \frac{453.6\ g}{\cancel{lb}} = \underline{350\ g}$$

given unit cancels — Requested unit remains in the numerator

Many problems, however, involve conversions where one simple relationship between given and requested units is not available. In these cases, one must travel from given to requested in two or more carefully planned steps. For each step in the calculation, a conversion factor is needed.

Example B-2 Conversion of Meters to Feet

Convert 61.7 m to ft.

PROCEDURE

The only relationship between English and metric systems given is between in. and cm, three steps and three corresponding conversions are needed. The unit map is shown below.

Given → Requested

$$m \xrightarrow{(1)} cm \xrightarrow{(2)} in \xrightarrow{(3)} ft$$

SOLUTION

$$61.7\ \cancel{m} \times \overset{(1)}{\frac{10^2\ \cancel{cm}}{\cancel{m}}} \times \overset{(2)}{\frac{1\ \cancel{in.}}{2.54\ \cancel{cm}}} \times \overset{(3)}{\frac{1\ ft}{12\ \cancel{in.}}} = \underline{202\ ft}$$

(1) converts m to cm; (2) converts cm to in.; (3) converts in. to ft

Temperature is a measure of the heat intensity of the substance. Temperature is measured with a device called a **thermometer**. The **Fahrenheit** [t(F)], **Celsius** [t(C)], and **Kelvin** [T(K)] temperature scales relate to each other as follows.

$$t(F) = [t(C) \times 1.8] + 32$$

$$t(C) = \frac{[t(F) - 32]}{1.8}$$

$$T(K) = [t(C) + 273]$$

NEW TERMS

Celsius
Conversion factor
Dimensional analysis
Fahrenheit
Factor-label method
Kelvin
Mass
Temperature
Thermometer
Unit Factor
Unit map
Weight

SELF-TEST

B-1 Multiple Choice

___ 1. The metric prefix *milli* relates to the basic unit by:

(a) 10^3 (b) 10^2 (c) 10^{-6} (d) 10^{-3} (e) 1 cm^3

___ 2. Which of the following relationships is false?

(a) 1 mg = 10^6 kg
(b) 100 cm = 1 m
(c) 10^3 L = 1 kL
(d) 1 cg = 10^{-2} g

___ 3. Which of the following is an exact relationship?

(a) 1 yd = 3 ft
(b) 1 lb = 453.6 g
(c) 2.205 lb = 1 kg
(d) 1 km = 0.6215 mile

___ 4. One milliliter is the same as:

(a) 1 dm^3 (b) 1 m^2 (c) 1 mm^3 (d) 1 cm^2 (e) 1 cm^3

___ 5. Jupiter is a planet with a larger gravity than Earth. On Jupiter a person:

(a) weighs less.
(b) has more mass.
(c) weighs more.
(d) has less mass.

___ 6. If the Fahrenheit temperature increases by 36 degrees, the Celsius temperature increases by:

(a) 65 degrees
(b) 20 degrees
(c) 52 degrees
(d) 33 degrees
(e) 15 degrees

B-2 Matching

___ 1 m	(a) 2.205 kg	
___ 1 ft	(b) 0.4536 kg	
___ 1 L	(c) 1 dm^3	
___ 1 kg	(d) 10^3 kL	
___ 1 lb	(e) 10^3 mm	
	(f) 30.5 in.	(i) 453.6 kg
	(g) 1 m^3	(j) 30.5 cm
	(h) 10^6 mg	

B-3 Conversion Factors.

Write a relationship in factor form that accomplishes the following conversions.

e.g.: cm to in. 1 in./2.54 cm

1. cm to m ____________________
2. oz to lb ____________________
3. g to lb ____________________
4. L to qt ____________________
5. in. to cm ____________________

B-4 Conversions.

Make the following conversions. Use the factor-label method (dimensional analysis).

1. 14,850 in. to mi

2. 958 mm to km

3. 495 in. to cm

4. 0.811 L to qt

5. 18.8 oz to g

6. 37.0 barrels to kL (42.0 gallons = 1 barrel)

7. 195 m to yd

8. 85 mi/hr to m/min

9. 127°F to °C and K

10. -78°C to °F

Review Section C

CHAPTER SUMMARY SELF-TEST

C-1 Matching

___ Accuracy

___ Exponent

___ Mass

___ Thermometer

___ Fahrenheit

___ Exact relationship

(a) A measured relationship between units in two different systems of measurement.

(b) The attraction of gravity for a substance

(c) A device used to measure temperature

(d) Refers to how close a measurement is to the true value

(e) The number that is raised to a power

(f) A defined relationship between units in the same system of measurement

(g) The amount of matter in a substance

(h) A scale of temperature where the freezing point of water is defined as zero

(i) The power to which a number is raised

(j) Refers to how many significant figures are in a measurement

(k) A temperature scale with 180 divisions between the boiling point and freezing point of water

C-2 Problems

1. Add the following numbers. Express the answer to the proper decimal place.
   0.321, 4.24×10^{-2}, 4×10^{-4}

2. What is 0.062×10^{8} divided by 0.17×10^{-3}? Express the answer to the proper number of significant figures.

3. Convert 0.273 kg to mg.

4. Convert 17.5 lb to g.

5. Convert 23.0 mm/sec to in./hr.

6. Convert $100^{o}F$ to ^{o}C and to K.

Answers to Self-Tests

A-1 Multiple Choice

1. **c** The more precise measurement has the most significant figures.
2. **c** The first two zeros are not significant.
3. **d** 29.6 cm
4. **a** 26.7 Round off 26.696 to 26.7.
5. **c** 7.0×10^{-11}

A-2 Significant Figures

1.

157 - 3	0.02 - 1	0.209 - 3	120.0 - 4
103 - 3	0.50 - 2	1200 - 2	1006 - 4

2. (a) 0.1585 = $\underline{0.16}$ (b) 8520 = $\underline{8500}$
(c) 21.40411 = $\underline{21.4}$ (d) 0.36 = $\underline{0.4}$
(e) 36 = $\underline{36}$

A-3 Scientific Notation

1. (a) 4.2×10^{-3} (b) 4.79×10^{5}
   (c) 4.0×10^{-4} (d) 3.7×10^{-8}
   (e) 1.56×10^{-5}

2. (a) 0.00483 (b) 37,700
   (c) 0.00091 (d) 9,150,000

3. (a) $14.6 \times 10^{-3} = 0.0146$

$$\begin{array}{r} 0.0146 \\ 0.141 \\ 237 \times 10^{-5} = \underline{0.00237} \\ 0.15797 = \underline{0.158} \end{array}$$

(b) $0.00320 \times 10^{-3} = 3.20 \times 10^{-6}$

$(18.3 \times 10^{-6}) \times (3.20 \times 10^{-6}) = 58.56 \times 10^{-12} = 5.86 \times 10^{-11}$

(c) $$\frac{392 \times 10^{-22}}{0.022 \times 10^{24}} = \frac{3.92 \times 10^{-20}}{2.2 \times 10^{22}} = 1.8 \times 10^{-42}$$

(d) $$\frac{(0.0631 \times 10^{6}) \times (1.009 \times 10^{8})}{(7.71 \times 10^{-4})} =$$

$$\frac{(0.0631 \times 1.009)}{7.71} \times \frac{(10^{6} \times 10^{8})}{10^{-4}} = 0.0082578 \times 10^{18} = 8.26 \times 10^{15}$$

B-1 Multiple Choice

1. **d** 10^{-3}
2. **a** 1 mg = 10^{6} kg
3. **a** 1 yd = 3 ft
4. **e** 1 cm^3
5. **c** weighs more
6. **b** 20 degrees

   $$(36 \text{ ~~F div~~.} \times \frac{100 \text{ C div.}}{180 \text{ ~~F div.~~}} = 20 \text{ C div.})$$

B-2 Matching

e 1 m = 10^{3} mm

j 1 ft = 30.5 cm

c 1 L = 1 dm^3

h 1 kg = 10^{6} mg

b 1 lb = 0.4536 kg

B-3 Conversion Factors

1. 1 m/10^2 cm or 10^{-2} m/cm
2. 1 lb/16 oz
3. 1 lb/453.6 g
4. 1.057 qt/L
5. 2.54 cm/in.

B-4 Conversions

1. in. ⟶ ft ⟶ yd ⟶ mi

$$14{,}850\ \text{in.} \times \frac{1\ \text{ft}}{12\ \text{in.}} \times \frac{1\ \text{yd}}{3\ \text{ft}} \times \frac{1\ \text{mi}}{1760\ \text{yd}} = = \underline{0.2344\ \text{mi}}$$

2. mm ⟶ m ⟶ km

$$958\ \text{mm} \times \frac{1\ \text{m}}{10^3\ \text{mm}} \times \frac{10^{-3}\ \text{km}}{\text{m}} = 958 \times 10^{-6}\ \text{km} = \underline{9.58 \times 10^{-4}\ \text{km}}$$

3. in. ⟶ cm

$$495\ \text{in.} \times \frac{2.54\ \text{cm}}{\text{in.}} = \underline{1.26 \times 10^3\ \text{cm}}$$

4. L ⟶ qt

$$0.811\ \text{L} \times \frac{1.057\ \text{qt}}{\text{L}} = \underline{0.857\ \text{qt}}$$

5. oz ⟶ lb ⟶ g

$$18.8\ \text{oz} \times \frac{1\ \text{lb}}{16\ \text{oz}} \times \frac{453.6\ \text{g}}{\text{lb}} = \underline{533\ \text{g}}$$

6. barrels ⟶ gal ⟶ qt ⟶ kL

$$37.0\ \text{barrel} \times \frac{42\ \text{gal}}{\text{barrel}} \times \frac{4\ \text{qt}}{\text{gal}} \times \frac{1\ \text{L}}{1.057\ \text{qt}} \times \frac{10^{-3}\ \text{kL}}{\text{L}} = \underline{5.88\ \text{kL}}$$

7. m ⟶ cm ⟶ in. ⟶ ft ⟶ yd

$$195\ \text{m} \times \frac{10^2\ \text{cm}}{\text{m}} \times \frac{1\ \text{in.}}{2.54\ \text{cm}} \times \frac{1\ \text{ft}}{12\ \text{in.}} \times \frac{1\ \text{yd}}{3\ \text{ft}} = \underline{213\ \text{yd}}$$

8. mi/hr ⟶ km/hr ⟶ m/hr ⟶ m/min

$$\frac{85\ \cancel{mi}}{\cancel{hr}} \times \frac{1.609\ \cancel{km}}{\cancel{mi}} \times \frac{1\ m}{10^{-3}\ \cancel{km}} \times \frac{1\ \cancel{hr}}{60\ min} = \underline{2.3 \times 10^{3}\ m/min}$$

9. $$t(C) = \frac{[t(F) - 32]}{1.8} = \frac{(127 - 32)}{1.8} = \frac{95}{1.8} = 53^{o}C \quad T(K) = 53 + 273 = \underline{326\ K}$$

10. $t(F) = 1.8[t(C)] + 32 = 1.8(-78.0) + 32 = -140 + 32 = \underline{-108^{o}F}$

C-1 Matching

d	Accuracy refers to how close a measurement is to the true value.
i	An exponent is the power to which a number is raised.
g	Mass is the amount of matter in a substance.
c	A device used to measure temperature.
k	Fahrenheit is a temperature scale with 180 divisions between the boiling and freezing points of water.
f	A defined relationship between units in one system of measurement.

C-2 Problems

1.

$$\begin{array}{lr} & 0.321 \\ 4.24 \times 10^{-2} = & 0.0424 \\ 4 \times 10^{-4} = & \underline{0.0004} \\ & 0.3638 = \underline{0.364} \end{array}$$

2. $$\frac{0.0662 \times 10^{8}}{0.17 \times 10^{-3}} = \frac{6.62 \times 10^{6}}{1.7 \times 10^{-4}} = \underline{3.9 \times 10^{10}}$$

3. $$0.273\ \cancel{kg} \times \frac{1\ \cancel{g}}{10^{-3}\ \cancel{kg}} \times \frac{10^{3}\ mg}{\cancel{g}} = \underline{2.73 \times 10^{5}\ mg}$$

4. $$17.5\ \cancel{lb} \times \frac{453.6\ g}{\cancel{lb}} = \underline{7.94 \times 10^{3}\ g}$$

5. $$\frac{23.0\ \cancel{mm}}{\cancel{sec}} \times \frac{60\ \cancel{sec}}{\cancel{min}} \times \frac{60\ \cancel{min}}{hr} \times \frac{1\ \cancel{cm}}{10\ \cancel{mm}} \times \frac{1\ in.}{2.54\ \cancel{cm}} = \underline{3.26 \times 10^{3}\ in./hr}$$

6. $$t(C) = \frac{100 - 32}{1.8} = 38^{o}C \quad T(K) = 273 + 38 = \underline{311\ K}$$

Solutions to Black Text Problems

1-60 (a) $7.8 \times 10^3\ \cancel{m} \times \frac{1\ km}{10^3\ \cancel{m}} = \underline{7.8\ km}$ $\quad 7.8\ \cancel{km} \times \frac{1\ mi}{1.609\ \cancel{km}} = \underline{4.8\ mi}$

$4.8\ \cancel{mi} \times \frac{5280\ ft}{\cancel{mi}} = \underline{2.6 \times 10^4}\ ft$

(b) $0.450\ \cancel{mi} \times \frac{5280\ ft}{\cancel{mi}} = \underline{2380\ ft}$ $\quad 0.450\ \cancel{mi} \times \frac{1.609\ km}{\cancel{mi}} = \underline{0.724\ km}$

$0.724\ \cancel{km} \times \frac{10^3\ m}{\cancel{km}} = \underline{724\ m}$

(c) $8.98 \times 10^3\ \cancel{ft} \times \frac{1\ mi}{5280\ \cancel{ft}} = \underline{1.70\ mi}$ $\quad 1.70\ \cancel{mi} \times \frac{1.609\ km}{\cancel{mi}} = \underline{2.74\ km}$

$2.74\ \cancel{km} \times \frac{10^3\ m}{\cancel{km}} = \underline{2740\ m}$

(d) $6.78\ \cancel{km} \times \frac{1\ mi}{1.609\ \cancel{km}} = \underline{4.21\ mi}$ $\quad 4.21\ \cancel{mi} \times \frac{5280\ ft}{\cancel{mi}} = \underline{2.22 \times 10^4}\ ft$

$6.78\ \cancel{km} \times \frac{10^3\ m}{\cancel{km}} = \underline{6780\ m}$

1-61 (a) $6.78\ \cancel{gal} \times \frac{3.785\ L}{\cancel{gal}} = \underline{25.7\ L}$ $\quad 25.7\ \cancel{L} \times \frac{1.057\ qt}{\cancel{L}} = \underline{27.2\ qt}$

(b) $670\ \cancel{qt} \times \frac{1\ L}{1.057\ \cancel{qt}} = \underline{630\ L}$ $\quad 670\ \cancel{qt} \times \frac{1\ gal}{4\ \cancel{qt}} = \underline{170\ gal}$

(c) $7.68 \times 10^3\ \cancel{L} \times \frac{1.057\ qt}{\cancel{L}} = \underline{8.12 \times 10^3}\ qt$ $\quad 8.12 \times 10^3\ \cancel{qt} \times \frac{1\ gal}{4\ \cancel{qt}} = \underline{2.03 \times 10^3}\ gal$

1-63 $122\ \cancel{lb} \times \frac{453.6\ \cancel{g}}{\cancel{lb}} \times \frac{1\ kg}{10^3\ \cancel{g}} = \underline{55.3\ kg}$

1-65 $28.0\ \cancel{m} \times \frac{10^2\ \cancel{cm}}{\cancel{m}} \times \frac{1\ \cancel{in.}}{2.54\ \cancel{cm}} \times \frac{1\ \cancel{ft}}{12\ \cancel{in.}} \times \frac{1\ yd}{3\ \cancel{ft}} = \underline{30.6\ yd}$ (New punter is needed.)

1-67 $0.375\ \cancel{qt} \times \frac{1\ L}{1.057\ \cancel{qt}} = \underline{0.355\ L}$

1-69 6 ft 10 1/2 in. = 82.5 in. $82.5\ \cancel{in.} \times \frac{2.54\ \cancel{cm}}{\cancel{in.}} \times \frac{1\ m}{10^2\ \cancel{cm}} = \underline{2.10\ m}$

$212\ \cancel{lb} \times \frac{1\ kg}{2.205\ \cancel{lb}} = \underline{96.1\ kg}$

1-70 $55.0\ \cancel{L} \times \frac{1.057\ \cancel{qt}}{\cancel{L}} \times \frac{1\ gal}{4\ \cancel{qt}} = \underline{14.5\ gal}$

1-72 $0.200\ \text{gal} \times \frac{4\ \text{qt}}{\text{gal}} = 0.800\ \text{qt}$ $\qquad 0.800\ \text{qt} \times \frac{1\ \text{L}}{1.057\ \text{qt}} \times \frac{1\ \text{mL}}{10^{-3}\ \text{L}} = \underline{757\ \text{mL}}$

There is slightly more in a "fifth" than in 750 mL.

1-74 $\frac{65.0\ \text{mi}}{\text{hr}} \times \frac{1.609\ \text{km}}{\text{mi}} = \underline{105\ \text{km/hr}}$

1-78 $\frac{\$0.899}{\text{gal}} \times \frac{1\ \text{gal}}{4\ \text{qt}} \times \frac{1.057\ \text{qt}}{\text{L}} = \$0.238/\text{L}$ $\qquad 80.0\ \text{L} \times \frac{\$0.238}{\text{L}} = \underline{\$19.04}$

1-79 $551\ \text{mi} \times \frac{1\ \text{gal}}{21.0\ \text{mi}} \times \frac{\$0.899}{\text{gal}} = \underline{\$23.59}$

$482\ \text{km} \times \frac{1\ \text{mi}}{1.609\ \text{km}} \times \frac{1\ \text{gal}}{21.0\ \text{mi}} \times \frac{\$0.899}{\text{gal}} = \underline{\$12.82}$

1-80 $\$45.00 \times \frac{1\ \text{gal}}{\$0.899} \times \frac{21.0\ \text{mi}}{\text{gal}} \times \frac{1.609\ \text{km}}{\text{mi}} = \underline{1690\ \text{km}}$

1-83 $\$2.50 \times \frac{1\ \text{lb}}{\$0.95} \times \frac{145\ \text{nails}}{\text{lb}} = \underline{382\ \text{nails}}$

1-84 $5670\ \text{nails} \times \frac{1\ \text{lb}}{185\ \text{nails}} \times \frac{\$0.92}{\text{lb}} = \underline{\$28.20}$

1-85 $350\ \text{km} \times \frac{1\ \text{mi}}{1.609\ \text{km}} \times \frac{1\ \text{gal}}{24.5\ \text{mi}} \times \frac{\$1.22}{\text{gal}} = \underline{\$10.83}$

1-88 $442\ \text{mi} \times \frac{1.609\ \text{km}}{\text{mi}} \times \frac{1\text{hr}}{215\ \text{km}} = \underline{3.31\ \text{hr}}$

1-89 (a) $\$6.50 \times \frac{1.12\ \text{Euro}}{\$} = \underline{7.28\ \text{Euro}}$ $\qquad$ (b) $12.65\ \text{Euro} \times \frac{1\ \$}{1.12\ \text{Euro}} = \underline{\$11.29}$

1-90 $\frac{0.695\ \text{pd}}{\$} \times \frac{1\ \$}{1.12\ \text{Euro}} = \frac{0.621\ \text{pds}}{\text{Euro}}$

$25{,}500\ \text{Euro} \times \frac{0.621\ \text{pds}}{\text{Euro}} = \underline{15{,}800\ \text{pds}}$

1-93 $4.0 \times 10^8\ \text{mi} \times \frac{5280\ \text{ft}}{\text{mi}} \times \frac{12\ \text{in.}}{\text{ft}} \times \frac{2.54\ \text{cm}}{\text{in.}} \times \frac{1\ \text{s}}{3.0 \times 10^{10}\ \text{cm}} = \underline{2100\ \text{s}}$

$2100\ \text{s} \times \frac{1\ \text{min}}{60\ \text{s}} \times \frac{1\ \text{hr}}{60\ \text{min}} = \underline{0.58\ \text{hr}}$

1-94 $t(^{\circ}\text{F}) = 1.8(300^{\circ}\text{C}) + 32 = 520 + 32 = \underline{572^{\circ}\text{F}}$

1-95 $t(^{\circ}\text{C}) = \frac{76 - 32}{1.8} = \frac{44}{1.8} = \underline{24^{\circ}\text{C}}$

1-97 $t(^{o}F) = [(-39) \times 1.8] + 32 = \underline{-38^{o}F}$

1-99 $t(^{o}F) = [(35.0 \times 1.8) + 32.0] = \underline{95.0^{o}F}$

1-103 Since $t(^{o}C) = t(^{o}F)$ substitute $t(^{o}C)$ for $t(^{o}F)$ and set the two equations equal.

$$[t(^{o}C) \times 1.8] + 32 = \frac{t(^{o}C) - 32}{1.8} \qquad (1.8)^2 t(^{o}C) - t(^{o}C) = -32 - 32(1.8) \qquad t(^{o}C) = \underline{-40^{o}C}$$

1-106 $5.34 \times 10^{10}\ \cancel{ng} \times \frac{10^{-9}\ \cancel{g}}{\cancel{ng}} \times \frac{1\ lb}{453.6\ \cancel{g}} = \underline{0.118\ lb}$

1-108 $1.00\ \cancel{kg} \times \frac{10^3\ \cancel{g}}{\cancel{kg}} \times \frac{1\ \cancel{tr\ lb}}{373\ \cancel{g}} \times \frac{12\ \cancel{oz}}{\cancel{tr\ lb}} \times \frac{\$320}{\cancel{oz}} = \underline{\$10{,}300}$

1-109 $\frac{247\ lb}{82.3\ doz} = \underline{3.00\ lb/doz} \qquad \frac{82.3\ doz}{247\ lb} = \underline{0.333\ doz/lb}$

1-110 $12.0\ \cancel{fur} \times \frac{1\ \cancel{mi}}{8\ \cancel{fur}} \times \frac{5280\ \cancel{ft}}{\cancel{mi}} \times \frac{12\ \cancel{in.}}{\cancel{ft}} \times \frac{1\ hand}{4\ \cancel{in.}} = \underline{2.38 \times 10^4\ hands}$

1-112 $0.500\ \cancel{lb} \times \frac{453.6\ \cancel{g}}{\cancel{lb}} \times \frac{10^3\ \cancel{mg}}{\cancel{g}} \times \frac{1\ \cancel{cig.}}{11.0\ \cancel{mg}} \times \frac{1\ pkg}{20\ \cancel{cig.}} = \underline{1030\ pkgs}$

$1030\ \cancel{pkgs} \times \frac{1\ \cancel{day}}{3\ \cancel{pkg}} \times \frac{1\ yr}{365\ \cancel{day}} = \underline{1.41\ years}$

1-113 $1\ in. = 2.54\ cm \qquad 1\ in.^3 = 16.4\ cm^3 = 16.4\ mL$

$306\ \cancel{in.^3} \times \frac{16.4\ \cancel{mL}}{\cancel{in.^3}} \times \frac{1\ L}{10^3\ \cancel{mL}} = \underline{5.02\ L}$

1-115 $5.4 \times 10^7\ ^{o}F \qquad 3.0 \times 10^{7}\ ^{o}C + 273 = \underline{3.0 \times 10^7\ K}$

1-117 $1\ ft \times 30.0\ ft \times 50.0\ ft = 1500\ \cancel{ft^3\ snow} \times \frac{0.100\ ft^3\ water}{\cancel{ft^3\ snow}} = 150\ ft^3\ water$

$150\ \cancel{ft^3} \times \frac{62.0\ lb}{\cancel{ft^3}} = \underline{9300\ lb} \qquad 9300\ \cancel{lb} \times \frac{1\ ton}{2000\ \cancel{lb}} = \underline{4.65\ ton}$

2

Matter, Changes, and Energy

Review Section A *The Types of Matter and Its Properties*

OUTLINE

2-1 Types of Matter - Elements and Compounds

1. Elements
2. Compounds

OBJECTIVES

Distinguish between the two types of pure matter - elements and compounds.

2-2 Physical Properties of Matter

1. The physical states of matter — *Describe the three states of matter and give examples of each.*

2. Physical changes and pure substances — *Give examples of the physical properties of a substance and the physical changes it undergoes.*

2-3 A Physical Property: Density

1. Calculation of density
2. Density as a conversion factor — *Calculate the density of a substance and use it as a conversion factor between mass and volume.*

3. Specific gravity

2-4 Chemical Properties of Matter

1. Chemical properties
2. Chemical changes — *Give examples of the chemical properties of a substance and the chemical changes it undergoes.*

3. Conservation of matter — *Discuss the significance of the law of conservation of matter*

SUMMARY OF SECTIONS 2-1 THROUGH 2-4

Questions: *What are simplest forms of matter? How do we distinguish among these and other more complex types of matter? What are some of the ways we characterize a portion of matter? What do we mean when we say that "matter is conserved?"*

All **matter** in the universe is made up of less than 90 **elements. Compounds** are combinations of the various elements. Elements and compounds can be distinguished by their properties. For example, all matter exists in one of three physical states: **solid, liquid,** or **gas**. The physical state is an example of a **physical property**, and the change of a substance from one physical state (**phase**) to another is an example of a **physical change**. The **melting point** of a solid (the **freezing point** of a liquid) and the **boiling point** of a liquid (the **condensation point** of a gas) are all distinct physical properties. All **pure substances** (elements and compounds) have distinguishing physical properties.

Density is an important physical property of homogeneous matter that relates mass to a specified volume. For solids and liquids density is usually expressed in grams per milliliter (g/mL). For gases, density is expressed in grams per liter (g/L). Density can be used as a conversion factor which converts mass to an equivalent volume or vice versa. An example follows.

Example A-1 Density as a Conversion Factor

What is the volume in mL of 0.23 lb of table salt? (density = 2.16 g/mL)

PROCEDURE

A two-step conversion is necessary as shown by the following unit map.

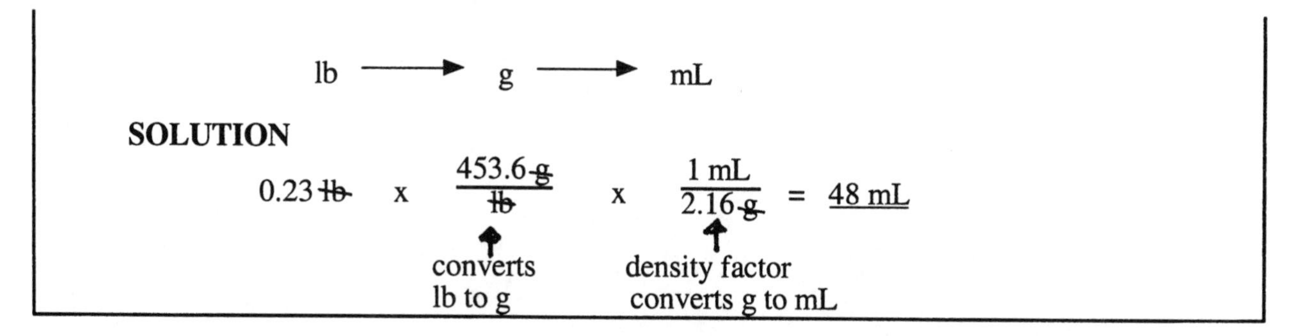

The density of a liquid can also be expressed as a **specific gravity.** Specific gravity is a comparison of the density of a substance to that of water at the same temperature. Since the density of water is essentially "one" g/mL, specific gravity has the same numerical value as density but is expressed without units.

All substances have **chemical properties** as well as physical properties. Chemical properties relate to the profound changes (called **chemical changes**) that the substance undergoes. Pure substances also have distinguishing chemical properties. In any chemical process the **law of conservation of mass** can be demonstrated.

NEW TERMS

Boiling point
Chemical change
Chemical property
Compound
Condensation point
Density
Element
Freezing point
Gas
Liquid
Melting point
Physical change
Physical property
Physical state
Property
Pure substance
Specific gravity
Solid
Substance

SELF-TEST

A-1 Multiple Choice

___ 1. Which of the following is an example of an element?

(a) water
(b) carbon
(c) concrete
(d) carbon dioxide

___ 2. Which of the following distinguishes an element from a compound?

(a) An element can be broken down into compounds.
(b) A compound has variable properties.
(c) An element cannot be broken down by chemical means.
(d) An element has definite properties.

___ 3. Which of the following describes the liquid state?

(a) It fills the entire container.
(b) It has definite dimensions.
(c) It fills the lower part of the container.
(d) It has a definite shape and volume.

___ 4. Which of the following is not a physical property of sodium?

(a) It is a solid.
(b) It is soft.
(c) It is shiny.
(d) It forms a compound in the presence of chlorine.
(e) It melts at 98 ^{o}C.

___ 5. Which of the following is not a chemical property of oxygen?

(a) It is a colorless gas.
(b) It supports combustion (burning).
(c) It reacts with almost all other elements.
(d) It is used in animal metabolism.
(e) It reacts with iron to form rust.

___ 6. Which of the following is an example of a physical change?

(a) burning
(b) decaying
(c) boiling
(d) tarnishing
(e) decomposing

___ 7. Which of the following is a unit of density?

(a) g/m^2 (b) lb/ft (c) qt/lb (d) lb/gal (e) m/L

___ 8. Solid X floats in liquid A but sinks in liquid B. The order of densities of the three substances is therefore:

(a) A > B > X
(b) A > X > B
(c) B > X > A
(d) X > B > A
(e) X > A > B

___ 9. A 55-g quantity of a substance has a volume of 10.0 mL. Its specific gravity is:

(a) 5.5 g/mL
(b) 0.55 g/mL
(c) 0.18
(d) 1.8
(e) 5.5

A-2 Matching

_____	density	(a) Cannot be decomposed into simpler substances
_____	solid	(b) Same as freezing point.
_____	compound	(c) Has a definite volume but not a definite shape
_____	melting point	(d) Observation does not require a change
_____	pure substance	(e) Temperature at which condensation occurs
_____	physical property	(f) The ratio of the volume to the mass

(g) Has definite dimensions

(h) Made up of two or more elements

(i) Has variable properties

(j) Requires the observation of a change

(k) The ratio of the mass to the volume

(l) Element or compound

A-3 Problems

1. The volume of a 65.5-g sample of a pure substance is 70.8 mL. What is its density?

2. A sample of metal has rectangular sides which measure 20.0 cm, 15.0 cm, and 35.0 cm. The sample has a mass of 84.6 kg. What is its density?

3. If the density of a substance is 2.70 g/mL, what is the mass of 1.12 L of the substance?

4. If the density of a substance is 0.954 g/mL, what is the volume of a 1.00-kg sample?

Review Section B *The Composition of Mixtures*

OUTLINE

2-5 Mixtures of Elements and Compounds

1. Heterogeneous mixtures
   a. Phases
   b. Filtration
2. Homogeneous mixtures
   a. Solutions
   b. Distillation
   c. Alloys
3. Solutions versus pure liquids
   a. Properties
   b. Density and the hydrometer

OBJECTIVES

Classify a sample of matter as either heterogeneous or homogeneous.

Describe how a solid can be separated from a liquid in a laboratory.

Describe how components of a solution can be separated in a laboratory.

Calculate the percent of a component from the masses of elements present or use percent as a conversion factor.

Distinguish between the properties of solutions and pure substances.

SUMMARY OF SECTION 2-5

Questions: *What is the nature of a specific mixture? How can we tell if a substance is pure or a mixture? How can some mixtures be separated?*

Much of what we see around us in nature is a mixture of pure substances, at least to some extent. Intimate mixtures in which the pure substances are mixed at the basic particle level are known as **homogeneous mixtures** or **solutions**. Examples of homogeneous mixtures are the metal **alloys** that have so many uses. Alloys are usually composed of a mixture of metals. Homogeneous matter exists in one phase. Pure substances themselves are examples of homogeneous matter (when they occur in one phase). **Heterogeneous matter**, on the other hand, exists in two or more phases with identifiable boundaries between phases. The classification of matter from heterogeneous mixtures down to the most basic substances (the elements) can be illustrated as follows.

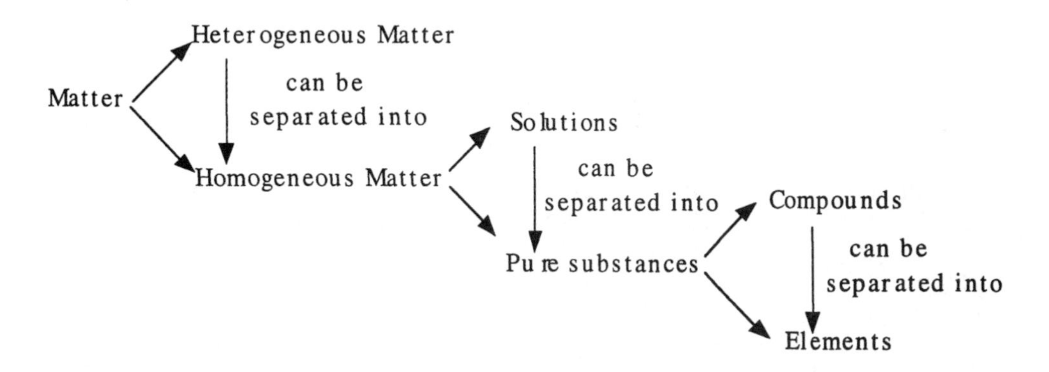

In the laboratory, heterogeneous mixtures of a solid in a liquid can be separated by **filtration**. Solutions may be separated into components by **distillation**. It is not always possible to distinguish a mixture, especially a solution, from a pure substance by casual observation. Often, one must closely examine properties such as melting point and boiling point to tell the difference. The densities of solutions (measured with a **hydrometer**) also vary according to the proportions of the components. Pure substances have distinct and unchanging properties. A mixture, on the other hand, melts and boils at different temperatures than would any of its pure components.

NEW TERMS

Alloy
Distillation
Filtration
Heterogeneous
Homogeneous
Hydrometer
Phase
Solution

SELF-TEST

B-1 Multiple Choice

___ 1. Which of the following is an example of a homogeneous mixture?

(a) gasohol
(b) oxygen gas
(c) table salt
(d) oil and vinegar

___ 2. Which of the following is an example of heterogeneous matter?

(a) air
(b) aluminum
(c) foamy shaving cream
(d) sugar

___ 3. Which of the following is not true about a solution?

(a) It is homogeneous.
(b) It has definite properties.
(c) It exists in one phase.
(d) It can often be separated into its components by distillation.

___ 4. Which of the following is an alloy?

(a) battery acid (b) bronze
(c) chromium (d) sodium sulfide

___ 5. A certain alloy of aluminum contains 3.5% magnesium. What mass of the alloy can be formed from 2.34 kg of magnesium?

(a) 82 kg (b) 8.2 kg
(c) 2600 kg (d) 67 kg

B-2 Matching

_____ one solid phase, definite melting point, can be decomposed further

_____ a liquid and a solid phase, definite melting point, can be decomposed further

_____ one liquid phase, melting point range, can be separated into other substances

_____ two solid phases, melting point range

_____ one liquid phase, sharp boiling point, cannot be decomposed further

(a) ice water (heterogeneous - pure)

(b) sodium chloride (homogeneous - pure)

(c) bromine (homogeneous - pure)

(d) sugar mixed with salt (heterogeneous - mixture)

(e) sugar water solution (homogeneous - mixture)

Review Section C *Energy and Matter*

OUTLINE	OBJECTIVES
2-6 Energy Changes in Chemical Reactions	
1. Energy	
2. Forms of energy	*Name and describe the various forms of energy.*
3. Conservation of energy	*Discuss the significance of the law of conservation of energy*
4. Energy and changes	*Classify a chemical or physical change as being either exothermic or endothermic.*
5. Types of energy	*Distinguish between kinetic and potential energy.*

2-7 Temperature Change and Specific Heat

1. Units of heat energy — *Define the calorie and the joule.*

2. Specific heat — *Calculate the specific heat of a substance given appropriate experimental data and use it as a conversion factor among heat, temperature change, and mass.*

SUMMARY OF SECTIONS 2-6 AND 2-7

***Questions:** What does energy have to do with chemical reactions? How do we describe and quantify energy? How is energy conserved? How does heat relate to temperature change?*

Chemistry is also concerned with the energy that accompanies chemical changes. **Energy** (the ability to do work), however, comes in various forms, each of which can be converted into the others. In these changes energy is neither created nor destroyed, which is known as the law of **conservation of energy.** When chemical energy is transformed into heat energy in a chemical reaction the reaction is said to be **exothermic**. When heat energy is changed into chemical energy, the reaction is said to be **endothermic**. There are also two types of energy: **potential energy** (energy of position) and **kinetic energy** (energy of motion).

The **specific heat** of a homogeneous substance relates how much heat (in **calories** or **joules**) it takes to change one gram of the substance by one degree Celsius. Water is the standard which has a specific heat of 1 cal/g $\cdot$ ^{o}C.

Example C-1 Specific Heat and Temperature Change

How many calories does it take to change 10.0 g of water a total of 5.2 Celsius degrees?

PROCEDURE

$$\text{Sp ht} = \frac{\text{cal}}{\text{g} \cdot {}^{o}\text{C}}$$

Solving algebraically for *cal* we obtain

$$\text{cal} = \text{Sp ht} \times \text{g} \times {}^{o}\text{C}(\Delta T)$$

SOLUTION

$$\frac{1.00 \text{ cal}}{\cancel{\text{g}} \cdot \cancel{{}^{o}\text{C}}} \times 10.0 \cancel{\text{g}} \times 5.2 \cancel{{}^{o}\text{C}} = \underline{52 \text{ cal}}$$

In the final section, we observed two very important laws relating to chemical reactions. Mass and energy are neither created nor destroyed in a chemical reaction. Much of the quantitative relationships that we will study later are based on the law of **conservation of mass**. However, according to Einstein's law there is a relationship between mass and energy. Fortunately though, the interconversion that occurs in chemical reactions is so small that it can be ignored entirely.

NEW TERMS

Calorie
Chemical energy
Conservation of energy
Conservation of mass
Endothermic
Energy
Exothermic
Heat energy
Joule
Kinetic energy
Potential energy
Specific Heat

SELF-TEST

C-1 Multiple Choice

___ 1. Which of the following is an example of potential energy?

(a) a running halfback
(b) a bowling ball at the top of the stairs
(c) water flowing over a falls
(d) a moving train

___ 2. When wood burns, which two forms of energy are released?

(a) heat, chemical
(b) electrical, heat
(c) heat, light
(d) light, chemical
(e) light, mechanical

___ 3. The ____________ energy of gasoline is converted in an automobile engine into ____________ energy of the moving tires.

(a) heat, electrical
(b) chemical, mechanical
(c) light, heat
(d) chemical, light
(e) light, mechanical

___ 4. An exothermic chemical reaction results when:

(a) electrical energy is changed into heat energy.
(b) heat energy is changed into chemical energy.
(c) heat energy is changed into mechanical energy.
(d) chemical energy is changed into heat energy.

___ 5. Which of the following are possible units of specific heat?

(a) $J/^{o}C$
(b) $g/cal \cdot {}^{o}C$
(c) $J/g \cdot K$
(d) $K/g \cdot cal$

___ 6. How many grams of water can be heated by 50 calories a total of 2.0 Celsius degrees?

(a) 25 g (b) 50 g (c) 100 g (d) 200 g (e) 75 g

C-2 Problems

1. It takes 36.0 calories to heat a 10.0-g sample of a metal from $0.0^{\circ}C$ to $31.0^{\circ}C$. What is the specific heat of the metal?

2. The specific heat of a substance is $0.523\ J/g \cdot {}^{\circ}C$. How many joules are liberated when 65.0 g of the substance cools a total of 12.0 Celsius degrees?

Review Section D

CHAPTER SUMMARY SELF-TEST

D-1 Matching

_____	Potential energy	(a) Has a definite volume and a definite shape
_____	A form of energy	(b) A chemical reaction that gives off heat
_____	Liquid	(c) Electricity
_____	Element	(d) A pure substance
_____	Property	(e) Contains two or more phases
_____	Exothermic reaction	(f) A unique, observable characteristic or trait
_____	Phase	(g) Energy of position
_____	Distillation	(h) A chemical reaction that absorbs heat
_____	Solution	(i) A laboratory process used to separate the components of a solution

(j) Has a definite volume and can be poured

(k) A homogeneous mixture

(l) A laboratory procedure used to separate a liquid from a solid phase

(m) Energy of motion

(n) Any physical state with uniform properties and identifiable boundaries

D-2 Problems

1. The density of a metal is 2.73 g/mL. What is the volume in liters occupied by 225 kg of the metal?

2. The specific heat of a substance is 0.275 cal/g · ^{o}C. What mass in grams is heated a total of 10.0 Celsius degrees by 50.0 calories?

Answers to Self-Tests

A-1 Multiple Choice

1. **b** The others are all compounds. Carbon is listed among the elements in the front cover.
2. **c** An element is the most basic substance that cannot be decomposed by chemical means.
3. **c** It fills the lower part of the container.
4. **d** It forms a compound in the presence of chlorine.
5. **a** It is a colorless gas. Color is a physical property.
6. **c** Boiling
7. **d** lb/gal
8. **b** A>X>B
9. **c** 5.5

A-2 Matching

k Density is the ratio of the mass to the volume.

g A solid has definite dimensions.

h A compound is made up of two or more elements.

b The melting point is the same as the freezing point.

l A pure substance is an element or a compound.

d The observation of a physical property does not require a change.

A-3 Problems

1. 65.6 g ~ 70.8 mL

   ? g ~ 1.00 mL $\qquad \dfrac{65.5\ g}{70.8\ mL} = \underline{0.925\ g/mL}$

2. Volume = 20.0 cm x 15.0 cm x 35.0 cm = 10,500 cm^3 = 10,500 mL

   Mass = 84.6 kg = 84,600 g

   $$\text{Density} = \frac{84600\ g}{10500\ mL} = \underline{8.06\ g/mL}$$

3. **L → mL → g**

   $$1.12\,\cancel{L} \times \frac{10^3\,\cancel{mL}}{\cancel{L}} \times \frac{2.70\ g}{\cancel{mL}} = \underline{3.02 \times 10^3\ g}$$

4. **kg → g → mL**

   $$1.00\,\cancel{kg} \times \frac{1\,\cancel{g}}{10^{-3}\,\cancel{kg}} \times \frac{1\ mL}{0.954\,\cancel{g}} = \underline{1.05 \times 10^3\ mL}$$

B-1 Multiple Choice

1. **a** Gasohol. Oxygen is a pure substance (element). Table salt is a pure substance (compound). Oil and vinegar form a heterogeneous mixture.

2. **c** Foamy shaving cream is a mixture of air (gas) and liquid. Air is a homogeneous mixture and aluminum (element) and sugar (compound) are pure substances.

3. **b** The properties of a solution depend upon the proportions of components.

4. **b** Bronze is an example of an alloy.

5. **d** $2.34\ \cancel{kg\ magnesium} \times \dfrac{100\ kg\ alloy}{3.5\ \cancel{kg\ magnesium}} = 67\ kg\ alloy$

B-2 Matching

b Sodium chloride is one solid phase that can be decomposed into elements. It has a definite melting point.

a Ice water is composed of a solid and liquid phase (heterogeneous) but both phases are the same pure compounds.

e A sugar water solution is one homogeneous liquid phase but is a mixture that can be separated into other substances.

f A mixture of sugar and salt is a heterogeneous mixture with two identifiable solid phases.

c Bromine is an element that is composed of a single liquid phase with a sharp boiling point. It cannot be separated into more basic components.

C-1 Multiple Choice

1. **b** A bowling ball at the top of the stairs.

2. **c** Heat and light.

3. **b** Chemical, mechanical. (The compounds of gasoline store chemical energy which is transformed into mechanical energy to move the gears, drive shaft, and tires.)

4. **d** Chemical energy is changed into heat energy.

5. **c** J/g · K

6. **a** $$\frac{50\,\cancel{\text{cal}} \times \frac{1\,\text{g} \cdot \cancel{^{\circ}\text{C}}}{\cancel{\text{cal}}}}{2.0\,\cancel{^{\circ}\text{C}}} = \underline{25\text{ g}}$$

C-2 Problems

1. $$\frac{36.0\text{ cal}}{10.0\text{ g} \times 31.0\,^{\circ}\text{C}} = \underline{0.116\text{ cal/g} \cdot {^{\circ}\text{C}}}$$

2. $$\frac{0.523\text{ J}}{\cancel{\text{g}} \cdot \cancel{^{\circ}\text{C}}} \times 65.0\,\cancel{\text{g}} \times 12.0\,\cancel{^{\circ}\text{C}} = \underline{408\text{ J}}$$

D-1 Matching

g	Potential energy is energy of position.
c	Electricity is a form of energy.
j	A liquid has a definite volume and can be poured.
d	An element is a pure substance.
f	A property is a unique, observable characteristic or trait.
b	An exothermic reaction is a reaction that gives off heat.
n	A phase is a physical state with uniform properties and identifiable boundaries.
i	Distillation is a laboratory process used to separate the components of a solution.
k	A solution is a homogeneous mixture.

D-2 Problems

1. $$225\,\cancel{\text{kg}} \times \frac{1\,\cancel{\text{g}}}{10^{-3}\,\cancel{\text{kg}}} \times \frac{1\,\cancel{\text{mL}}}{2.73\,\cancel{\text{g}}} \times \frac{10^{-3}\,\text{L}}{\cancel{\text{mL}}} = \underline{82.4\text{ L}}$$

2. $$\text{mass} = \frac{\text{cal}}{\text{sp ht} \times \Delta T} = \frac{50.0\,\cancel{\text{cal}}}{0.275\,\frac{\cancel{\text{cal}}}{\text{g}\cdot\cancel{^\circ\text{C}}} \times 10.0\,\cancel{^\circ\text{C}}} = \underline{18.2\text{ g}}$$

Solutions to Black Text Problems

2-19 $\frac{208\text{ g}}{80.0\text{ mL}} = \underline{2.60\text{ g/mL}}$

2-21 $\frac{1064\text{ g}}{657\text{ mL}} = \underline{1.62\text{ g/mL}}$ (carbon tetrachloride)

2-24 $671\,\cancel{\text{mL}} \times \frac{2.16\text{ g}}{\cancel{\text{mL}}} = \underline{1450\text{ g}}$

2-25 $1.00\,\cancel{\text{L}} \times \frac{1\,\cancel{\text{mL}}}{10^{-3}\,\cancel{\text{L}}} \times \frac{0.67\text{ g}}{\cancel{\text{mL}}} = \underline{670\text{ g}}$

2-26 $1.00\,\cancel{\text{gal}} \times \frac{3.785\,\cancel{\text{L}}}{\cancel{\text{gal}}} \times \frac{\cancel{\text{mL}}}{10^{-3}\,\cancel{\text{L}}} \times \frac{0.67\,\cancel{\text{g}}}{\cancel{\text{mL}}} \times \frac{1\text{ lb}}{453.6\,\cancel{\text{g}}} = \underline{5.6\text{ lb}}$

2-28 $1.00\ \cancel{kg} \times \frac{10^3\ \cancel{g}}{\cancel{kg}} \times \frac{1.00\ mL}{1.60\ \cancel{g}} = \underline{625\ mL}$

2-29 Vol. = 92.45 - 14.00 = 78.45 mL

density = 136.5 g/78.45 mL = $\underline{1.74\ g/mL}$ (magnesium)

2-30 $1.05\ \cancel{lb} \times \frac{453.6\ g}{\cancel{lb}} = 476\ g$ $\quad 476\ g/10^3\ mL = \underline{0.476\ g/mL}$ Yes, it floats.

2-31 155 g/163 mL = 0.951 g/mL

$4.56\ \cancel{kg} \times \frac{10^3\ \cancel{g}}{\cancel{kg}} \times \frac{1.00\ mL}{0.951\ \cancel{g}} = \underline{4790\ mL}$

Pumice floats in water but sinks in alcohol.

2-34 $3.00\ cm \times 8.50\ cm \times 6.00\ cm = 153\ cm^3 = 153\ mL$

$153\ \cancel{mL} \times \frac{13.6\ g}{\cancel{mL}} = \underline{2080\ g}$

2-35 mass of liquid = 143.5 - 32.5 = 111.0 g $\quad$ 111.0 g/125 mL = $\underline{0.888\ g/mL}$

2-37 Water: $1.00\ \cancel{L} \times \frac{1\ \cancel{mL}}{10^{-3}\ \cancel{L}} \times \frac{1.00\ g}{\cancel{mL}} = 1000\ g$

Gasoline: $1.00\ \cancel{L} \times \frac{1\ \cancel{mL}}{10^{-3}\ \cancel{L}} \times \frac{0.67\ g}{\cancel{mL}} = 670\ g$

One liter of water has a greater mass.

2-38 $5.65\ \cancel{oz} \times \frac{1\ \cancel{lb}}{16\ \cancel{oz}} \times \frac{453.6\ g}{\cancel{lb}} = 160\ g$ $\quad 33.3 - 25.0 = 8.3\ mL$

160 g/8.3 mL = $\underline{19\ g/mL}$ It'a gold.

2-41 One needs a conversion factor between mL (cm^3) and ft^3.

$$\left(\frac{2.54\ cm}{in.}\right)^3 = \frac{16.4\ cm^3}{in.^3} = \frac{16.4\ mL}{in.^3} \qquad \left(\frac{12\ in.}{ft}\right)^3 = \frac{1728\ in.^3}{ft^3}$$

$$\frac{1.00\ \cancel{g}}{\cancel{mL}} \times \frac{1\ lb}{453.6\ \cancel{g}} \times \frac{16.4\ \cancel{mL}}{\cancel{in.^3}} \times \frac{1728\ \cancel{in.^3}}{ft^3} = \underline{62.5\ lb/ft^3}$$

2-42 $4.5\ \cancel{mL} \times \frac{2.0 \times 10^7\ \cancel{g}}{\cancel{mL}} \times \frac{1\ lb}{453.6\ \cancel{g}} = \underline{2.0 \times 10^5\ lb\ (100\ tons)}$

2-57 Mass of mixture = 85 + 942 = 1027 g

$\frac{85\ \cancel{g}}{1027\ \cancel{g}} \times 100\% = \underline{8.3\%\ tin}$

2-59 $255\ \cancel{kg\ \text{"nickels"}} \times \frac{25\ kg\ nickel}{100\ \cancel{kg\ \text{"nickels"}}} = \underline{64\ kg\ nickel}$

2-61 $122\ \cancel{\text{lb iron}} \times \dfrac{100\text{ lb duriorn}}{86\ \cancel{\text{lb iron}}} = \underline{140\text{ lb duriorn}}$

2-70 $\dfrac{73.2\text{ J}}{10.0\text{ g} \cdot 8.58\ ^{o}\text{C}} = \underline{0.853\text{ J/g} \cdot {}^{o}\text{C}}$

2-71 $\dfrac{56.6\text{ cal}}{365\text{ g} \cdot 5.0\ ^{o}\text{C}} = \underline{0.031\text{ cal/g} \cdot {}^{o}\text{C (gold)}}$

2-73 $^{o}\text{C} = \dfrac{\text{cal}}{\text{sp. heat x g}} = \dfrac{150\ \cancel{\text{cal}}}{0.092\dfrac{\cancel{\text{cal}}}{\cancel{\text{g}} \cdot {}^{o}\text{C}} \times 50.0\ \cancel{\text{g}}} = 33\ ^{o}\text{C rise}$

$t(^{o}\text{C}) = 25 + 33 = 58\ ^{o}\text{C}$

This compares to a 3.0 ^{o}C rise in temperature for 50.0 g of water.

2-75 heat (J) = sp heat x g x ^{o}C $\qquad \dfrac{0.895\text{ J}}{\cancel{\text{g}} \cdot \cancel{^{o}\text{C}}} \times 43.5\ \cancel{\text{g}} \times 13\ \cancel{^{o}\text{C}} = \underline{506\text{ J}}$

2-76 58 -25 = 33 ^{o}C rise in temperature

$\text{g} = \dfrac{\text{cal}}{\text{sp heat} \cdot {}^{o}\text{C}} = \dfrac{16.0\ \cancel{\text{cal}}}{0.106\dfrac{\cancel{\text{cal}}}{\text{g} \cdot \cancel{^{o}\text{C}}} \cdot 33\ \cancel{^{o}\text{C}}} = \underline{4.6\text{ g}}$

2-79 $^{o}\text{C} = \dfrac{\text{J}}{\text{g x sp heat}} \qquad \dfrac{50.0\ \cancel{\text{J}}}{12.0\ \cancel{\text{g}} \times 0.444\dfrac{\cancel{\text{J}}}{\cancel{\text{g}} \cdot {}^{o}\text{C}}} = 9.38\ ^{o}\text{C rise (iron)}$

$\dfrac{50.0\ \cancel{\text{J}}}{12.0\ \cancel{\text{g}} \times 0.13\dfrac{\cancel{\text{J}}}{\cancel{\text{g}} \cdot {}^{o}\text{C}}} = 32\ ^{o}\text{C rise (gold)}$

$\dfrac{50.0\ \cancel{\text{J}}}{12.0\ \cancel{\text{g}} \times 4.18\dfrac{\cancel{\text{J}}}{\cancel{\text{g}} \cdot {}^{o}\text{C}}} = 0.997\ ^{o}\text{C rise (water)}$

2-81 $^{o}\text{C} = \dfrac{\text{cal}}{\text{g x sp heat}} \qquad \dfrac{1000\ \cancel{\text{cal}}}{50.0\ \cancel{\text{g}} \times 1.00\dfrac{\cancel{\text{cal}}}{\cancel{\text{g}} \cdot {}^{o}\text{C}}} = 20\ ^{o}\text{C rise}$

$t(^{o}\text{C}) = 20 + 25 = \underline{45}\ ^{o}\text{C}$

2-83 $\dfrac{0.895\text{ J}}{\cancel{\text{g}} \cdot \cancel{^{o}\text{C}}} \times 50.0\ \cancel{\text{g}} \times 65\ \cancel{^{o}\text{C}} = \underline{2910\text{ J}}$

2-87 Heat gained by water $100\ \cancel{\text{mL}} \times \dfrac{1.00\text{ g}}{\cancel{\text{mL}}} \times (28.7 - 25.0\ \cancel{^{o}\text{C}}) \times \dfrac{4.184\text{ J}}{\cancel{\text{g}} \cdot \cancel{^{o}\text{C}}} = 1550\text{ J}$

Heat gained by lead $\cancel{\text{g lead}} \times (42.8 - 28.7\ \cancel{^{o}\text{C}}) \times \dfrac{0.128\text{ J}}{\cancel{\text{g}} \cdot \cancel{^{o}\text{C}}} = 1550\text{ J}$

g lead = $\underline{860\text{ g}}$

2-88 Heat lost by metal = heat gained by water

$$100.0\ \text{g} \times 68.7\ ^{\circ}\text{C} \times \text{sp ht.} = 100.0\ \cancel{\text{g}} \times 6.3\ \cancel{^{\circ}\text{C}} \times \frac{1.00\ \text{cal}}{\cancel{\text{g}} \cdot \cancel{^{\circ}\text{C}}}$$

specific heat = $\underline{0.092\ \text{cal/g}} \cdot {}^{\circ}\text{C}$ The metal is <u>copper</u>.

2-89 Density of A = 0.86 g/mL; density of B = 0.89 g/mL
Liquid A floats on liquid B.

2-91

$$100\ \cancel{\text{mL}} \times \frac{1.06\ \text{g}}{\cancel{\text{mL}}} = 106\ \text{g solution}$$

$$106\ \cancel{\text{g solution}} \times \frac{14\ \text{g sugar}}{100\ \cancel{\text{g solution}}} = \underline{15\ \text{g of sugar}}$$

2-92 volume of metal = 24.96 - 18.22 = 6.74 mL

62.485 g/6.74 mL = <u>9.27 g/mL</u>

2-94

$$305\ \cancel{\text{mL}} \times \frac{1.00\ \text{g}}{\cancel{\text{mL}}} = 305\ \text{g water}$$

$$\frac{10.0\ \text{g}}{(305 + 10.0)\text{g}} \times 100\% = \underline{3.17\%\ \text{salt}}$$

2-96

$$50.0\ \cancel{\text{mL gold}} \times \frac{19.3\ \text{g}}{\cancel{\text{mL gold}}} = 965\ \text{g gold}$$

$$50.0\ \cancel{\text{mL alum.}} \times \frac{2.70\ \text{g}}{\cancel{\text{mL alum.}}} = 135\ \text{g alum.} \qquad \frac{965\ \text{g gold}}{(965 + 135)\text{g alloy}} \times 100\% = \underline{87.7\%\ \text{gold}}$$

2-98

$$\frac{215\ \text{J}}{25.0\ \text{g} \times 66^{\circ}\text{C}} = 0.13\ \frac{\text{J}}{\text{g} \cdot {}^{\circ}\text{C}}\ \text{(gold)} \qquad 25.0\ \cancel{\text{g gold}} \times \frac{1\ \text{mL}}{19.3\ \cancel{\text{g gold}}} = \underline{1.30\ \text{mL}}$$

2-101

$$250 \times 10^{3}\ \cancel{\text{g}} \times (125 - 95\ \cancel{^{\circ}\text{C}}) \times \frac{0.444\ \text{J}}{\cancel{\text{g}} \cdot \cancel{^{\circ}\text{C}}} = 3.33 \times 10^{6}\ \text{J needed by engine}$$

$$\cancel{\text{g}}\ \text{water} \times (95.0 - 25.0\ \cancel{^{\circ}\text{C}}) \times \frac{4.184\ \text{J}}{\cancel{\text{g}} \cdot \cancel{^{\circ}\text{C}}} = 3.33 \times 10^{6}\ \text{J}$$

$$\text{g water} = 1.14 \times 10^{4}\ \cancel{\text{g}} \times \frac{1.00\ \cancel{\text{mL}}}{\cancel{\text{g}}} \times \frac{1\ \text{L}}{10^{3}\ \cancel{\text{mL}}} = \underline{11.4\ \text{L}}$$

3

Elements, Compounds, and Their Composition

Review Section A *Elements and Compounds and Their Composition*

OUTLINE

3-1 Names and Symbols of the Elements
1. The origin of some names
2. The symbol of an element

OBJECTIVES

List the names and symbols of some common elements.

3-2 Composition of an Element and Atomic Theory

1. The atomic theory	*List the conclusions of Dalton's atomic theory.*
2. The composition of an atom	*Describe the fundamental particles of an element.*

3-3 Elements and Compounds Composed of Molecules, and Their Formulas

1. The covalent bond	*Describe the function of the covalent bond in a molecule.*
2. Molecular compounds	*Describe the fundamental particle of a molecular compound.*
3. Formulas	*List the elements and the number of atoms of each element in a compound from the formula.*
4. Structural formulas	

3-4 The Electrical Nature of Matter: Compounds Composed of Ions

1. Electrostatic charges	
2. Ions	*Define the terms "ion," "cation," and "anion."*
3. Formula units	*Write the correct formula for a compound composed of specified ions.*

SUMMARY OF SECTIONS 3-1 THROUGH 3-4

Questions: *What are the origins of some of the elements' names? How did the atomic concept come to be accepted? How does a compound differ from an element? What distinguishes one compound from another? Why are some compounds hard with high melting points? Why are these different from the compounds that are gases, liquids, or soft solids?*

Of the 114 known elements, less than 90 occur naturally on Earth. Only ten, however, account for over 99% of the crust of the earth. The names of the elements have a variety of sources, but all are designated by a **symbol**. Symbols may have one, two, or, in a few cases, three letters.

Elements are composed of basic particles called atoms. An **atom** is the smallest particle of an element that has the characteristics of that element and which can enter into chemical reactions. Even before powerful microscopes could create images of atoms, scientists had presumed that they existed. In fact, since the early 1800's, the atomic theory of John Dalton has been the accepted way to explain the nature of matter.

We now look at how atoms are found in the two pure forms of matter, elements and compounds. Elements are composed of like atoms that occur as either individual atoms as found in a noble gas or as **molecules**. The atoms of molecules are intimately bound to each other by a force called a **covalent bond**. **Molecular compounds** are composed of molecules that contain atoms of different elements. The **formula** of a compound expresses the number of atoms of each element comprising the particular molecule. **Structural formulas** show more detail such as the sequence of the bonded atoms. The formula of ethyl alcohol is as follows.

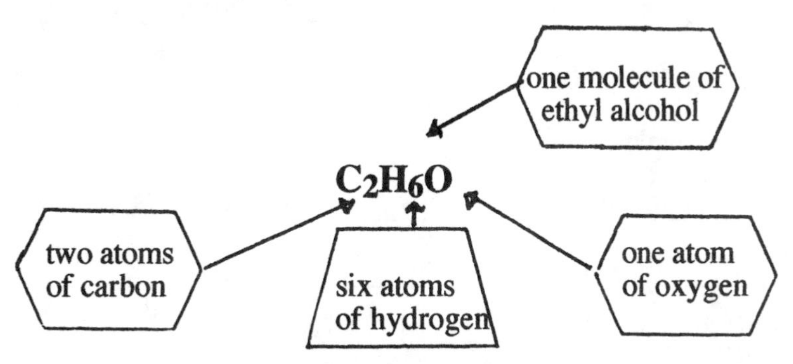

Another important category of compounds exists, known as **ionic compounds**. They are composed of **ions** rather than discrete molecules. Ions are atoms or groups of atoms that have a net electrical charge. The charge on an atom (or group of atoms forming a **polyatomic ion**), is represented in the upper right-hand corner of the symbol or symbols (e.g., S^{2-}, Ba^{2+}, $SO_3{}^{2-}$).

The ions in ionic compounds are held together by attractive **electrostatic forces**. Electrostatic forces include attractions between unlike charges and repulsions between like charges. The formula of an ionic compound represents one **formula unit** which expresses the ratio of **cations** and **anions** present in the compound. For example, the formula of potash indicates the following.

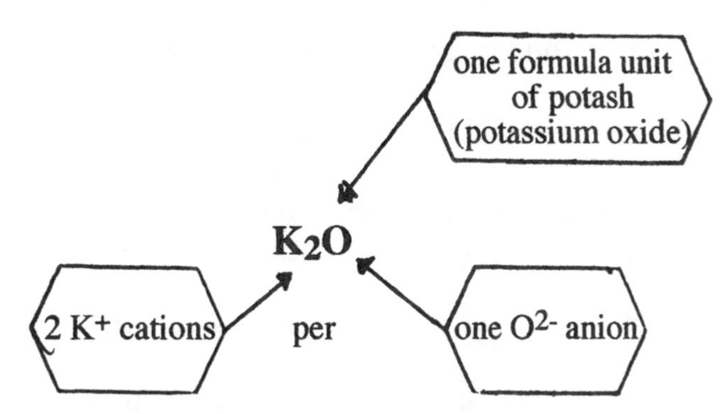

Eventually, we will discuss why ionic compounds form in some cases and why molecular compounds form in others.

NEW TERMS

Anion	Ion
Atom	Ionic compound
Cation	Molecular compound
Covalent bond	Molecule
Electrostatic force	Polyatomic ion
Formula	Structural formula
Formula unit	Symbol

SELF-TEST

A-1 Fill in the blanks.

Write the chemical symbols for:

________ nitrogen ________ magnesium ________ lead

________ aluminum ________ sodium ________ chlorine

Write the name of the element for:

___________ P ___________ Fe ___________ S

___________ Ca ___________ K ___________ B

A-2 Multiple Choice

___ 1. Which of the following formulas represents molecules of an element?

(a) Ni (b) N_2 (c) H_2O (d) Ne (e) NaCl

___ 2. Which of the following formulas represents molecules of a compound?

(a) Mn (b) P_4 (c) PF_3 (d) Br_2 (e) K

___ 3. Which of the following is a polyatomic anion?

(a) NH_4^+ (b) SO_3 (c) F^- (d) SO_3^{2-}

___ 4. An ionic compound contains one Ni^{2+} ion and two ClO_3^- ions. Which of the following is the proper representation of the formula?

(a) $NiCl_2O_6$ (b) $Ni_2Cl_2O_6$ (c) $Ni(ClO_3)_2$ (d) $(Ni)(ClO_3)_2$

___ 5. How many atoms of carbon are in one formula unit of $Fe_2(C_2O_4)_3$?

(a) 6 (b) 2 (c) 3 (d) 4 (e) 12

___ 6. A certain formula of an ionic compound contains two K^+ cations and one anion. The anion could be:

(a) Br^- (b) S^{2-} (c) N^{3-} (d) C^{4-}

___ 7. A certain formula of an ionic compound contains three CrO_4^{2-} ions and. two cations. The cation could be:

(a) Na^+ (b) Mg^{2+} (c) Al^{3+} (d) Mn^{4+}

Review Section B *The Atom and Its Composition*

OUTLINE

3-5 Composition of the Atom

1. The nuclear model
2. The particles in the atom

3-6 Atomic Number, Mass Number, and Atomic Weight

1. The nuclear particles
2. Isotopes
3. Isotopic mass and atomic mass
4. The charge on an ion

OBJECTIVES

Describe the nuclear atom, including the name, location, mass (in amu), and electrical charge of the three primary particles in the atom.

Distinguish between the atomic number and the mass number of an atom.

Give the number of protons, neutrons, and electrons in a specified neutral isotope.

Calculate the atomic mass of an element given the percent distribution and isotopic masses of its isotopes.

Determine the number of electrons and protons in a specified monatomic ion.

SUMMARY OF SECTIONS 3-5 AND 3-6

Questions: *Can an atom be broken down into even more fundamental particles? How do the atoms of one element differ from those of another? What determines the mass of an atom? How does an atom acquire a charge?*

At first it was assumed that atoms were hard spheres, but around the turn of the century experiments proved that the atom was considerably more complex. The experiments of Thomson and Rutherford, among others, led to the nuclear model of the atom. In this model, the atom is composed of three particles: **electrons, protons**, and **neutrons**. The neutrons and protons (called **nucleons**) exist in a small, dense core of the atom called the **nucleus**. Protons carry a positive charge and the nucleus contains nearly all of the mass of the atom. The comparatively small, negatively charged electrons exist in the vast, mostly empty space of the atom outside of the nucleus.

The atoms of a particular element are characterized by the **atomic number,** which is the number of protons (or the total positive charge) in the nucleus. The atoms of the same element may have different **mass numbers**, however; the mass number is the total number of protons and neutrons. Atoms of the same element with different mass numbers are known as **isotopes** of that element. Neutral atoms have the same number of electrons as protons. For example, consider the isotope ${}^{40}_{18}Ar$.

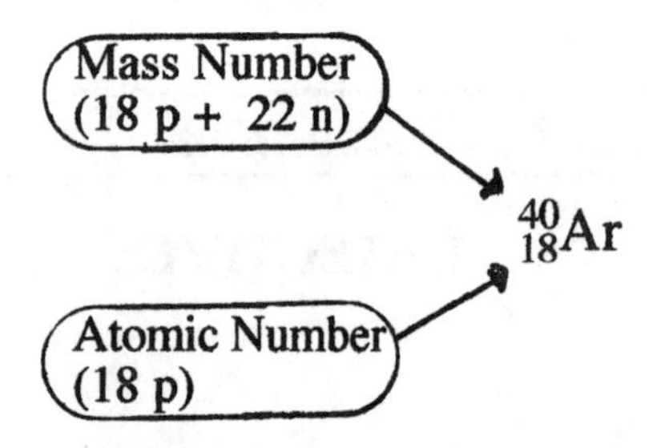

The **isotopic mass** of an isotope is obtained by comparing the mass of the isotope to the mass of $^{12}_{6}C$ the standard, which is defined as having a mass of exactly 12 **atomic mass units (amu)**. Elements generally occur in nature as mixtures of isotopes. The **atomic mass** of an element is the weighted average of the isotopic masses of the naturally occurring isotopes.

An electrical charge on an atom results when there are more or less electrons present than protons. For example, an ion of sulfur exists that has 18 electrons and 16 protons and thus has a charge of -2.

NEW TERMS

Atomic mass
Atomic mass unit (amu)
Atomic number
Electron
Isotope
Isotopic mass
Mass number
Neutron
Nucleon
Nucleus
Proton

SELF-TEST

B-1 Multiple Choice

___ 1. Which of the following describes an electron?

(a) mass = 1 amu, charge = -1
(b) mass = 0 amu, charge = 0
(c) mass = 0 amu, charge = -1
(d) mass = 1 amu, charge = +1

___ 2. The nucleus of an atom contains:

(a) protons and electrons
(b) most of the mass and all of the positive charge
(c) neutrons and electrons
(d) most of the positive charge and all of the mass

___ 3. How many neutrons are in $^{239}_{94}Pu$?

(a) 94 (b) 239 (c) 333 (d) 145

___ 4. One isotope of sulfur is $^{32}_{16}S$. A second isotope of sulfur contains atoms with

(a) 16 protons and 17 neutrons
(b) 32 nucleons with 17 protons
(c) 16 neutrons with 17 protons
(d) 32 protons and 16 neutrons

___ 5. An element has an atomic mass of about one-third of the standard. The element is

(a) Cl (b) O (c) He (d) Li

___ 6. How many electrons are in $^{128}_{52}Te^{2-}$?

(a) 52 (b) 54 (c) 50 (d) 76 (e) 78

___ 7. A certain ion has a +2 charge and contains 23 electrons. This is an ion of what element?

(a) Mn (b) V (c) Co (d) Sc

C-2 Matching

___ proton
___ neutron
___ nucleon
___ atomic number
___ mass number
___ neutral atom
___ atomic mass

(a) The number of neutrons in an atom

(b) Has a -1 charge and a mass of about 1 amu

(c) The number of nucleons in an atom

(d) The mass of an atom compared to ^{12}C

(e) Has a +1 charge and a mass of about 1 amu

(f) An atom with equal numbers of electrons and protons

(g) Has a 0 charge and a mass of about 1 amu

(h) The total number of neutrons and electrons in an atom

(i) Has either a +1 or 0 charge and a mass of about 1 amu

(j) The number of protons in an atom

(k) An atom with equal numbers of protons and neutrons

(l) The mass of a naturally occurring element compared to ^{12}C

B-3 Problems

1. $^{89}_{39}Y$ has ______ protons, ______ neutrons, and ______ electrons. The mass number is ______. The atomic number is ______.

2. ^{91}Zr has ______ protons, ______ neutrons, and ______ electrons. The mass number is ______. The atomic number is ______.

3. $^{197}_{78}?$ has ______ protons, ______ neutrons, and ______ electrons. The symbol of the element is ______.

4. The naturally occurring isotopic mixture of an element has a mass 2.58 times that of the standard, ^{12}C. What is the atomic mass of the element and what is the element?

5. If the isotopic mass of ^{12}C were defined as exactly five, state the approximate isotopic masses of the following.

   (a) ^{14}N (b) ^{238}U

6. Magnesium occurs in nature as a mixture of three isotopes: ^{24}Mg isotopic mass = 23.99 amu, ^{25}Mg, iso. mass = 24.99, and ^{26}Mg, iso. mass = 25.98 amu. If magnesium is 79.0% ^{24}Mg, 10.0% ^{25}Mg, and 11.0% ^{26}Mg, what is the atomic mass of magnesium?

7. Iridium (Ir) occurs in nature as a mixture of two isotopes: ^{191}Ir (iso. mass = 191.04 amu) and ^{193}Ir (iso. mass = 193.04 amu). Using the atomic mass of iridium from the periodic table, determine the percent of each isotope present.

B-4 Matching.

___ A monatomic anion

___ A monatomic cation

___ A molecule

___ A polyatomic ion

(a) A sulfur atom with chemical bonds to two oxygen atoms containing a total of 32 electrons

(b) A strontium atom with 36 electrons

(c) An oxygen atom with 8 electrons

(d) An iodine atom with 54 electrons

(e) A chlorine atom bound to an oxygen atom containing a total of 26 electrons

Review Section C

CHAPTER SUMMARY SELF-TEST

C-1 Matching

___ a proton

___ molecule of an element

___ an isotope with 13 neutrons

___ an isotope of sodium

___ formula of a molecular compound

___ an anion

___ an electron

___ formula of an ionic compound

(a) NH_4^+

(b) F_2

(c) Na^+

(d) $Ca(ClO)_2$

(e) $HClO_4$

(f) ${}^{1}_{1}H^+$

(g) ${}^{24}_{11}X$

(h) a neutral particle with a mass of about 1 amu

(i) a negatively charged particle in an atom

(j) I^-

(k) ${}^{23}_{12}X$

(l) Xe

Answers to Self-Tests

A-1 Fill in the Blanks

N - nitrogen, Mg - magnesium, Pb - lead, Al - aluminum, Na - sodium, Cl - chlorine

phosphorus - P, iron - Fe, sulfur - S, calcium - Ca, potassium - K, boron - B

A-2 Multiple Choice

1. **b** N_2
2. **c** PF_3
3. **d** SO_3^{2-}
4. **c** $Ni(ClO_3)_2$
5. **a** 6
6. **b** S^{2-} $(2K^+)(S^{2-}) = K_2S$
7. **c** Al^{3+} $(2Al^{3+})(3CrO_4^{2-}) = Al_2(CrO_4)_3$

B-1 Multiple Choice

1. **c** mass = 0 amu, charge = -1
2. **b** most of the mass and all of the positive charge
3. **d** 145
4. **a** 16 protons and 17 neutrons
5. **c** He
6. **b** 54
7. **a** Mn

B-2 Matching

e A proton has a +1 charge and a mass of about 1 amu.

g A neutron has 0 charge and a mass of about 1 amu.

i A nucleon is either a proton or a neutron.

j The number of protons in an atom is the atomic number.

c The number of nucleons in an atom is the mass number.

f A neutral atom contains equal numbers of protons and electrons.

l Atomic mass is the mass of a naturally occurring element compared to ^{12}C.

B-3 Problems

1. $^{89}_{39}Y$ has 39 protons, 50 neutrons, and 39 electrons. The mass number is 89. The atomic number is 39.

2. ^{91}Zr has 40 protons, 51 neutrons, and 40 electrons. The mass number is 91. The atomic number is 40.

3. $^{197}_{78}?$ has 78 protons, 119 neutrons, and 78 electrons. The symbol of the element is Pt.

4. ^{12}C = 12.0000 amu
   2.58 x 12.000 amu = <u>30.96</u> amu. The element must be phosphorus (P).

5. (a) ^{14}N has a mass $\frac{14}{12}$ that of ^{12}C. If ^{12}C = 5.00, $^{14}N = \frac{14}{12} \times 5.00 = \underline{5.83}$

   (b) ^{238}U has a mass $\frac{238}{12}$ that of ^{12}C. If ^{12}C = 5.00,

   $^{238}U = \frac{238}{12} \times 5.00 = \underline{99.2}$

6. Calculate the weighted average of all isotopes.

   ^{24}Mg: 0.790 x 23.99 = 18.95 amu
   ^{25}Mg: 0.100 x 24.99 = 2.50 amu
   ^{26}Mg: 0.110 x 25.98 = <u>2.85 amu</u>
   24.30 amu = <u>24.3 amu</u>

7. At. wt. of Ir = 192.22 amu
   Let X = decimal fraction of ^{191}Ir
   Then (1 - X) = decimal fraction of ^{193}Ir
   191.04 X = mass due to ^{191}Ir, 193.04 (1 - X) = mass due to ^{193}Ir
   191.04 X + 193.04 (1 - X) = 192.22
   -2.00 X = -0.82
   X = 0.41   0.41 x 100% = 41% ^{191}Ir 100% - 41% = 59% ^{193}Ir

B-4 Matching

d An iodine atom with 54 electrons would be an anion with a -1 charge.

b A strontium atom with 36 electrons would be a cation with a +2 charge.

a SO_2 would be a neutral molecule with 32 electrons. (Two oxygens and one sulfur would have a total positive charge of +32 in their nuclei.)

e A chlorine atom bound to a oxygen atom with a total of 26 electrons. (The total positive charge in a neutral oxygen and chlorine is +25.)

C-1 Matching

f A proton is the same as $^{1}_{1}H^{+}$.

b F_2 represents a molecule of an element.

g $^{24}_{11}X$ has 24 - 11 = 13 neutrons

k $^{23}_{12}X$ is an isotope of sodium.

e	$HClO_4$ represents the formula of a compound.
j	I^- is an anion.
i	An electron is a negatively charged particle in an atom.
d	$Ca(ClO)_2$ is the formula of an ionic compound.

Solutions to Black Text Problems

3-58 5.81 x 12.000 = 69.7 amu The element is Ga.

3-60 0.505 x 78.92 = 39.85
0.495 x 80.92 = 40.06
79.91 = 79.9 amu

3-61 ^{28}Si 0.9221 x 27.98 = 25.80
^{29}Si 0.0470 x 28.98 = 1.362
^{30}Si 0.0309 x 29.97 = 0.926
28.088 = 26.09 amu

3-63 Let X = decimal fraction of ^{35}Cl and Y =decimal fraction of ^{37}Cl. Since there are two isotopes present, X + Y = 1, Y = 1 - X.
(X x 35) + (Y x 37) = 35.5
(X x 35) + [(1 - X) x 37] = 35.5 X = 0.75 (75% ^{35}Cl) Y = 0.25 (25% ^{37}Cl)

3-70 Let x = mass no. of I and y = mass no. of Tl Then (1) x + y = 340 or x = 340 - y
Also, (2) $x = \frac{2}{3}y - 10$ Substituting for x from (1) and solving for y
y = 210 amu (Tl) and x = 340 - 210 = 130 amu (I)

3-72 121.8 (Sb) Sb^{3+} has 51-3 = 48 electrons
^{121}Sb has 121 - 51 = 70 neutrons ^{123}Sb has 123 - 51 = 72 neutrons
^{121}Sb =57.9% due to neutrons; ^{123}Sb 58.5% due to neutrons

3-73 0.602 x 196 amu = 118 neutrons (196 - 118= 78 protons) Element is platinum.
[platinum (Pt)] 78 - 2 = 76 electrons for Pt^{2+}

3-77 (a) H: $\frac{1.008}{12.00}$ x 8.000 = 0.672 (b) N: $\frac{14.01}{12.00}$ x 8.000 = 9.34
(c) Na: $\frac{22.99}{12.00}$ x 8.000 = 15.3 (d) Ca: $\frac{40.08}{12.00}$ x 8.000 = 26.72

3-78 $\frac{43.3}{10.0}$ = 4.33 times as heavy as ^{12}C 4.33 x 12.0 amu = 52.0 amu The element is Cr.

4

The Periodic Table and Chemical Nomenclature

Review Section A *Relationships Among the Elements and the Periodic Table*

OUTLINE

4-1 Grouping the Elements: Metals and Nonmetals, Noble Metals and Active Metals

1. Two categories of elements

OBJECTIVES

Describe the differences between a metal and a nonmetal.

2. Two categories of metals — *Describe the differences between active and noble metals.*

4-2 The Periodic Table

1. The origin of the table — *Give a brief description of the origin of the periodic table and how it was constructed.*
2. The Periodic Law

4-3 Periods and Groups in the Periodic Table

1. Periods — *Locate the first seven periods of elements on the periodic table.*
2. Groups
3. Groups of groups — *Describe the four major element categories and locate them on the periodic table. Locate the alkali metals, alkaline earth metals, and the halogens on the table.*

4-4 Putting the Periodic Table to Use

1. Metals and nonmetals — *Locate the metals, nonmetals, and metalloids on the periodic table.*
2. Physical states — *Locate the elements that are gases and liquids at room temperature on the periodic table.*
3. Molecular elements — *Locate the diatomic elements on the periodic table.*

SUMMARY OF SECTIONS 4-1 THROUGH 4-4

Questions: *Can elements be grouped according to some common properties? How does the periodic table display groups of related elements? Are there other convenient groupings of elements in the periodic table?*

One of the most important classifications of the elements is that of **metals** and **nonmetals**. A third class of elements (**metalloids**) is sometimes employed to describe elements with intermediate behavior. Metals may be subdivided further as either chemically reactive, such as sodium, or chemically inert such as gold. In fact, there are many other metals and nonmetals that are chemically similar. The **periodic table** shows these relationships in the form of a chart of the elements. This chart displays the elements by increasing atomic number with families of elements falling into vertical columns. This is known as the **periodic law**. The table was first displayed in this fashion by Mendeleev and Meyer around 1870. In any given **period** (a horizontal row of elements ending with a noble gas), nonmetals are found toward the end of the period. If a **group** (a vertical column of elements) contains nonmetals, they are found at the top.

The groups of elements can be classified into four main categories.

1. **Representative elements (Groups IA -VIIA)**
   The elements in several of the groups of representative elements have enough common properties that they are entitled to special names. These include the **alkali metals** (Group IA), the **alkaline earth metals** (Group IIA), the **chalcogens** (Group VIA), and the **halogens** (Group VIIA).

2. **Noble gases (Group VIIIA)**
   These are the nonmetallic gases at the end of each period. The heavier of these (Kr, Xe, and Rn) form a few compounds but generally these elements are chemically unreactive.

3. **Transition metals (Groups IIIB - IIB)**
   These elements are all metals and contain many of the familiar structural and coinage metals such as iron, chromium, nickel, silver, and gold.

4. **Inner transition metals**
   There are two categories of these elements containing 14 elements each - the **lanthanides** (beginning with cerium), and the **actinides** (beginning with thorium).

At **room temperature** (25°C) all three physical states are found among the elements with the solid state being the most common by far. The rest are gases except for two liquids - one a metal (mercury), and the other, a nonmetal (bromine). The gases are to the right and top in the periodic table. Some of the nonmetals exist in nature as molecules. All elements of Group VIIA as well as nitrogen, oxygen, and hydrogen exist as diatomic elements . The most common form of phosphorus is P_4, while the most common form of sulfur is S_8.

NEW TERMS

Actinide
Alkali metal
Alkaline earth metal
Group
Halogen
Inner transition metal
Lanthanide
Metalloid
Metal
Noble gas
Nonmetal
Period
Periodic law
Representative element
Room temperature
Transition metal

SELF-TEST

A-1 Multiple Choice

___ 1. Which of the following elements exists (at room temperature) as a liquid?

(a) Ce (b) Hf (c) Br (d) N (e) Ne

___ 2. Which of the following is a halogen?

(a) H (b) Li (c) Ca (d) I (e) Ar

___ 3. Which of the following elements appears at the end of the third period?

(a) Cl (b) K (c) Zn (d) Ne (e) Ar

___ 4. Which of the following elements is a metalloid?

(a) Sn (#50)
(b) Sb (#51)
(c) C (#6)
(d) Mg (#24)
(e) Zn (#30)

___ 5. How many elements are in the third period?

(a) 8 (b) 18 (c) 2 (d) 12 (e) 32

___ 6. Which of the following elements is a representative element?

(a) Zn (b) Se (c) Sc (d) Kr (e) Ce

___ 7. Which of the following groups does not contain a metal?

(a) IIIB
(b) IA
(c) VIA
(d) VIIA
(e) VIIB

___ 8. Which of the following is a representative element metal?

(a) Si (b) Fe (c) Ga (d) Cu (e) U

A-2 Matching

___	lanthanide	(a) Ru	(g) H
___	representative element gas at room temperature	(b) K	(h) Br
___	noble gas	(c) Sr	(i) Kr
___	an alkali metal	(d) Sm	(j) C
___	transition metal	(e) As	(k) W
___	an actinide	(f) Sn	(l) U

Review Section B *The Formulas and Names of Compounds*

OUTLINE / OBJECTIVES

4-5 Naming and Writing Formulas of Metal-Nonmetal Binary Compounds

1. Metals that form one charge — *Name binary compounds with common metals that form only one charge.*
2. Metals that form more than one charge — *Apply the Stock method for naming compounds with metals that form more than one charge and write formulas from the names.*

4-6 Naming and Writing Formulas of Compounds with Polyatomic Ions

1. Formulas and names of some polyatomic ions. — *Write the names and formulas of some common polyatomic ions.*

4-7 Naming Nonmetal-Nonmetal Binary Compounds

1. The order of presentation and the order of naming the two nonmetals
2. Greek prefixes — *Use the Greek prefixes to name nonmetal-nonmetal binary compounds and write formulas given their names.*

4-8 Naming Acids

1. Binary acids
2. Oxyacids — *Name binary acids and oxyacids and write formulas from the names.*

SUMMARY OF SECTIONS 4-5 THROUGH 4-8

Question: *How does the periodic table aid us in naming compounds?*

Before proceeding into other aspects of chemistry, we take some time in this chapter to systematize the naming of compounds (**chemical nomenclature**). In the first three sections, we focus on the nomenclature of metal compounds that are either **binary** (two element) or contain a polyatomic ion.

A potential problem we encounter is that some metals form a variety of compounds with the same nonmetal. For example, the following oxides of manganese are known: MnO, Mn_2O_3, MnO_2, and Mn_2O_7. Obviously, we need some system to distinguish among these compounds. The system used (the **Stock method**) is to indicate the positive charge on the metal (or the positive charge that it would have if it were an ion).

The naming of **salts** is summarized as follows.

1. Metal-nonmetal binary compounds

   (a) metals that have only one charge (IA, IIA, and Al)
   Example: $CaBr_2$ calcium bromide
   (Calcium forms only a +2 charge.)

   (b) metals that can have more than one charge (most other metals)
   Example: Cr_2O_3 chromium(III) oxide
   (The Roman numeral III indicates that Cr has a +3 charge.)

   CrO chromium(II) oxide
   (In this case, the Cr has a +2 charge.)

2. Metal-polyatomic ions (**oxyanions**)

   (a) metals that have one charge
   Example: $Sr_3(PO_4)_2$ strontium phosphate
   (To arrive at the formula we recall that the total positive charge on the metal must cancel the total negative charge on the anions.
   Thus, $(Sr^{2+} \times \mathbf{3}) + (PO_4^{3-} \times \mathbf{2}) = 0$)

   (b) metals that form more than one charge
   Example: $Ti_2(SO_4)_3$ titanium(III) sulfate
   (By solving the following algebra equation, we can arrive at the charge on the metal. $(Tl^{x} \times \mathbf{2}) + (SO_4^{2-} \times \mathbf{3}) = 0$ $x = \underline{+3}$)

The formula can also be determined by the "cross charge" method. That is, the number of each respective charge becomes the subscript of the other ion as follows.

barium chloride	$Ba^{(2+)}$ ⟷ $Cl^{(1-)}$	$BaCl_2$
iron(III) sulfate	$Fe^{(3+)}$ ⟷ $(SO_4)^{(2-)}$	$Fe_2(SO_4)_3$

Some unique problems are encountered with the naming of nonmetal-nonmetal compounds. For example, which element is written and named first? With a few exceptions for several common compounds, we write and name the element that is closest to being a metal first. The Stock system is not generally used because of certain ambiguities that can result. Instead we make use of Greek prefixes.

Most anions combined with hydrogen form a class of compounds called **acids**. We will see in a later chapter why these compounds are so special that they have their own nomenclature.

The naming of nonmetal-nonmetal compounds is summarized as follows:

1. Nonmetal-nonmetal binary compounds
   Example: N_2O_3 — dinitrogen trioxide
   [Name the most metallic element first then the other element in the same manner as an anion. Use Greek prefixes to indicate number of atoms present.]

2. Acids

(a) **binary acids**
   Example: H_2Te — hydrotelluric acid
   [If parent anion ends in *ide* (e.g., telluride), add *hydro* to root plus *ic* ending.]

(b) **oxyacids**
   Example: HNO_3 — nitric acid
   HNO_2 — nitrous acid
   [If parent anion ends in *ate* (i.e, nitrate), add *ic* ending. If parent anion ends in *ite* (i.e., nitrite) add *ous* ending.]

NEW TERMS

Acid (binary and oxy-)	Oxyanion
Chemical nomenclature	Salt
Classical method	Stock method

SELF-TEST

B-1 Writing Formulas and Names

1. Write the formulas of the following ions including the charges.

(a) carbonate ______________	(f) chlorate ______________
(b) bromide ______________	(g) nitrite ______________
(c) bisulfate ______________	(h) nitride ______________
(d) sulfide ______________	(i) nitrate ______________
(e) acetate ______________	(j) ammonium ______________

2. Write the correct formulas in the boxes.

	HSO_3^-	CrO_4^{2-}	PO_4^{3-}
Ba^{2+}	$Ba(HSO_3)_2$		
Fe(III)			
NH_4^+			

3. Write the correct names for the compounds in problem 2.

	HSO_3^-	CrO_4^{2-}	PO_4^{3-}
Ba^{2+}	______________	______________	______________
Fe(III)	______________	______________	______________
NH_4^+	______________	______________	______________

4. Of the following sets of two elements, choose which would be written and named first in a binary compound formed by the two elements:

(a) Si and Cl ____________ (b) Xe and O ____________

(c) O and S ____________ (d) I and Te ____________

(e) F and H ____________ (f) N and P ____________

(g) Cl and Br ____________ (h) Cl and N ____________

5. Write the number that corresponds to each prefix:

(a) di ____________ (b) penta ____________

(c) mono ____________ (d) hepta ____________

(e) tetra ____________ (f) hexa ____________

6. Write the formulas and the names of the acids formed by the following anions:

	Formula	Name
(a) $Cr_2O_7^{2-}$	__________	____________________
(b) HSO_3^-	__________	____________________
(c) PO_4^{3-}	__________	____________________
(d) Br^-	__________	____________________

Review Section C

CHAPTER SUMMARY SELF-TEST

C-1 Problems

1. Write the correct names for:

(a) CaO ______________________________

(b) BaI_2 ______________________________

(c) Al_2S_3 ______________________________

(d) VO_2 ______________________________

(e) $FeCl_3$ ______________________________

(f) Au_2S_3 ______________________________

(g) $LiClO_4$ ______________________________

(h) $Zr(OH)_4$ ______________________________

(i) $(NH_4)_2Se$ ______________________________

(j) $Ni(HCO_3)_2$ ______________________________

(k) ICl_3 ______________________________

(l) AsF_3 ______________________________

(m) I_2O_7 ______________________________

(n) HI ______________________________

(o) HClO ______________________

(p) H_2CrO_4 ______________________

2. Write formulas for the following:

(a) rubidium oxide ____________

(b) aluminum iodide ____________

(c) beryllium selenide ____________

(d) manganese(II) oxide ____________

(e) manganese(VII) oxide ____________

(f) platinum(IV) chloride ____________

(g) potassium oxalate ____________

(h) cesium bisulfate ____________

(i) radium cyanide ____________

(j) cadmium(II) phosphate ____________

(k) cobalt(II) nitrate ____________

(l) tetraphosphorus hexoxide ____________

(m) dibromine trioxide ____________

(n) disulfur dichloride ____________

(o) hydrofluoric acid ____________

(p) sulfuric acid ____________

(q) carbonic acid ____________

3. The following compounds have names that are not usually accepted. Write the correct name and explain the error.

Formula	Wrong Name	Correct Name
K_2CrO_4	dipotassium chromate	______________
Reason: ______________		
$Ca(HSO_4)_2$	calcium(II) bisulfate	______________
Reason: ______________		
H_2SO_4	hydrogen sulfate	______________
Reason: ______________		
PbO	lead oxide	______________
Reason: ______________		
SO_3	sulfur(VI) oxide	______________
Reason: ______________		
$Sr(NO_3)_2$	strontium nitrite	______________
Reason: ______________		

Answers to Self-Tests

A-1 Multiple Choice

1. **c** Br (exists in nature as Br_2)
2. **d** I
3. **e** Ar
4. **b** Sb (#51)
5. **a** 8
6. **b** Se
7. **d** VIIA
8. **c** Ga

A-2 Matching

d	Samarium (Sm) is a lanthanide.
g	Hydrogen (H) is a representative element gas.
i	Krypton (Kr) is a noble gas.
b	Potassium (K) is a representative element metal.
a	Ruthenium (Ru) is a transition metal.
l	Uranium (U) is an actinide.

B-1 Writing Formulas and Names

1.

(a) CO_3^{2-} (f) ClO_3^-

(b) Br^- (g) NO_2^-

(c) HSO_4^- (h) N^{3-}

(d) S^{2-} (i) NO_3^-

(e) $C_2H_3O_2^-$ (j) NH_4^+

2.

	$BaCrO_4$	$Ba_3(PO_4)_2$
$Fe(HSO_3)_3$	$Fe_2(CrO_4)_3$	$FePO_4$
NH_4HSO_3	$(NH_4)_2CrO_4$	$(NH_4)_3PO_4$

3.

barium bisulfate	barium chromate	barium phosphate
iron(III) bisulfite	iron(III) chromate	iron(III) phosphate
ammonium bisulfite	ammonium chromate	ammonium phosphate

4.

(a) Si (b) Xe
(c) S (d) Te
(e) H (f) P
(g) Br (h) N

5.

(a) di - 2 (b) penta - 5
(c) mono - 1 (d) hepta - 7
(e) tetra - 4 (f) hexa - 6

6.

(a) $H_2Cr_2O_7$ dichromic acid (b) H_2SO_3 sulfurous acid

(c) H_3PO_4 phosphoric acid (d) HBr hydrobromic acid

C-1 Problems

1.

(a) calcium oxide (b) barium iodide
(c) aluminum sulfide (d) vanadium(IV) oxide
(e) iron(III) chloride (f) gold(III) sulfide
(g) lithium perchlorate (h) zirconium(IV) hydroxide
(i) ammonium selenide (j) nickel(II) bicarbonate
(k) iodine trichloride (l) arsenic trifluoride
(m) diiodine heptoxide (n) hydroiodic acid
(o) hypochlorous acid (p) chromic acid

2.

(a) Rb_2O (b) AlI_3
(c) BeSe (d) MnO
(e) Mn_2O_7 (f) $PtCl_4$
(g) $K_2C_2O_4$ (h) $CsHSO_4$
(i) $Ra(CN)_2$ (j) $Cd_3(PO_4)_2$
(k) $Co(NO_3)_2$ (l) P_4O_6
(m) Br_2O_3 (n) S_2Cl_2
(o) HF (p) H_2SO_4
(q) H_2CO_3

3.

(a) K_2CrO_4 potassium chromate Since chromate has a -2 charge and potassium is always +1, the Greek prefix is unnecessary.

(b) $Ca(HSO_4)_2$ calcium bisulfate Since calcium has only a +2 charge, the (II) is unnecessary.

(c) H_2SO_4 sulfuric acid Oxyacids have only an acid name.

(d) PbO	lead(II) oxide	Lead forms ions with more than one charge so the Stock method is generally used to indicate charge
(e) SO_3	sulfur trioxide	Greek prefixes are used in nonmetal - nonmetal binary compounds rather than the Stock method.
(f) $Sr(NO_3)_2$	strontium nitrate	The anion is a nitrate not a nitrite.

5

Modern Atomic Theory

Review Section A *The Energy of the Electron in the Atom*

OUTLINE / OBJECTIVES

5-1 The Emission Spectra of the Elements

1. Light energy — *Describe the relationship between the wavelength of light and energy.*
2. Light spectra — *Distinguish between the properties of continuous and discrete spectra.*

5-2 A Model for the Electrons in the Atom

1. The Bohr model
2. Energy states in a hydrogen atom — *List the important assumptions of Bohr's model for the hydrogen atom.*

5-3 Modern Atomic Theory - A Closer Look at Energy Levels

1. Orbitals and shells — *Describe the types of orbitals that are present in the first three shells in an atom.*
2. Shapes of orbitals — *Describe the shapes of s, p, and d orbitals*
3. Subshells — *List the types of orbitals in a specific subshell*

SUMMARY OF SECTIONS 5-1 THROUGH 5-3

Questions: *What is the nature of light? How does one color differ from another? How does an atom emit light? How are the electrons in atoms arranged? What is meant by orbitals, shells, and subshells?*

Light is a form of energy that interacts with matter. The light coming to us from the sun contains all of the colors of the rainbow and is thus known as a **continuous spectrum**. Light from a hot, glowing, gaseous element, however, emits specific colors of light and is known as a **discrete spectrum**. The light that we perceive in the **visible spectrum** is bordered by light with **wavelengths** slightly longer than red (the **infrared**) and slightly shorter than violet (the **ultraviolet**). It was the orderly spectrum of hydrogen, the lightest element, that attracted the most curiosity of scientists in the early part of the 20th century, however.

In 1913, Niels Bohr explained the origin of the discrete spectrum of hydrogen by describing the energy states available to the one electron in hydrogen. His model was analogous to the planets revolving around the sun except that the electron can exist in only definite (**quantized**) **energy levels** with each level designated by a **principal quantum number (*n*)**. When energy in the form of heat is supplied to the atom, the electron jumps from the lowest energy state (the **ground state**) to a higher energy state (an **excited state**). When the electron falls back down to a lower energy state, it re-emits the energy in the form of light. Since the difference in energy between the higher and lower energy levels is discrete, the emitted light has a discrete energy. In the visible region of the spectrum, hydrogen emits four colors: violet, indigo, green, and red. Even though modern theories use a more complicated model of the atom, Bohr's model remains useful to the chemist because it presents an easily-visualized picture.

The modern view of the atom, called the **wave mechanical model**, provides more detail about the energy and other properties of the electron in the hydrogen atom. In this model, the electron is not viewed as a hard particle with a definite energy and position but instead as having a probability of existing in a region of space known as an **orbital**. Different types of orbitals exist with different shapes. For example, ***s*** type orbitals have spherical shapes that include most of the electron density. In the second shell (n = 2), ***p*** type orbitals exist which have two "lobes" of electron density lying along a three-coordinate axis. There are three *p* type orbitals in the second and successive shells. The third and higher energy shells also have ***d*** type orbitals and the fourth and higher shells have ***f*** type orbitals. The latter two types have complex shapes. If the principal energy level is known as a **shell**, the orbitals of each type within a shell together make up what is known as a **subshell**. For example the second shell (n = 2) contains two subshells, the 2*s* (one orbital) and the 2*p* (three orbitals). The n = 3 shell has "*s*", "*p*", and "*d*" subshells, which are listed in order of increasing energy. This information is summarized in the following table. Notice that the capacity of each shell is $2n^2$ electrons.

Table 5-A

Shell	1	2	3	4
Subshell	*s*	*s* *p*	*s* *p* *d*	*s* *p* *d* *f*
Subshell capacity	2	2 6	2 6 10	2 6 10 14
Shell capacity	2	8	18	32

NEW TERMS

Continuous spectrum
Discrete spectrum
Excited state
Ground state
Infrared light
Orbital
Principal quantum number
Shell (Principal energy level)
Spectrum
Subshell
Ultraviolet
Visible light
Wavelength
Wave mechanics

SELF-TEST

A-1 Multiple Choice

____ 1. Which of the following is an important postulate of the Bohr model of the atom?

(a) electrons exist in indefinite orbits around the nucleus
(b) the energy of the light is directly proportional to its wavelength
(c) atoms emit light when electrons jump to higher energy orbits
(d) the discrete spectrum of hydrogen is a property of the nucleus
(e) atoms emit discrete wavelengths of light because there is a discrete difference in energy between orbits

____ 2. How many orbitals are in the 4*f* subshell?

(a) 14 (b) 7 (c) 6 (d) 10 (e) 4

____ 3. How many total orbitals are in the $n = 2$ shell?

(a) 2 (b) 3 (c) 4 (d) 8 (e) 6

____ 4. How many "lobes" does one *p* orbital have?

(a) one (b) none (c) two (d) four (e) three

____ 5. What is the capacity of the $n = 3$ shell?

(a) 2 (b) 8 (c) 18 (d) 10 (e) 32

____ 6. What subshells are in the $n = 3$ shell?

(a) *s, p, d*
(b) *s, p*
(c) *s, p, f*
(d) *p, d, f*
(e) *s, d, f*

____ 7. What is the electron capacity of the *d* subshell?

(a) 2 (b) 6 (c) 10 (d) 8 (e) 14

____ 8. Which of the following subshells does not exist?

(a) 7*s* (b) 2*d* (c) 2*p* (d) 5*f* (e) 6*d*

____ 9. Which of the following subshells is the highest in energy?

(a) 6*p* (b) 5*d* (c) 6*s* (d) 4*f* (e) 5*f*

A-2. Matching

____ Orbital

____ Excited state

____ Ultraviolet light

____ Shell

____ Discrete spectra

____ Subshell

____ Wavelength

(a) When an electron occupies an energy level higher than the lowest available energy level

(b) A principal energy level in an atom

(c) Light that is lower in energy than the visible

(d) The rainbow

(e) A region of space in which one or two electrons are found

(f) The distance between the lowest and highest point on a wave

(g) Bands of light of definite energies

(h) When all electrons are in the lowest available energy states

(i) All of the orbitals of one type within a shell

(j) Light with shorter wavelengths than the visible

(k) The distance between two equivalent points on a wave

A-3. Orbital Shape

1. Sketch the shapes of all orbitals in the $n = 2$ shell.

Review Section B *The Distribution of Electrons in Atoms*

OUTLINE	OBJECTIVES
5-4 Electron Configurations of the Elements	
1.The Aufbau principle	*List the normal order of filling of the first ten subshells.*
2. A periodic property	*Write the electron configuration for the atoms of a specified element.*

5-5 Orbital Diagrams of the Elements

1. The Pauli exclusion principle
2. Hund's rule

Write orbital diagrams for the outer electrons of a specified element.

SUMMARY OF SECTIONS 5-4 AND 5-5

Questions: *How does theory deal with the electrons in the atoms of an element? How do these assignments of electrons correlate with the periodic table?*

We are now ready to discuss the **electron configuration** of the elements. This configuration summarizes the location of all the electrons in the atoms of an element (shell, subshell, and number of electrons in each subshell). The standard method for electron configuration is as follows.

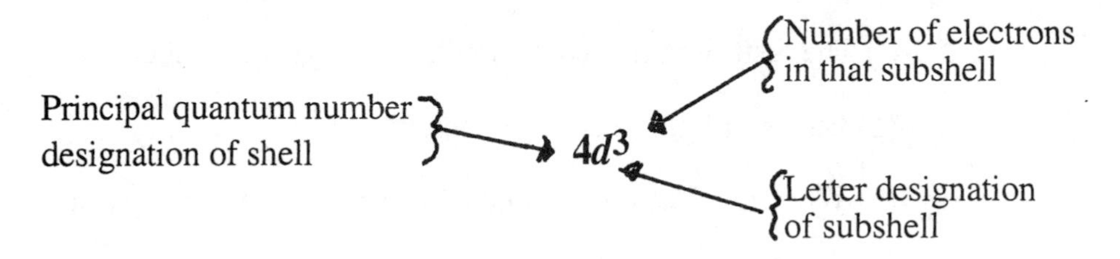

Electrons fill the subshells in the order of the lowest energy levels first. This is known as the **Aufbau principle**. The electron configuration of a few sample elements are as follows.

Element	Electron Configuration	Comments
Li	$1s^2 2s^1$	Two electrons fill the first shell, the third is in the next lowest level which is the 2*s* subshell.
Ne	$1s^2 2s^2 2p^6$	This element completes the filling of the two subshells of the $n = 2$ shell.
K	[Ar] $4s^1$	The [Ar] shorthand notation represents all of the electrons in the noble gas Ar. Notice that the 4*s* is lower in energy than the 3*d* so the 4*s* fills first.
Zn	[Ar] $4s^2 3d^{10}$	The 3*d* fills after the 4*s*. This completes filling of the $n = 3$ shell.
Kr	[Ar] $4s^2 3d^{10} 4p^6$	This completes the filling of the 4*p* subshell. Notice that noble gases have filled outer *ns* and *np* subshells (except He which doesn't have a 1*p* subshell).
La	[Xe] $6s^2 5d^1$	The periodic table tells us that this element has a d^1 configuration since it comes under Y which has $4d^1$.

Ce [Xe] $6s^2 5d^1 4f^1$ The next element after La has one electron in the $4f$ subshell.

When the orbitals in a subshell are represented as boxes, the illustration is known as an **orbital diagram** or **box diagram**. The orbitals present in the various subshells are thus illustrated as follows.

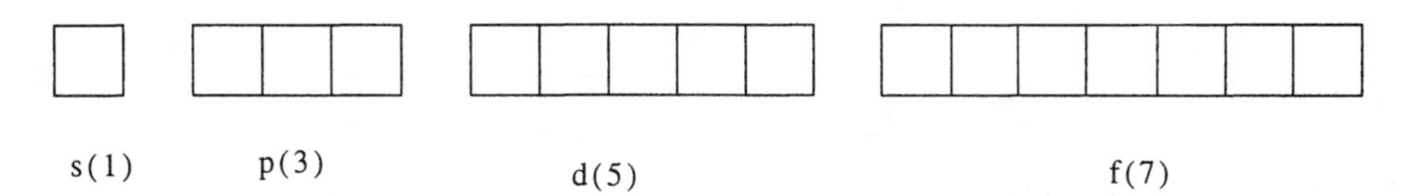

s(1) p(3) d(5) f(7)

Electrons in an atom have one additional important property known as electronic "spin." Spin arises because the electron has properties of a spinning particle. The spin of an electron is represented as an arrow that points either up or down illustrating the two possible spins. According to the **Pauli exclusion principle**, if an orbital is occupied by two electrons, they must have opposite or "paired" spins. The two "paired" electrons in an orbital are represented as one arrow pointing up and the other down indicating opposite spins.

According to **Hund's rule**, electrons occupy separate orbitals with their spins parallel (given a choice). Thus the orbital diagram of a nitrogen atom is represented as follows. Notice that, according to Hund's rule, the three electrons are in separate p orbitals with parallel spins leaving nitrogen with three unpaired electrons.

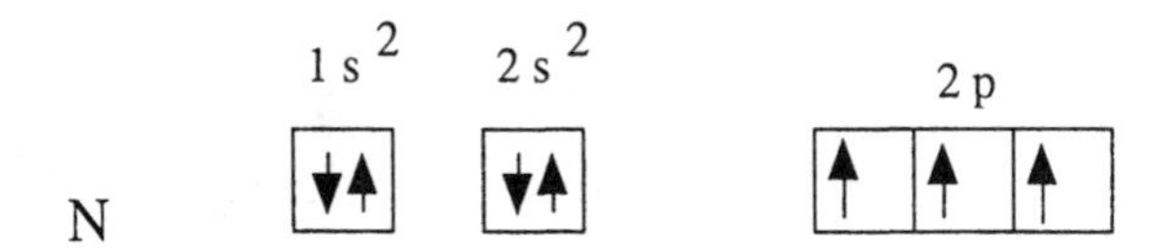

NEW TERMS

Aufbau principle	Hund's rule
Box diagram	Orbital diagram
Electron configuration	Pauli exclusion principle
Electron spin	Unpaired electrons

SELF-TEST

B-1. Multiple Choice

____ 1. Which of the following groups (vertical columns) in the periodic table have no unpaired electrons?

(a) IIB (b) IA
(c) IIIA (d) IVA
(e) IIIB

____ 2. How many unpaired electrons are in the atoms of group IVB?

(a) none (b) one (c) two (d) three (e) four

____ 3. Which of the following orbital diagrams is excluded by the Pauli exclusion principle?

(a)

↑↓		

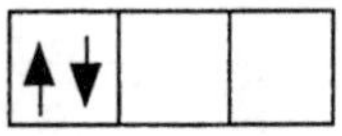

(b)

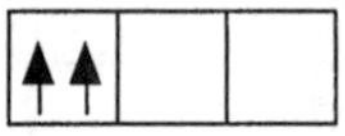

↑↑		

(c)

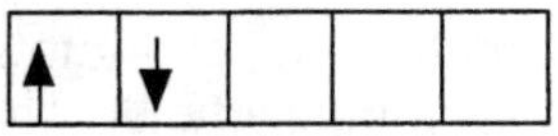

↑	↓			

(d)

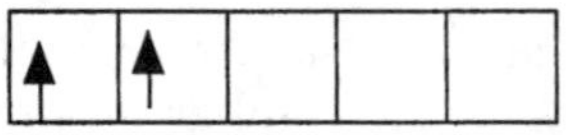

↑	↑			

____ 4. Which of the following orbital diagrams is excluded by Hund's rule?

(a)

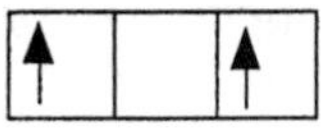

↑		↑

(b)

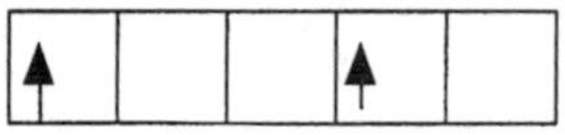

↑			↑	

(c)

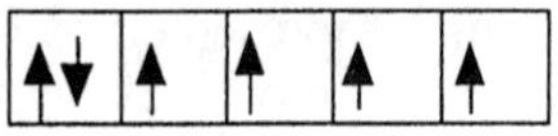

↑↓	↑	↑	↑	↑

(d)

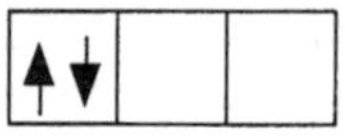

↑↓		

B-2. Matching

____ Aufbau principle

____ Hund's rule

____ Unpaired electron

____ Pauli exclusion principle

(a) Electrons occupy excited states first.

(b) Present when an orbital is singly occupied

(c) Electrons occupy the lowest possible energy state.

(d) Electrons occupy separate orbitals of the same energy with parallel spin

(e) Present when an orbital is doubly occupied

(f) Electrons cannot occupy the same orbital with the same spin.

B-3. Problems

Write the orbital diagram showing the arrangement of all electrons beyond the previous noble gas for the following. For example, the orbital diagram for Ca is

$4s^2$

(a) V

(b) Cl

(c) Ce

(d) Zn

Review Section C *Periodic Properties*

OUTLINE

5-6 Electron Configuration and the Periodic Table

1. General electron configurations
2. Using the periodic table

5-7 Periodic Trends

1. Atomic radii
2. Ionization energy

OBJECTIVES

Using only the periodic table, give the expected electron configurations of specified elements for all electrons beyond the previous noble gas.

Describe the general trends in first ionization energies and predict which metals may also form a +2 or +3 ion.

SUMMARY OF SECTIONS 5-6 AND 5-7

Questions: *How can we use the periodic table to determine electron configuration? What trends in atomic properties can be predicted directly from the periodic table?*

Many years after the periodic table was first displayed, we find that the vertical columns have the same subshell configuration but successively higher shells. Thus electron configuration is called a *periodic* property. We can put this to good use in establishing the electron configurations of the elements by locating the element on the table and then writing the electron configuration expected for an element in that position. The four categories of elements can then be illustrated by their general electron configurations as follows. [NG] represents a noble gas configuration.

Representative elements: [NG] ns^x or [NG] ns^2np^y

e.g., IA (alkali metals) [NG] ns^1 Rb: [Kr] $5s^1$

IVA [NG] ns^2np^2 Si: [Ne] $3s^23p^2$

Noble gases: [NG] ns^2np^6 e.g., Ar: [Ne] $3s^23p^6$

Transition metals: [NG] $ns^2\ (n\text{-}1)d^x$ e.g., Zr: [Kr] $5s^24d^2$

Inner transition metals: [NG] $ns^2\ (n\text{-}1)d^1\ (n\text{-}2)f^x$

e.g., Ce: [Xe] $6s^25d^14f^1$

Besides electron configuration, there are other properties of the atoms of the elements that vary in a periodic manner. The first property discussed that shows definite trends is the **atomic radius** of the atoms. The radius is the distance from the nucleus to the outermost electrons. There are a trend across a period and a trend down a group.

To understand these trends, one must appreciate how one electron affects the charge of the nucleus felt by another electron. Electrons in the same subshell do not **shield** each other effectively. As a result, the more electrons there are in an outer subshell, the more the nuclear charge is felt by all of these outer electrons and the smaller the atom. Thus elements generally become smaller as a subshell is filling. Going down a group, the outer electrons lie in a shell farther from the nucleus so atoms are generally larger.

Notice that the trend toward large atomic radii (down and to the left in the periodic table) is the same as the trend toward metallic character. This arrangement is no coincidence since **ionization energy** (the energy required to remove an electron from a gaseous atom) is related to the trends in the atomic radii. The smaller the atom, the more tightly the outer electrons are held and the harder they are to remove. Thus large atoms like metals lose an electron easily compared to smaller atoms like the nonmetals. This is a very important difference in the properties of the metals and nonmetals. We also noticed that it is comparatively very difficult to remove electrons from noble gas structures. For example, the energy required to form Mg^{3+} is very large compared to the energy required to form Mg^{2+}. This is because Mg has two outer electrons beyond a noble gas (i.e., [Ne] $3s^2$). Notice that the first two electrons removed lie outside the neon core. The third electron, however, must be removed from the neon core and is much more firmly attached.

NEW TERMS

Atomic radius
Ionization energy
Shielding

SELF-TEST

C-1 Problems

1. Using only the periodic table, write the electron configuration for

(a) N ________________ (b) Ga ____________________

(c) I ________________ (d) Pm ____________________

(e) Ta ________________

2. Using only the periodic table, write the symbol of the element whose atoms have the following electron configurations:

____ (a) [Ne] $3s^2$

____ (b) [Ar] $4s^2 3d^5$

____ (c) [Xe] $6s^2 4f^{14} 5d^{10}$

____ (d) [Rn] $7s^2 6d^1 5f^3$

3. Indicate the symbol of the appropriate element.

____ (a) the first element with a *d* electron

____ (b) the first element with an *f* electron

____ (c) the first element to have a filled *d* subshell

____ (d) the first element with a d electron and a filled 4*f*

____ (e) the first element that has the general configuration

$$ns^2(n\text{-}1)d^{10}np^3$$

____ (f) the atomic number of the first element with a filled 6*d* subshell

____ (g) the first element with a filled 5*p* subshell

C-2 Multiple Choice

____ 1. Which of the following elements has the general electron configuration ns^2np^5 (n is the principal quantum number)?

(a) P (b) Ar (c) S (d) Mn (e) Cl

____ 2. From the periodic table, predict which of the following elements has the smallest radius:

(a) Ga (b) P (c) O (d) S (e) N

____ 3. From the periodic table, predict which of the following elements has the lowest ionization energy:

(a) K (b) Br (c) Ca (d) Rb (e) I

____ 4. From the periodic table, predict which of the following cations would require the largest amount of energy to form:

(a) Ba^{3+} (b) Ga^{3+} (c) Cs^+ (d) Sr^{2+} (e) Ca^+

____ 5. Which of the following would have the smallest radius?

(a) Cl^- (b) S^{2-} (c) S (d) Na (e) Na^+

C-3 Matching:

If any of the following ions are energetically feasible (comparably easy to form), put a check. If not, select the reason from the possibilities listed.

Ion	Reason
___ Ba^{2+}	(a) Nonmetals do not easily form cations
___ O^{2+}	(b) An electron would have to be removed from a noble gas core.
___ Na^{2+}	(c) Metals to not easily form anions.
___ F^{2-}	(d) The outer *p* subshell cannot accommodate all the electrons.
___ Mg^{2-}	

Review Section D

CHAPTER SUMMARY SELF-TEST

D-1 Match the Elements

Using the periodic table, write the symbol of the appropriate element. The elements that correspond to a particular statement are listed below.

____ 1. The element that has one electron in addition to a complete $4s$ subshell

____ 2. The element that completes the filling of the $n = 3$ shell

____ 3. Two elements with two unpaired electrons in the $4p$ subshell

____ 4. The element with four electrons more than the noble gas xenon

____ 5. The element with the electron configuration: [Kr] $5s^2 4d^{10} 5p^3$

____ 6. The element with the electron configuration: [Ar] $4s^1 3d^5$

____ 7. The element with one electron in a p subshell and a full $4d$ subshell

____ 8. The first element with p electrons that also has no unpaired electrons

____ 9. The element that starts the filling of the $n = 4$ shell

____ 10. The first element that has six filled orbitals

____ 11. The first element that has three unpaired electrons in orbitals that lie at a 90° angle from each other

____ 12. The first element with paired electrons in a $3d$ orbital

Elements:

(a) Cr
(b) Ge
(c) Ne
(d) Mg
(e) Fe
(f) Sb
(g) Se
(h) Cu
(i) K
(j) Ce
(k) N
(l) In
(m) Sc

Answers to Self-Tests

A-1 Multiple Choice

1. **e** Atoms emit discrete wavelengths of light because there is a discrete difference in energy between orbits.

2. **b** 7

3. **c** 4 (one s and three p)

4. **c** two lobes on either side of a nucleus

5. **c** $2n^2 = 18$

6. **a** *s, p, d*

7. **c** 10

8. **b** 2*d*

9. **e** 5*f*

A-2 Matching

e A region of space in which one or two electrons are found

a An electron is in an excited state when it occupies an energy level higher than the lowest available energy level.

j Ultraviolet light has wavelengths shorter than light in the visible.

b A shell is a principal energy level in an atom.

g Discrete spectra are bands of light of definite energies.

i All of the orbitals of one type within a shell

l Wavelength of light is the distance between two equivalent points on a wave.

A-3 Orbital Shape

The shapes of the $n = 2$ orbitals (s and p)

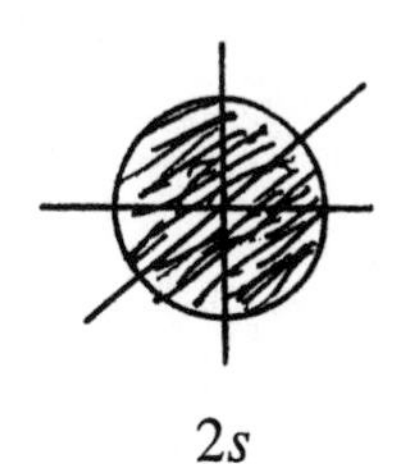

$2s$

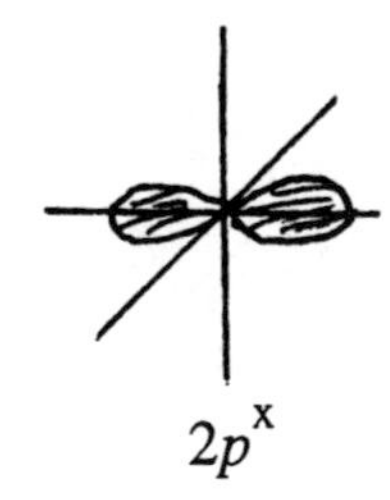

$2p^x$

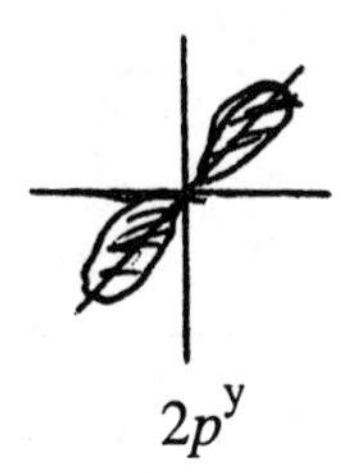

$2p^y$

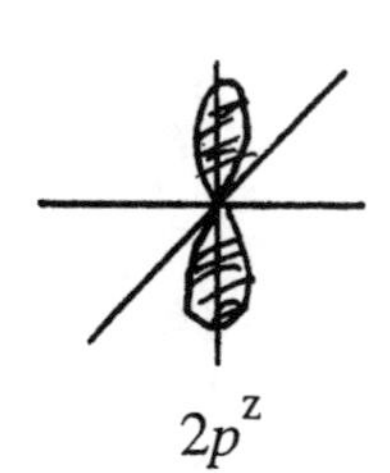

$2p^z$

B-1 Multiple Choice

1. **a** IIB
2. **c** two unpaired electrons in d orbitals
3. **b**

4. **d**

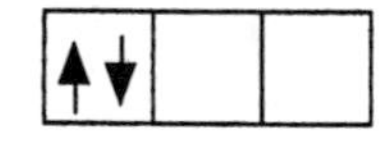

B-2 Matching

c The Aufbau principle states that electrons occupy the lowest possible energy state.

d Hund's rule states that electrons occupy separate orbitals of the same energy with parallel spins.

b Unpaired electrons are present when an orbital is singly occupied.

f The Pauli principle states that electrons in the same orbital must have opposite spins.

B-3 Problems

(a) V

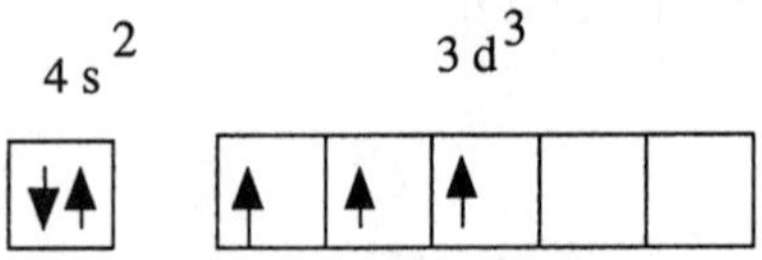

$4s^2$ $3d^3$

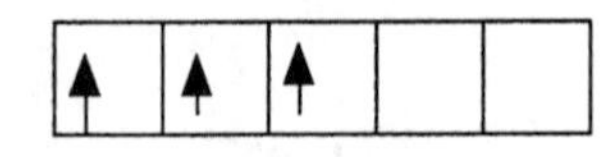

(b) Cl

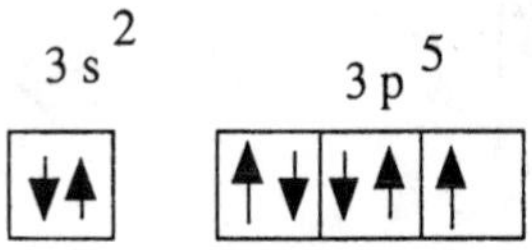

$3s^2$ $3p^5$

(c) Ce

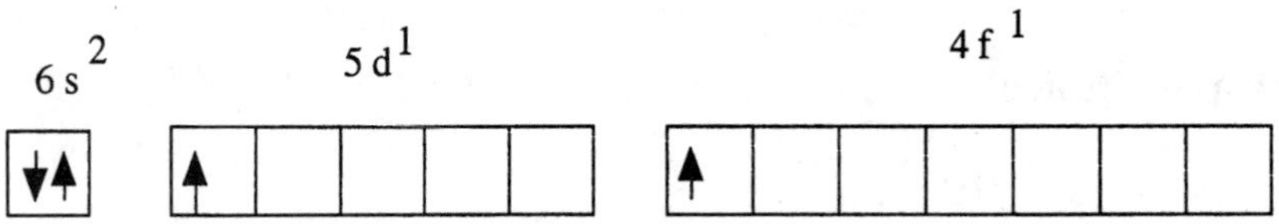

$6s^2$ $5d^1$ $4f^1$

(d) Zn

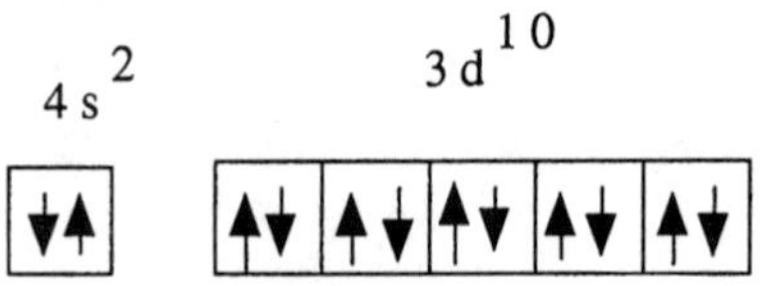

$4s^2$ $3d^{10}$

C-1 Problems

1. Electron configuration

(a) N [He] $2s^22p^3$

(b) Ga [Ar] $4s^23d^{10}4p^1$

(c) I [Kr] $5s^24d^{10}5p^5$

(d) Pm [Xe] $6s^25d^14f^4$

(e) Ta [Xe] $6s^24f^{14}5d^3$

2. Locate the elements

(a) [Ne] $3s^2$ Mg

(b) [Ar] $4s^23d^5$ Mn

(c) [Xe] $6s^24f^{14}5d^{10}$ Hg

(d) [Rn] $7s^26d^15f^3$ U

3. Indicate the element

(a) The first element with a *d* electron is Sc (#21).

(b) The first element with an *f* electron is Ce (#58).

(c) The first element to have a filled *d* subshell is Cu (#29).

(d) The first element with a d electron and a filled 4*f* is Lu (#71)

(e) The first element that has the general configuration $ns^2(n\text{-}1)d^{10}np^3$ is As (#33).

(f) The atomic number of the first element with a filled 6*d* subshell is element #112 or #111 if $7s^16d^{10}$.

(g) The first element with a filled 5*p* subshell is Xe (#54).

C-2 Multiple Choice

1. **e** Cl

2. **c** O

3. **d** Rb

4. **a** Ba^{3+} (The third electron would have to come from a filled inner subshell.)

5. **e** Na^+

C-3 Matching

Ba^{2+} is fine.

O^{2+} does not form because nonmetals do not form cations. (a)

Na^{2+} does not form because an electron would have to be removed from a noble gas core. (b)

F^{2-} does not form because the outer *p* subshell in fluorine cannot accommodate two extra electrons. (d)

S^{2-} is OK.

Mg^{2-} does not form because metals do not easily form anions. (c)

D-1 Match the Elements

1. Sc
2. Cu
3. Si and S
4. Ce
5. Sb
6. Cr
7. In
8. Ne
9. K
10. Mg
11. N
12. Fe

6

The Chemical Bond

Review Section A *Chemical Bonds and The Nature of Ionic Compounds*

OUTLINE

6-1 Bond Formation and the Representative Elements

1. Valence electrons and the octet rule
2. How representative elements attain an octet of electrons.

   Objective: List the three ways representative elements attain octets.

3. Lewis dot symbols of the representative elements.

   Objective: Write Lewis dot symbols for the atoms of any representative element.

6-2 Formation of Ions

1. Formation of metal cations
2. Formation of nonmetal anions

   Objective: Determine the charge acquired by a specified representative element by application of the octet rule.

6-3 Formulas of Binary Ionic Compounds

1. Formation of binary ionic compounds

   Objective: Write the formula of binary ionic compounds formed by representative elements.

2. The ionic bond and a crystal lattice

SUMMARY OF SECTIONS 6-1 THROUGH 6-3

Questions: *Why do some atoms bond with others and some do not? How do representative elements change their electron configuration? How can we predict whether an atom gains or loses electrons?*

One of the most fundamental topics of chemistry is how the atoms of one element bond to those of another. From a discussion of the structure of atoms in Chapter 3, we progressed to the arrangements of the electrons in those atoms in Chapter 5. That discussion gave us the theoretical basis for the periodic table, which was introduced in Chapter 4. With this background, we are now ready to discuss why the elements combine in the ratios and in the manner in which they do.

At the end of Chapter 5, we learned that metals can lose electrons comparatively easily and that nonmetals have a tendency to gain electrons. To illustrate these losses and gains of electrons, we make use of **Lewis dot symbols** where outer *s* and *p* electrons (known as **valence electrons**) are represented as dots. By use of Lewis symbols and through a knowledge of the electron configuration of the ions, we can see that many ions of the representative elements form as predicted by the octet rule. The **octet rule** states that these atoms gain, lose or share electrons so as to have access to eight valence electrons. Binary ionic compounds thus form between metals and nonmetals. The ions in the solid exist in a geometric arrangement known as a **lattice,** which is held together by **ionic bonds.** The formula of the compound is determined by the charges on the ions. A formula represents the ratio of cations to anions present so that there is no net charge, illustrated for the compound Tl_2O_3 [thallium (III) oxide] as follows:

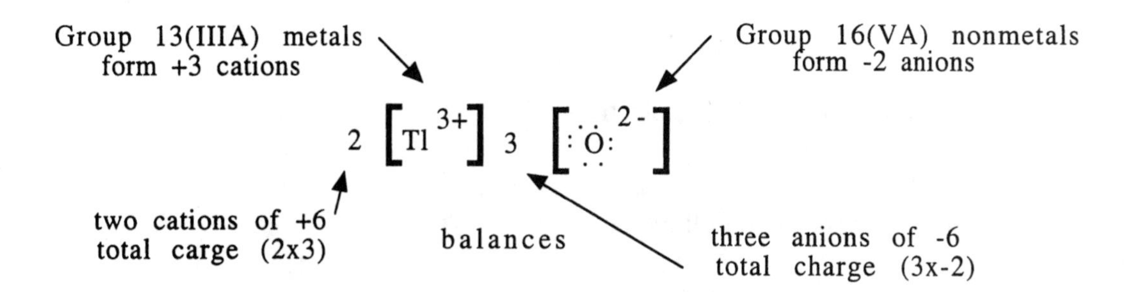

NEW TERMS

- Ionic bond
- Lattice
- Lewis dot symbol
- Octet rule
- Valence electron

SELF-TEST

A-1 Multiple Choice

___ 1. What is the expected charge on a Te ion?

(a) +2 (b) -2 (c) -1 (d) +3 (e) +6

___ 2. Which of the following pairs of atoms would not be expected to combine to form a binary ionic compound?

(a) B and H
(b) Na and H
(c) Li and Cl
(d) Rb and O
(e) K and S

___ 3. What is the expected charge on a Group IIA ion?

(a) +1 (b) +6 (c) +2 (d) -2 (e) -3

___ 4. Based on the charge on the metal ion, which of the following metal-nonmetal compounds would not be expected to be ionic?

(a) LiBr (b) CaO (c) Rb_2S (d) CrO (e) Mn_2O_7

___ 5. Which of the following ionic compounds would not be expected to exist based on the octet rule?

(a) LiI (b) CaS (c) BeF_2 (d) BaCl (e) Mg_3N_2

A-2 Matching

___ A nonmetal ion with the same electron configuration as Xe — (a) Te^-

(b) Sb^{3-}

___ A metal ion that does not follow the octet rule — (c) H^-

___ A metal ion with the same electron configuration as Xe — (d) Al^{3+}

(e) I^-

___ A possible -3 ion

(f) Pb^{2+}

___ An ion predicted for Group IIIA

(g) B^{3+}

___ A nonmetal ion that violates the octet rule — (h) La^{3+}

(i) N^{3-} (j) Br^- (k) Si^{4+} (l) S^{2-}

A-3 Problems

1. Write the ion expected to form from the following representative elements:

   (a) Sr (b) Ga

   (c) H (d) Se

   (e) I

2. Write the formula of the ionic compound formed between each pair:

   (a) Al and Se ________________________

   (b) Be and F ________________________

   (c) Ba and O ________________________

   (d) K and N ________________________

Review Section B *Chemical Bonds and the Nature of Molecular Compounds*

OUTLINE / OBJECTIVES

6-4 The Covalent Bond

1. The sharing of electrons — *Describe how the octet rule can be satisfied by covalent bonds.*
2. Lewis structures
3. The nature of a covalent bond
4. Multiple covalent bonds — *Distinguish among single, double, and triple bonds.*

6-5 Writing Lewis Structures

1. Rules for writing Lewis structures
2. Predicting bonds with the *6N + 2* rule
3. Examples of Lewis structures of molecules and polyatomic ions — *Demonstrate the octet rule by writing the Lewis structures of specified molecular compounds and polyatomic ions.*
4. Exceptions to the octet rule

6-6 Resonance Structures

1. The limitations of some Lewis structures
2. Resonance structures and hybrids — *Describe what is implied by resonance structures.*
3. Bonding in allotropes of oxygen — *Write equivalent resonance structures for specified molecules or ions.*

SUMMARY OF SECTIONS 6-4 THROUGH 6-6

Questions: *Why do certain elements combine to form ionic and others molecular compounds? How do the electron configurations of two elements help us predict formulas of molecular compounds? How do we represent the bonds in molecular compounds?*

We now turn the discussion to compounds where atoms share rather than gain or lose electrons. Such compounds are generally, but not always, between two nonmetals where neither element loses electrons easily. Bonds of this nature are known as **covalent bonds.** The octet rule predicts that representative elements (other than H) achieve an octet by having access to eight valence electrons. The octet is established through a combination of unshared electrons and shared electrons in a bond. This process is illustrated below with the **Lewis structure** of Br_2. (Two dots represents an **unshared pair** of outer electrons, a dash represents a shared pair of electrons in a covalent bond.)

Each Br has six unshared electrons but counts the two shared electrons in the bond for a total of eight.

Notice that there are a total of fourteen valence electrons (seven from each Br).

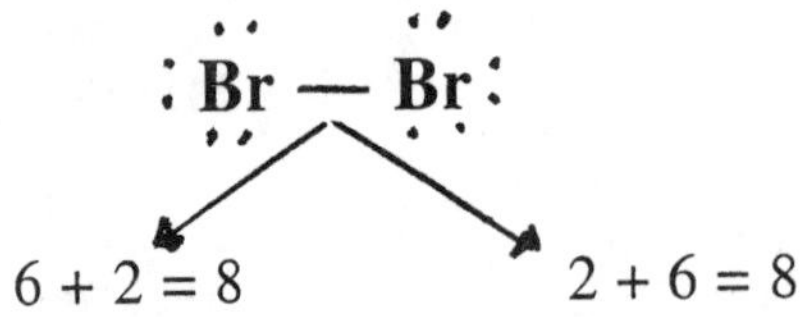

$$:\ddot{\underset{..}{Br}} - \ddot{\underset{..}{Br}}:$$

6 + 2 = 8　　　　2 + 6 = 8

In many of the representative elements, especially C, N, O, and S compounds, two pairs (a **double bond**) or three pairs (a **triple bond**) of electrons are shared between two atoms. For example, the Lewis structures of CO_2 and CO illustrate this as follows:

Double bonds

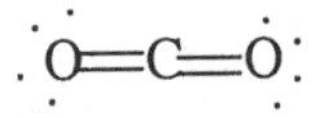

$$\ddot{O}{=}C{=}\ddot{O}$$

C has access to four shared pairs for a total of eight.
Each O has two unshared pairs and access to two shared pairs for a total of eight.

Triple bond

$$:C{\equiv}O:$$

Both C and O have an unshared pair and access to three shared pairs for a total of eight for each.

The charge and formula of polyatomic ions can also be rationalized by Lewis structures and the octet rule. For example, the structure of the NO_2^- ion is illustrated as follows. Notice that the NO_2^- ion has 18 valence electrons (five from N, six each from the two O's, and one extra electron to account for the -1 charge).

N has access to one unshared pair and three shared pairs

$$\left[\ddot{O}{=}\ddot{N}{-}\ddot{\underset{..}{O}}:\right]^-$$

This O has two unshared pairs and two shared pairs.

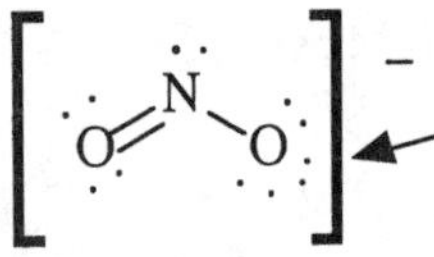

This O has three unshared pairs and one shared pair.

The octet rule is not an unbreakable law. Some stable molecules exist where an atom has access to less than eight electrons. In other cases, we can write Lewis structures that follow the octet rule (e.g., O_2 and BF_3) but we know that these structures do not agree with experimental observations. Also, nonmetals of the third period and down form many compounds where an atom has access to more than eight electrons. These compounds are not discussed in this text.

Writing Lewis structures of molecules and ions is an important and fundamental endeavor of general chemistry and should be mastered. Structures not only allow us to understand the ratio of atoms in a molecule or an ion but, in some cases, allow us to predict possible formulas. From the Lewis structure, it is also possible to predict the geometry of the molecule or ion, which is discussed in a later section.

Most students have very little difficulty writing correct Lewis structures with a knowledge of the guidelines listed in the text and a generous amount of practice. We will run through the rules for C_2H_4 (ethylene) as an example.

1. H and C are both nonmetals, so this is a molecular compound.

2. The number of valence or outer electrons is:

atom	number of atoms		number of valence electrons		total
C	2	x	4	=	8
H	4	x	1	=	4
					12

3. Arrange the atoms symmetrically.

H H C C H H

4. Form bonds between atoms
(There is always one less bond than there are atoms.)

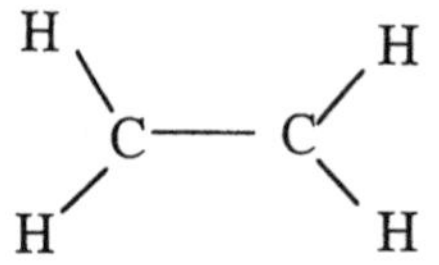

5. Add the remaining electrons
(12 - 10 = 2) to atoms with less
than eight remaining.

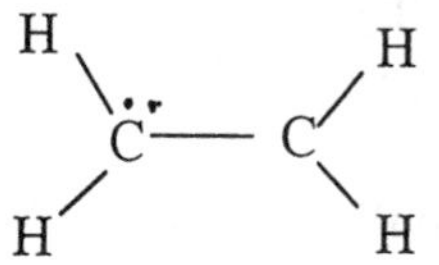

6. Check octets (two for H's).
Notice one C has only six electrons.
By making a double bond with the
other C, both C's have access to eight.

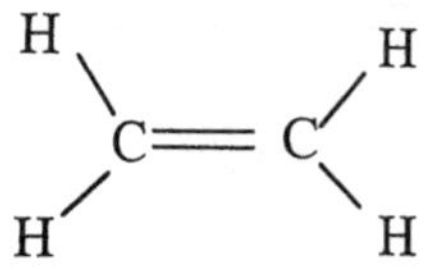

The *"6N + 2" rule* was introduced as a convenient way to check Lewis structures. N is equal to the number of atoms in the formula excluding hydrogen. If $6N + 2$ is equal to the number of valence electrons then only single bonds are present. If $(6N + 2)$ - (# val. electrons) = 2, then one double bond is present. If $(6N + 2)$ - (# val. electrons) = 4, then two double bonds or one triple bond is present. For example, in ozone (O_3), an **allotrope** of oxygen, $6N + 2 = 20$. Since there are 18 valence electrons (3 x 6), 20 - 18 = 2, indicating one double bond.

It is possible that a Lewis structure may present an incomplete picture. For example, consider one of the structures of the nitrite ion shown below. Each structure implies that one N-O bond is different from the other. In fact, they are both identical. To remedy this misconception, we can write the two equally possible Lewis structures which are called **resonance structures.** The actual structure is a **resonance hybrid** of the two shown. Both N-O bonds are thus predicted to be identical and between a double and a single bond. This corresponds to experimental observation.

NEW TERMS

Allotrope	Resonance hybrid
Covalent bond	Resonance structure
Double bond	Triple bond
Lewis structure	Unshared pair

SELF-TEST

B-1 Multiple Choice

___ 1. Which of the following two elements would be expected to form a covalent bond?

(a) Na, O (b) H, S
(c) K, Te (d) Mn, F
(e) Ba, S

___ 2. Based on the octet rule, which of the following compounds would not be expected to exist?

(a) NH_4 (b) HI (c) OF_2 (d) CF_4 (e) H_2Te

___ 3. How many total valence electrons are present in the H_2CO molecule?

(a) 16 (b) 24 (c) 12 (d) 14 (e) 18

___ 4. How many valence electrons are present in the NO^+ ion?

(a) 14 (b) 15 (c) 10 (d) 11 (e) 12

___ 5. The N_2 molecule contains which of the following bonds?

(a) an ionic bond (b) a single bond
(c) a double bond (d) a triple bond
(e) a quadruple bond

___ 6. Which of the following is not a resonance structure for $C_2O_4^{2-}$?

(a) 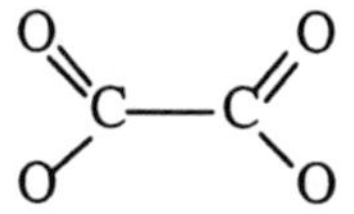 (b) 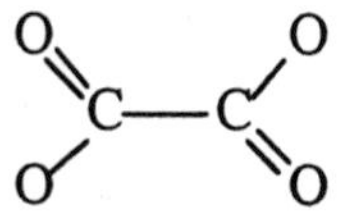 (c)

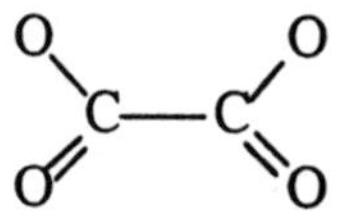

(d) 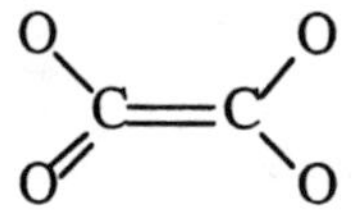 (e)

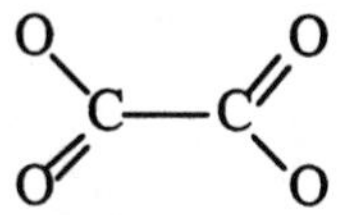

___ 7. Which of the following has a carbon-oxygen bond half-way between a single and a double bond?

(a) CO_3^{2-} (b) $H_3C\text{-}CO_2^-$ (c) CO_2 (d) CO

B-2 Problems

1. Write the simplest formulas of covalent compounds formed between the following two elements (single bonds only):

   (a) P and H ________________

   (b) F and I ________________

   (c) O and Br ________________

   (d) Se and H ________________

2. Using only the $6N + 2$ rule, complete the following table for the given molecules or ions:

Species	Single bonds	Number of double bonds	Triple bonds
SeO_3	______	______	______
NO^+	______	______	______
ClO_4^-	______	______	______
C_4H_8	______	______	______
CH_3NO_2	______	______	______

3. Write Lewis structures for the following. Where appropriate, write equivalent resonance structures.

(a) $AsCl_3$

(b) TeO_2

(c) NCO^- (C in the middle)

(d) Na_2SO_3

(e) HNO_3

(f) C_2H_2

Review Section C *The Distribution of Charge in Chemical Bonds*

OUTLINE	OBJECTIVES
6-7 Electronegativity and Polarity of Molecules	
1. The definition of electronegativity	*Give the general periodic trends in electronegativities.*
2. Polar bonds and dipoles	*Determine the direction of a bond dipole from the electronegativities of the elements.*
3. Electronegativity difference and ionic bonds	*Predict ionic bonding from differences in electronegativities.*

6-8 Geometry of Simple Molecules

1. VSEPR Theory — *Describe the basic premise of the VSEPR theory.*
2. The molecular geometry of a molecule
3. Predicting molecular geometry from a Lewis structure — *Using VSEPR theory, determine the molecular geometry of a compound from its Lewis structure.*

6-9 Polarity of Molecules

1. Molecular dipoles
2. Polarity and molecular geometry — *Predict which molecules are polar based on their molecular geometry.*

SUMMARY OF SECTIONS 6-7 THROUGH 6-9

Questions: *Are there bonds that are intermediate between ionic and covalent? How can Lewis structures be used to predict molecular geometry? How can molecular geometry be used to predict molecular polarity?*

Electronegativity is a periodic property of elements that is a measure of the attraction an atom has for electrons in a chemical bond. Nonmetals are more electronegative than metals with the most electronegative element, fluorine, having an electronegativity of 4.0. When different elements form a bond, the more electronegative atom acquires a partial negative charge leaving the other atom with a partial positive charge. The bond then contains a **dipole** (two poles) and is thus a **polar bond**. This is illustrated as follows:

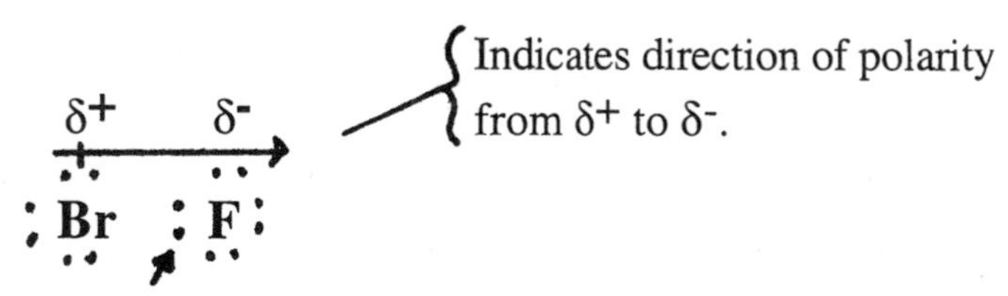

Pair of electrons in the bond are drawn closer to the F thus giving the F a partial negative charge (indicted by δ^-)

The existence of polar covalent bonds indicates that the dividing line between purely covalent bonds (equal sharing between elements of the same electronegativity) and ionic (no sharing between elements with a difference in electronegativity of 2.0 or greater) is not exact. The more polar a bond, the more ionic character it has. Thus a polar covalent bond represents the intermediate ground between the two extremes of an ionic bond and a **nonpolar** covalent bond (equal sharing) illustrated as follows:

Na^{+} $:\ddot{F}:^{-}$ $\qquad\qquad$ $:\ddot{F}—\ddot{F}:$

Ionic	Nonpolar Covalent
Large difference in electro-negativity between Na and F. *Complete electron exchange.*	No difference in electro-negativity between two F's. *Equal sharing of electrons.*

Even though a molecule may contain polar bonds, the molecule itself may not be polar. The polarity of the molecule depends on the geometry of the molecule. The **molecular geometry** can be determined from its Lewis structure and the application of **VSEPR theory.** This simple theory simply tells us that electron pairs on a central atom, either bonded or as lone pairs, tend to repel each other to the maximum extent. The three symmetrical geometries assumed by molecules containing no lone pairs are linear (two bonds), trigonal planar (three bonds), and tetrahedral (four bonds). When the molecule has lone pairs in place of bonded pairs, the molecular geometries are described as V-shaped (two bonds and one lone pair), trigonal pyramid (three bonds and one lone pair), and V-shaped (two bonds and two lone pairs).

A polar bond causes a bond dipole which is like a force with both magnitude and direction. If the forces are equal and are oriented in space so as to oppose each other, the dipoles cancel. If a molecule has one of the three symmetrical geometries and all terminal atoms are the same, the molecule is nonpolar. An example of a nonpolar molecule with three polar bonds is as follows:

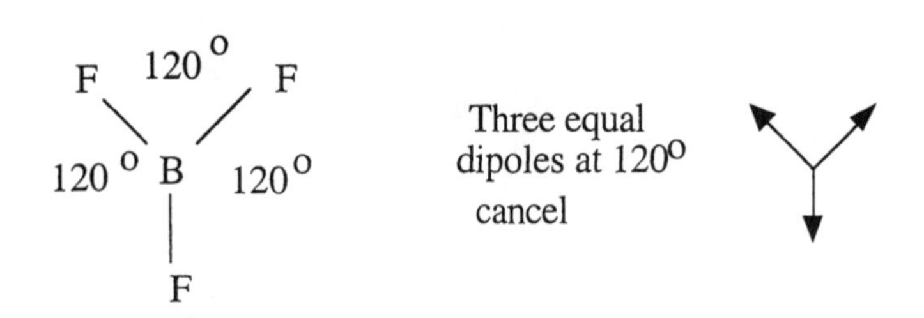

Because H_2O is a V-shaped molecule, it has a net **molecular dipole.** This very important fact accounts for many of the properties of water that will be discussed in Chapter 10.

NEW TERMS

Dipole	Nonpolar bond
Electronegativity	Polar covalent bond
Molecular dipole	VSEPR theory
Molecular geometry	

SELF-TEST

C-1 Multiple Choice

___ 1. Which of the following has the highest electronegativity?

(a) N (b) I (c) C (d) H (e) Na

___ 2. Which of the following two elements would have the greatest difference in electronegativity?

(a) Al, I (b) K, N (c) K, S (d) B, H (e) C, H

___ 3. Which of the following contains a polar covalent bond?

(a) H_2 (b) N_2 (c) KBr (d) H_2S (e) KF

___ 4. Which of the following elements always has the negative dipole in any polar covalent bond?

(a) Fr (b) At (c) F (d) H (e) O

___ 5. Which of the following molecules is polar?

(a) SO_3 (b) O_3 (c) N_2 (d) BCl_3 (e) SO_2

Review Section D

CHAPTER SUMMARY SELF-TEST

D-1 Formulas

Write the formula of the simplest compound formed by each pair of elements. Indicate whether the compound is primarily ionic or covalent.

	Formula	Ionic or Covalent
(a) Si and H	________	________
(b) Sr and Cl	________	________
(c) S and F	________	________
(d) Ga and O	________	________
(e) Mg and N	________	________
(f) B and S	________	________

D-2 Matching

___	A formula unit with both ionic and covalent bonds	(a) NO
		(b) Cl_2O
___	A formula unit with only ionic bonds	(c) SO_2
___	A molecule that cannot follow the octet rule	(d) Rb_2Se
		(e) N_2
___	A molecule containing the most electronegative element	(f) $KBrO_4$
___	A molecule with a nonpolar covalent bond	(g) CCl_4
		(h) SiF_4
___	A molecule with resonance structures	
		(i) CO

D-3 Bonding

Given the following five compounds with three atoms in one formula unit:

OF_2, K_2O, SeO_2, CO_2, and SrF_2

(a) The two formula units that contain ions ______ and ______

(b) The formula unit that has V-shaped molecules with an angle of about 120° ______

(c) The formula unit that has nonpolar molecules ______

(d) The formula unit that has V-shaped molecules with an angle of about 109° ______

Answers to Self-Tests

A-1 Multiple Choice

1. **b** -2, Te is in Group VIA which can gain two electrons.
2. **a** B and H. Both of these elements are nonmetals.
3. **c** +2. Alkaline earth metals can lose two electrons to form an octet.
4. **e** Mn_2O_7. The Mn would be a +7 ion, which is too high for an ion.

5. **d** BaCl Either the Ba would have a +1 charge or the Cl a -2 charge, neither of which has an octet of electrons.

A-2 Matching

e I^- is a nonmetal ion with the same electron configuration as Xe.

d Pb^{2+} is a metal ion that does not follow the octet rule.

g La^{3+} is a metal ion with the same electron configuration as Xe.

h N^{3-} is a possible -3 ion. (Sb is a metal.)

d Al^{3+} is an ion predicted for Group IIIA. (B is a nonmetal.)

a Te^- is a nonmetal ion that violates the octet rule.

A-3 Problems

1. (a) Sr^{2+} [Group IIA]

   (b) Ga^{3+} [Group IIIA]

   (c) H^-

   (d) Se^{2-} [Group VIA]

   (e) I^- [Group VIIA]

2. (a) Al^{3+} and Se^{2-} = Al_2Se_3

   (b) Be^{2+} and F^{1-} = BeF_2

   (c) Ba^{2+} and O^{2-} = BaO

   (d) K^{1+} and N^{3-} = K_3N

B-1 Multiple Choice

1. **b** H, S Both elements are nonmetals.
2. **a** NH_4 There would be nine electrons on the nitrogen.
3. **c** 12 [4 + 6 + (2 x 1) = 12]
4. **c** 10 (6 + 5 - 1 = 10)
5. **d** a triple bond

6. **d**

(d) 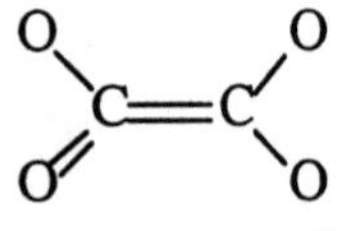 This is not a correct Lewis structure.

7. **b** $H_3C\text{-}CO_2^-$ The two resonance structures are:

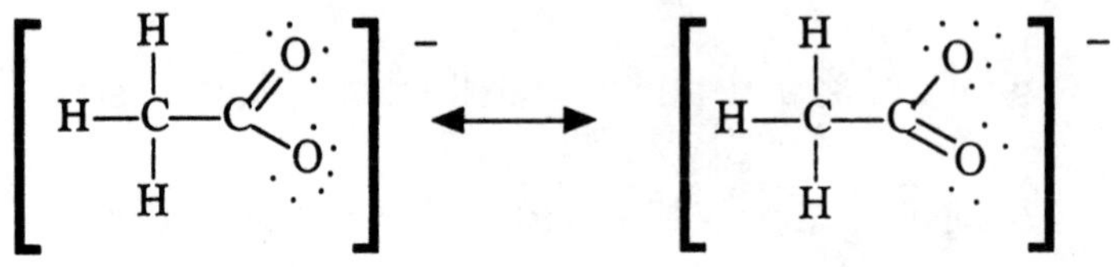

$$\left[H_3C-C(=O)-O^- \right]^- \longleftrightarrow \left[H_3C-C(-O^-)=O \right]^-$$

B-2 Problems

1.

(a) PH_3 H—P—H (with H below P)

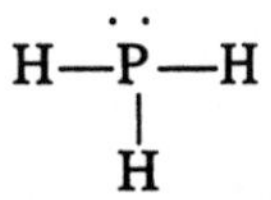

(b) IF :F—I:

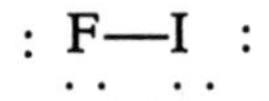

(c) Br_2O :Br—O—Br:

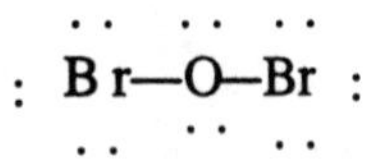

(d) H_2S H—Se—H

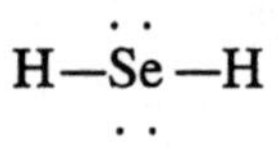

2. SeO_3 two single, one double

NO^+ one triple

ClO_4^- four single

C_4H_8 10 single, one double (There are 12 atoms so there are a total of 11 bonds.)

CH_3NO_2 five single, one double

3. (a) $AsCl_3$ Valence electrons

As	1 x 5 =	5
Cl	3 x 7 =	21
		26

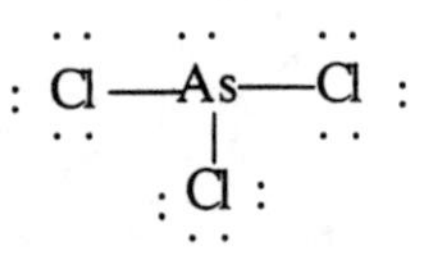

:Cl—As—Cl: (with :Cl: below As)

(b) TeO_2

	Valence electrons	
Te	1 x 6 =	6
O	2 x 6 =	12
		18

Two resonance structures

(c) NCO^- Valence electrons

C	1 x 4 =	4
N	1 x 5 =	5
O	1 x 6 =	6
Charge =		1
		16

$[:\ddot{N}=C=\ddot{O}:]^-$

(d) Na_2SO_3 This contains $2(Na^+)$ ions and a SO_3^{2-} ion.

For SO_3^{2-}

	Valence electrons	
S	1 x 6 =	6
O	3 x 6 =	18
Charge =		2
		26

$\left[:\ddot{O}-\ddot{S}-\ddot{O}: \right]^{2-}$ (with $:\ddot{O}:$ bonded below S)

(e) HNO_3 Valence electrons

H	1 x 1 =	1
N	1 x 5 =	5
O	3 x 6 =	18
		24

(The N is the central atom, and the H is attached to O.)
Two resonance structures.

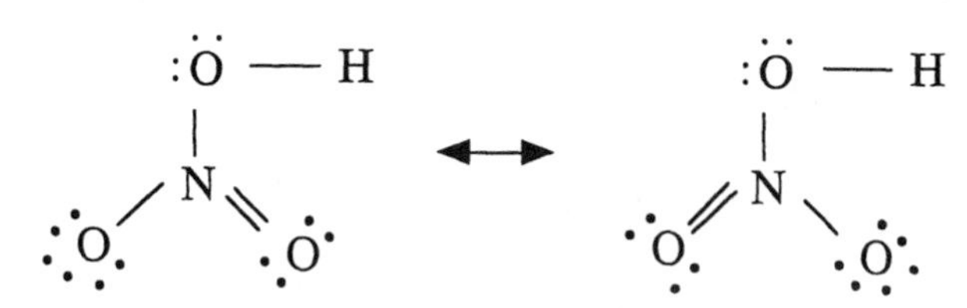

(f) C_2H_2

	Valence electrons	
H	2 x 1 =	2
C	2 x 4 =	8
		10

H—C≡C—H

C-1 Multiple Choice

1. **a** N
2. **b** K, N (0.8 and 3.0)
3. **d** H_2S
4. **c** F It is the most electronegative element.
5. **e** SO_2 The bond dipoles do not cancel because the molecule is V-shaped.

D-1 Formulas

(a) SiH_4 Covalent

(b) $SrCl_2$ Ionic

(c) SF_2 Covalent

(d) Ga_3O_2 Ionic

(e) Mg_3N_2 Ionic by metal-nonmetal criterion but molecular by electronegativity difference

(f) B_2S_3 Covalent

D-2 Matching

f $KBrO_4$ contains both ionic and covalent bonds (i.e., $K^+BrO_4^-$).

d Rb_2Se contains only an ionic bond (i.e., Rb^+ and Se^{2-}).

a NO cannot follow the octet rule because it has 11 valence electrons.

h SiF_4 is a molecule containing the most electronegative element fluorine.

e N_2 is a molecule with a nonpolar covalent bond.

c SO_2 is a molecule with two resonance structures.

D-3 Bonding

(a) K_2O ($2K^+, O^{2-}$) and SrF_2 ($Sr^{2+}, 2F^-$) contain only ions.

(b) SeO_2 has one lone pair on the central atom so it is V-shaped with an angle of about 120°.

(c) CO_2 is linear so it is nonpolar.

(d) OF_2 has two lone pairs on the central atom so it is V-shaped with an angle about 109°.

Solutions to Black Text Problems

6-94 Assume 100 g of compound. There is then 27.4 g Na, 14.3 g C, 57.1 g O, and 1.19 g H per 100 g.

$$27.4\ \cancel{\text{g Na}} \times \frac{1\ \text{mol Na}}{22.99\ \cancel{\text{g Na}}} = 1.19\ \text{mol Na} \qquad 14.3\ \cancel{\text{g C}} \times \frac{1\ \text{mol C}}{12.01\ \cancel{\text{g C}}} = 1.19\ \text{mol C}$$

$$57.1\ \cancel{\text{g O}} \times \frac{1\ \text{mol O}}{16.00\ \cancel{\text{g O}}} = 3.57\ \text{mol O} \qquad 1.19\ \cancel{\text{g H}} \times \frac{1\ \text{mol H}}{1.008\ \cancel{\text{g H}}} = 1.18\ \text{mol H}$$

$$\text{O: } \frac{3.57}{1.18} = 3.0 \qquad \text{Formula} = NaHCO_3 = Na^+\ HCO_3^-$$

H — O — C(=O)(—O:) $^-$

The geometry around the C is trigonal planar with the approximate H-O-C angle of 120°.

6-96 In 100 g of compound there are 19.8 g Ca, 1.00 g H, 31.7 g S, and 47.5 g O.

$$19.8\ \cancel{\text{g Ca}} \times \frac{1\text{mol Ca}}{40.08\ \cancel{\text{g Ca}}} = 0.494\ \text{mol Ca} \qquad 1.00\ \cancel{\text{g H}} \times \frac{1\ \text{mol H}}{1.008\ \cancel{\text{g H}}} = 0.992\ \text{mol H}$$

$$31.7\ \cancel{\text{g S}} \times \frac{1\text{mol S}}{32.07\ \cancel{\text{g S}}} = 0.988\ \text{mol S} \qquad 47.5\ \cancel{\text{g O}} \times \frac{1\text{mol O}}{16.00\ \cancel{\text{g O}}} = 2.97\ \text{mol O}$$

$$\text{H: } \frac{0.992}{0.494} = 2.0 \qquad \text{S: } \frac{0.988}{0.494} = 2.0 \qquad \text{O: } \frac{2.97}{0.494} = 6.0$$

Formula = $CaH_2S_2O_6$ = $Ca(HSO_3)_2$ calcium bisulfite or calcium hydrogen sulfite

H — O — S(—O:) — O: $^-$

The geometry around the S is trigonal pyramid. The H-O-S angle is about 109°.

7

Chemical Reactions

Review Section A *The Representation of Chemical Changes and Some Types of Changes*

OUTLINE / **OBJECTIVES**

7-1 The Chemical Equation

1. Representing reactions — *Describe the information represented by a chemical equation.*

2. Balancing equations — *Balance simple equations.*

7-2 Combustion, Combination, and Decomposition Reactions

1. Combustion
2. Combination
3. Decomposition

Classify appropriate reactions into one of the three types listed.

SUMMARY OF SECTIONS 7-1 AND 7-2

Questions: *How do chemists represent chemical changes? Why must equations be balanced? What are the characteristics of three simple groupings of reactions?*

Learning chemistry is much like learning a foreign language. The letters of the language are the symbols of the elements, and the letters combine into words that are the formulas of compounds. Finally, we see in this chapter, that the words combine into sentences which are the chemical equations. **Chemical equations** represent, not only what chemical changes occur, but *how much.* Before a chemical equation can tell how much, it must follow the law of conservation of mass. This means that the equation must be balanced with the same number of atoms on both sides of the equation. This is accomplished by appropriate **coefficients** placed before specific compounds in the equation. The implications of a typical **balanced equation** are illustrated as follows:

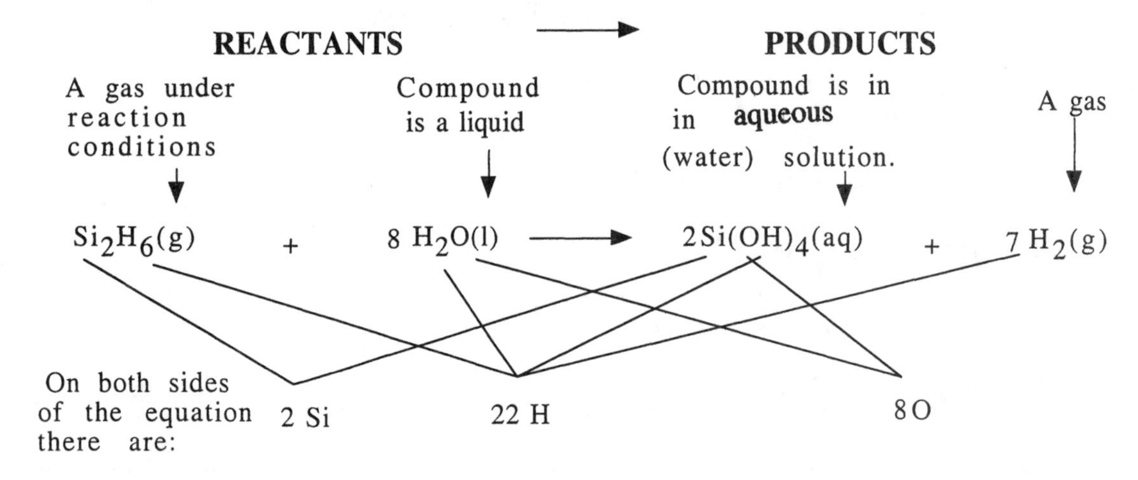

Consider how the above equation was balanced. Elements other than hydrogen or oxygen should be balanced first. Thus silicon is the first element balanced in the above equation by placing a coefficient of "2" before the $Si(OH)_4$. Hydrogen or oxygen is balanced next. Since oxygen appears in the fewer number of compounds, we next place an "8" in front of the H_2O to balance the eight oxygens in $2Si(OH)_4$. We now have 22 hydrogens on the left and eight hydrogens on the right in the $2Si(OH)_4$. By placing a coefficent of "7" in front of the H_2, we now have 22 hydrogens on the right. The equation is balanced.

There are several ways to categorize or classify chemical reactions. One simple way that includes a great many chemical reactions is described in this chapter. The five types of reactions described in this chapter do not include all chemical reactions; in later chapters, we will find there are other convenient classifications of chemical reactions that suit certain purposes. In any case, three of these five types and an example of each are as follows:

1. **Combustion reactions**

   Ex. $2C_2H_2(g) + 5O_2(g) \rightarrow 4CO_2(g) + 2H_2O(l)$

2. **Combination reactions** ($A + B \rightarrow C$)

   Ex. $LiOH(aq) + CO_2(g) \rightarrow LiHCO_3(aq)$

3. **Decomposition reactions** ($C \rightarrow A + B$)

   Ex. $2Au_2O_3(s) \rightarrow 4Au(s) + 3O_2(g)$

We can recognize these three types quite easily. Oxygen is involved as a reactant in any combustion reaction. A combination reaction results in one product and a decomposition reaction is just the opposite. That is, it has only one reactant.

NEW TERMS

Balanced equation	Combustion reaction
Chemical equation	Decomposition reaction
Coefficient	Product
Combination reaction	Reactant

SELF-TEST

A-1 Balancing Equations

1. Balance the following:

   (a) $UO_2 + HF \rightarrow UF_4 + H_2O$

   (b) $HCl + MgCO_3 \rightarrow MgCl_2 + CO_2 + H_2O$

   (c) $NaNO_3 \rightarrow NaNO_2 + O_2$

   (d) $Cu + HNO_3 \rightarrow Cu(NO_3)_2 + NO_2 + H_2O$

   (e) $Si_2H_6 + O_2 \rightarrow SiO_2 + H_2O$

   (f) $P_4O_{10} + H_2O \rightarrow H_3PO_4$

   (g) $UO_2 + CaH_2 \rightarrow U + CaO + H_2$

2. Change the following into formulas including all information and balance. Use the following symbols: solid (s), liquid (l), gas (g), aqueous solution (aq), yields or produces ($\rightarrow$), and heat ($\xrightarrow{\Delta}$).

   (a) Liquid alcohol (C_2H_6O) undergoes combustion to produce gaseous carbon dioxide and water.

(b) Solid iron(III) hydroxide decomposes to iron(III) oxide and water when heated.

(c) Sulfur (S_8) combines with fluorine to produce sulfur hexafluoride gas.

Review Section B *The Role of Ions in Water and Some Ways the Ions Interact*

OUTLINE	OBJECTIVES
7-3 The Formation of Ions in Water	
1. The composition of a solution	*Define the following terms - soluble, insoluble, solute, solvent, and solution.*
2. Ions in solution from ionic compounds	
3. Ions in solution from strong acids	*Write equations illustrating the solution of soluble ionic compounds and strong acids in water.*
7-4 Single-Replacement Reactions	
1. Substitution of metals in solution	*Write a balanced equation illustrating a single-replacement reaction.*
2. Molecular, total ionic, and net ionic equations	*Write balanced molecular, total ionic, and net ionic equations for specified single-replacement reactions.*
3. The activity series	*Use the table of the activity series to predict single-replacement reactions.*
7-5 Double-Replacement Reactions - Precipitation	
1. Solubility rules for some ionic compounds	*Determine whether a specified ionic compound is soluble or not from a table.*
2. The formation of a precipitate from two soluble compounds	*Use the table of solubilities to predict the occurrence of precipitation reactions and write the balanced molecular, total ionic, and net ionic equations illustrating the reactions.*
7-6 Double-Replacement Reactions - Neutralization	
1. Acids and bases	*Identify the strong acids and bases and the ions they form in aqueous solution.*

2. Salts — *Identify a salt as a product of a neutralization reaction.*

3. The formation of a salt and water — *Write the balanced molecular, total ionic, and net ionic equations for neutralizations of strong acids and bases.*

SUMMARY OF SECTIONS 7-3 THROUGH 7-6

Questions: *What happens when ionic compounds and strong acids dissolve in water? Why do certain metals dissolve in acid? How can we tell which ionic compounds do not dissolve in water? How do solids form in water? How are acids neutralized?*

When ionic compounds dissolve in water, the water molecules disperse the ions into a homogeneous mixture known as a **solution**. The solution of a typical ionic compound that is **soluble** in water is represented as follows:

$$Na_2CrO_4(s) \xrightarrow{H_2O} 2Na^+(aq) + CrO_4{}^{2-}(aq)$$

Strong acids, although molecular compounds in the pure state, are also ionized by water to produce ions. The solution of a typical strong acid in water is represented as follows:

$$HClO_4(aq) \xrightarrow{H_2O} H^+(aq) + ClO_4{}^-(aq)$$

Compounds with low solubilities are known as **insoluble**. Regardless of whether a compound is considered soluble or insoluble, there is a limit to how much of any compound dissolves in water. At a certain temperature the limit is known as the compound's **solubility**.

Ions are involved in many chemical reactions that occur in aqueous solution. In this chapter we focused on two types of reactions: single-replacement and double-replacement reactions.

Single replacement reactions ($A + BC \rightarrow AC + B$)

In ancient times, it was thought that lead could be transmuted into gold by immersing a lead bar in a certain aqueous solution. Actually, what was happening was a single-replacement reaction involving the exchange of Pb^{2+} ions for Au^{3+} ions and solid gold for solid lead. In other words, that certain solution contained gold ions in solution. This reaction is illustrated by the following **molecular equation**:

$$3Pb(s) + 2Au(NO_3)_3(aq) \rightarrow 3Pb(NO_3)_2(aq) + 2Au(s)$$

Reactions can also be illustrated by equations showing the neutral ionic compounds as separate cations and anions in solution. This is known as the **total ionic equation** and is illustrated as follows:

$$3Pb(s) + 2Au^{3+}(aq) + 6NO_3{}^-(aq) \rightarrow 3Pb^{2+}(aq) + 6NO_3{}^-(aq) + 2Au(s)$$

The previous equation can be simplified by subtracting out of the equation the ions that are identical on both sides of the equation. These are known as the **spectator ions** and the resulting equation is known as the **net ionic equation**. The net ionic equation allows us to focus on the species that actually changed in the reaction. By subtracting $6NO_3^-$ ions from both sides of the equation we have:

$$3Pb(s) + 2Au^{3+}(aq) \rightarrow 3Pb^{2+}(aq) + 2Au(s)$$

Metals have different abilities to replace ions of other metals from solutions. With simple experiments, an order can be established which is known as the **activity series**. (See Table 7-2 in the text.) A specific metal will spontaneously react with all of the metal ions underneath that metal in the table. (Notice that Pb is higher than Au^{3+} so the reaction shown above is spontaneous.) In this table, H_2 is treated as a metal and H^+ (from strong acids) as its metal ion.

Double replacement reactions (AB + CD $\rightarrow$ AD + CB)

Another type of reaction involves two ionic compounds in water. If solutions of two soluble compounds are mixed, a chemical reaction may occur if the cation in one solution forms an insoluble compound with the anion in the other solution. Table 7-3 in the text can be used to determine whether some familiar ionic compounds are insoluble. When the solutions are mixed, the ions join together to form the solid state known as a **precipitate**. Such reactions are thus known as **precipitation reactions** and are one type of double-replacement reaction. An example of a precipitation reaction follows.

Example B-1 The Equations Illustrating a Precipitation Reaction.

Write the balanced molecular, total ionic, and net ionic equations when solutions of sodium chloride and silver nitrate are mixed.

PROCEDURE

If the reactants are NaCl and $AgNO_3$, an exchange of ions produces possible products AgCl and $NaNO_3$. From Table 7-3 in the text, we note that AgCl is insoluble.

Solution #1		Solution #2		Solution #3 (#1 + #2)
	+		*produces*	
[NaCl(aq)]		[$AgNO_3$(aq)]		[AgCl(s) + $NaNO_3$(aq)]
Na^+ (aq)		Ag^+ (aq)	$\longrightarrow$	AgCl (s)
Cl^- (aq)		NO_3^- (aq)		Na^+ (aq) + NO_3^- (aq)

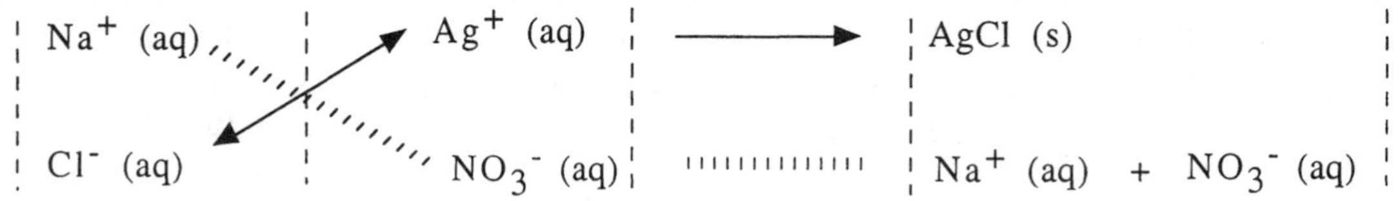

This chemical reaction can be illustrated by equations written in three different forms.

SOLUTION

Molecular equation: $NaCl(aq) + AgNO_3(aq) \rightarrow AgCl(s) + NaNO_3(aq)$

Total ionic equation: $Na^+(aq) + Cl^-(aq) + Ag^+(aq) + NO_3^-(aq) \rightarrow AgCl(s) + Na^+(aq) + NO_3^-(aq)$

Net ionic equation: $Ag^+(aq) + Cl^-(aq) \rightarrow AgCl(s)$

In another example, consider what happens when solutions of $(NH_4)_2S$ and $Pb(NO_3)_2$ are mixed. The compounds formed by an exchange of ions are NH_4NO_3 and PbS. Table 7-3 tells us that all nitrates are soluble but most sulfides (including Pb^{2+}) are insoluble. Therefore, the following reaction occurs:

$$\underset{\text{(soluble)}}{(NH_4)_2S(aq)} + \underset{\text{(soluble)}}{Pb(NO_3)_2(aq)} \rightarrow \underset{\text{(insoluble)}}{PbS(s)} + \underset{\text{(soluble)}}{2NH_4NO_3(aq)}$$

A second type of double-replacement reaction is possible where one of the products is a molecular compound rather than an insoluble ionic compound. **Neutralization reactions** are examples of this latter type of reaction. As mentioned earlier, strong acids are completely ionized by water to form $H^+(aq)$ ions. Another unique type of compound in water is known as a base. Bases obtain their unique character from the production of hydroxide (OH^-) ions in aqueous solution. The common **strong bases** (e.g., NaOH), which are ionic solids when pure, dissolve to form the same ions in solution.

$$NaOH(s) \xrightarrow{H_2O} Na^+(aq) + OH^-(aq)$$

When strong acids and strong bases are mixed, a neutralization reaction occurs. In a neutralization, the cation (of the acid) and the anion (of the base) combine to form the molecular compound water. The spectator ions usually remain in solution forming what is known as a **salt.** An example of a neutralization reaction between a strong acid and a strong base follows.

Example B-2 The Equations Illustrating a Neutralization Reaction

Write the balanced molecular, total ionic, and net ionic equations when solutions of nitric acid and sodium hydroxide are mixed.

PROCEDURE

Solution #1		Solution #2		Solution #3 (#1 + #2)
$[HNO_3(aq)]$	+	$[NaOH(aq)]$	*produces*	$[H_2O + NaNO_3(aq)]$
NO_3^- (aq)		OH^- (aq)	$\longrightarrow$	H_2O
H^+ (aq)		Na^+ (aq)		Na^+ (aq) + NO_3^- (aq)

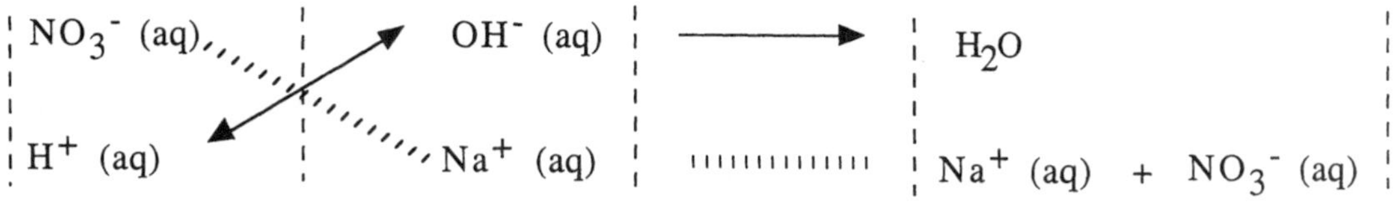

SOLUTION

Molecular equation: $HNO_3(aq) + NaOH(aq) \rightarrow H_2O + NaNO_3(aq)$

Total ionic equation: $H^+(aq) + NO_3^-(aq) + Na^+(aq) + OH^-(aq) \rightarrow H_2O + Na^+(aq) + NO_3^-(aq)$

Net ionic equation:	$H^+(aq) + OH^-(aq) \rightarrow H_2O$ (The net ionic equations for all neutralizations between strong acids and bases are the same.)

The molecular, total ionic, and net ionic equations for another example of a neutralization follow.

$$\textit{Acid} + \textit{Base} \rightarrow \textit{Salt} + \textit{Water}$$

Molecular: $H_2SO_4(aq) + 2KOH(aq) \rightarrow K_2SO_4(aq) + 2H_2O$

Total ionic: $2H^+(aq) + SO_4^{2-}(aq) + 2K^+(aq) + 2OH^-(aq) \rightarrow 2K^+(aq) + SO_4^{2-}(aq) + 2H_20$

Net ionic: $H^+(aq) + OH^-(aq) \rightarrow H_2O$

NEW TERMS

Acid
Activity series
Base
Double-replacement reaction
Insoluble
Molecular equation
Net ionic equation
Neutralization reaction
Precipitate
Precipitation reaction
Salt
Single-replacement reaction
Soluble
Solute
Solution
Solvent
Spectator ion
Strong acid
Strong base
Total ionic equation

SELF-TEST

B-1 Problems

1. Write equations illustrating the solution of the following compounds in water.

   (a) $Sr(ClO_3)_2$

   (b) KNO_2

   (c) $(NH_4)_2SO_3$

(d) HI

2. Write the total ionic and net ionic equations for:

(a) $Sn(s) + 2AgNO_3(aq) \rightarrow Sn(NO_3)_2(aq) + 2Ag(s)$

(b) $K_2SO_3(aq) + Sr(ClO_3)_2(aq) \rightarrow SrSO_3(s) + 2KClO_3(aq)$

(c) $2(NH_4)_3PO_4(aq) + 3CoCl_2(aq) \rightarrow Co_3(PO_4)_2(s) + 6NH_4Cl(aq)$

(d) $HClO_4(aq) + LiOH(aq) \rightarrow LiClO_4(aq) + H_2O(l)$

3. Write the balanced molecular, total ionic, and net ionic equations illustrating the following:

(a) When cobalt metal is placed in a hydrobromic acid solution, hydrogen gas is evolved and a solution of $CoBr_3$ is formed.

(b) When an aqueous solution of ammonium sulfate is mixed with an aqueous solution of lead(II) acetate, a precipitate of lead(II) sulfate forms.

(c) When a solution of iron(III) chlorate is mixed with a solution of potassium hydroxide, a precipitate of iron(III) hydroxide forms.

(d) When solutions of strontium hydroxide and trifluoroacetic acid ($HC_2F_3O_2$) are mixed, a neutralization occurs. (Trifluoroacetic acid is a strong acid.)

Review Section C

CHAPTER SUMMARY SELF-TEST

C-1 Types of Reactions

Balance the following equations. Classify the reactions as Combination (CO), Decomposition (D), Combustion (CB), Single-Replacement (SR), Double-Replacement Precipitation (DRP), or Double-Replacement Neutralization (DRN).

1. $Na(s) + O_2(g) \rightarrow Na_2O_2(s)$
2. $H_2SO_4(aq) + Ca(OH)_2(aq) \rightarrow CaSO_4(s) + H_2O(l)$
3. $AlI_3(aq) + AgNO_3(aq) \rightarrow AgI(s) + Al(NO_3)_3(aq)$
4. $Pt(SO_4)_2(aq) + Sn(s) \rightarrow Pt(s) + SnSO_4(aq)$
5. $B_4H_{10}(g) + O_2(g) \rightarrow B_2O_3(s) + H_2O(l)$
6. $H_3PO_3(aq) \rightarrow H_2O(l) + P_4O_6(s)$
7. $C_6H_{12}O_6(s) \rightarrow C(s) + H_2O\ (l)$
8. $Mg(s) + N_2(g) \rightarrow Mg_3N_2(s)$
9. $CaCO_3(s) + H_2O(l) + CO_2(g) \rightarrow Ca(HCO_3)_2(s)$
10. $Cl_2O_3(g) + H_2O(l) \rightarrow HClO_2(aq)$
11. $HCl(aq) + Mg(OH)_2\ (s) \rightarrow MgCl_2(aq) + H_2O(l)$
12. $Sc(s) + AgNO_3(aq) \rightarrow Sc(NO_3)_3(aq) + Ag(s)$
13. $Ag_2O(s) \rightarrow Ag(s) + O_2(g)$
14. $NH_3(g) + O_2(g) \rightarrow NO(g) + H_2O(l)$

C-2 Net Ionic Equations

Write the balanced net ionic equations for any single-replacement or double-replacement reactions in Problem C-1.

Answers to Self-Tests

A-1 Balancing Equations

1. (a) $UO_2 + 4HF \rightarrow UF_4 + 2H_2O$

   (b) $2HCl + MgCO_3 \rightarrow MgCl_2 + CO_2 + H_2O$

   (c) $2NaNO_3 \rightarrow 2NaNO_2 + O_2$

   (d) $Cu + 4HNO_3 \rightarrow Cu(NO_3)_2 + 2NO_2 + 2H_2O$

   (e) $2Si_2H_6 + 7O_2 \rightarrow 4SiO_2 + 6H_2O$

   (f) $P_4O_{10} + 6H_2O \rightarrow 4H_3PO_4$

   (g) $UO_2 + 2CaH_2 \rightarrow U + 2CaO + H_2$

2. (a) $C_2H_6O(l) + 3O_2(g) \rightarrow 2CO_2(g) + 3H_2O(l)$

   (b) $2Fe(OH)_3(s) \xrightarrow{\Delta} Fe_2O_3(s) + 3H_2O(g)$

   (c) $S_8(s) + 24F_2(g) \rightarrow 8SF_6(g)$

B-1 Problems

1. (a) $Sr(ClO_3)_2 \xrightarrow{H_2O} Sr^{2+}(aq) + 2ClO_3^-(aq)$

   (b) $KNO_2 \xrightarrow{H_2O} K^+(aq) + NO_2^-(aq)$

   (c) $(NH_4)_2SO_3 \xrightarrow{H_2O} 2NH_4^+(aq) + SO_3^{2-}(aq)$

   (d) $HI(aq) \xrightarrow{H_2O} H^+(aq) + I^-(aq)$

2. (a) $Sn(s) + 2Ag^+(aq) + 2NO_3^-(aq) \rightarrow Sn^{2+}(aq) + 2NO_3^-(aq) + 2Ag(s)$

$Sn(s) + 2Ag^+(aq) \rightarrow Sn^{2+}(aq) + 2Ag(s)$

(b) $2K^+(aq) + SO_3^{2-}(aq) + Sr^{2+}(aq) + 2ClO_3^-(aq) \rightarrow$

$SrSO_3(s) + 2K^+(aq) + 2ClO_3^-(aq)$

$Sr^{2+}(aq) + SO_3^{2-}(aq) \rightarrow SrSO_3(s)$

(c) $6NH_4^+(aq) + 2PO_4^{3-}(aq) + 3Co^{2+}(aq) + 6Cl^-(aq) \rightarrow$

$Co_3(PO_4)_2(s) + 6NH_4^+(aq) + 6Cl^-(aq)$

$3Co^{2+}(aq) + 2PO_4^{3-}(aq) \rightarrow Co_3(PO_4)_2(s)$

(d) $H^+(aq) + ClO_4^-(aq) + Li^+(aq) + OH^-(aq) \rightarrow$

$Li^+(aq) + ClO_4^-(aq) + H_2O(l)$

$H^+(aq) + OH^-(aq) \rightarrow H_2O(l)$

3. (a) $2Co(s) + 6HBr(aq) \rightarrow 2CoBr_3(aq) + 3H_2(g)$

$2Co(s) + 6H^+(aq) + 6Br^-(aq) \rightarrow Co^{3+}(aq) + 6Br^-(aq) + 3H_2(g)$

$2Co(s) + 6H^+(aq) \rightarrow Co^{3+}(aq) + 3H_2(g)$

(b) $(NH_4)_2SO_4(aq) + Pb(C_2H_3O_2)_2(aq) \rightarrow$

$PbSO_4(s) + 2NH_4C_2H_3O_2(aq)$

$2NH_4^+(aq) + SO_4^{2-}(aq) + Pb^{2+}(aq) + 2C_2H_3O_2^-(aq) \rightarrow$

$PbSO_4(s) + 2NH_4^+(aq) + 2C_2H_3O_2^-(aq)$

$Pb^{2+}(aq) + SO_4^{2-}(aq) \rightarrow PbSO_4(s)$

(c) $Fe(ClO_3)_3(aq) + 3KOH(aq) \rightarrow Fe(OH)_3(s) + 3KClO_3(aq)$

$Fe^{3+}(aq) + 3ClO_3^-(aq) + 3K^+(aq) + 3OH^-(aq) \rightarrow$

$Fe(OH)_3(s) + 3K^+(aq) + 3ClO_3^-(aq)$

$Fe^{3+}(aq) + 3OH^-(aq) \rightarrow Fe(OH)_3(s)$

(d) $2HC_2F_3O_2(aq) + Sr(OH)_2(aq) \rightarrow Sr(C_2F_3O_2)_2(aq) + 2H_2O(l)$

$2H^+(aq) + 2C_2F_3O_2^-(aq) + Sr^{2+}(aq) + 2OH^-(aq) \rightarrow$
$Sr^{2+}(aq) + 2C_2F_3O_2^-(aq) + 2H_2O(l)$

$H^+(aq) + OH^-(aq) \rightarrow H_2O(l)$

C-1 Types of Reactions

1. $2Na(s) + O_2(g) \rightarrow Na_2O_2(s)$ **(CO and CB)**

2. $H_2SO_4(aq) + Ca(OH)_2(aq) \rightarrow CaSO_4(s) + 2H_2O(l)$
(DRP and DRN)

3. $AlI_3(aq) + 3AgNO_3(aq) \rightarrow 3AgI(s) + Al(NO_3)_3(aq)$ **(DRP)**

4. $Pt(SO_4)_2(aq) + 2Sn(s) \rightarrow Pt(s) + 2SnSO_4(aq)$ **(SR)**

5. $2B_4H_{10}(g) + 11O_2(g) \rightarrow 4B_2O_3(s) + 10H_2O(l)$ **(CB)**

6. $4H_3PO_3(aq) \rightarrow 6H_2O(l) + P_4O_6(s)$ **(D)**

7. $C_6H_{12}O_6(s) \rightarrow 6C(s) + 6H_2O\ (l)$ **(D)**

8. $3Mg(s) + N_2(g) \rightarrow Mg_3N_2(s)$ **(CO)**

9. $CaCO_3(s) + H_2O(l) + CO_2(g) \rightarrow Ca(HCO_3)_2(s)$ **(CO)**

10. $Cl_2O_3(g) + H_2O(l) \rightarrow 2HClO_2(aq)$ **(CO)**

11. $2HCl(aq) + Mg(OH)_2\ (s) \rightarrow MgCl_2(aq) + H_2O(l)$ **(DRN)**

12. $Sc(s) + 3AgNO_3(aq) \rightarrow Sc(NO_3)_3(aq) + 3Ag(s)$ **(SR)**

13. $2Ag_2O(s) \rightarrow 4Ag(s) + O_2(g)$ **(D)**

14. $4NH_3(g) + 5O_2(g) \rightarrow 4NO(g) + 6H_2O(l)$ **(CB)**

C-2 Net Ionic Equations

2. $2H^{+}(aq) + SO_4^{2-}(aq) + Ca^{2+}(aq) + 2OH^{-}(aq) \longrightarrow CaSO_4(s) + H_2O(l)$

3. $I^{-}(aq) + Ag^{+}(aq) \longrightarrow AgI(s)$

4. $Pt^{4+}(aq) + 2Sn(s) \longrightarrow Pt(s) + 2Sn^{2+}(aq)$

11. $2H^{+}(aq) + Mg(OH)_2\,(s) \longrightarrow Mg^{2+}(aq) + H_2O(l)$

12. $Sc(s) + 3Ag^{+}(aq) \longrightarrow Sc^{3+}(aq) + 3Ag(s)$

8

Quantitative Relationships in Chemistry

Review Section A *The Measurement of Masses of Elements and Compounds*

OUTLINE / OBJECTIVES

8-1 Relative Masses of Elements

1. Counting by weighing — *Describe how large numbers of items can be counted by weighing.*

8-2 The Mole and the Molar Masses of the Elements

1. The mole as a number — *Compare the mole to other counting units.*
2. The mole as a mass — *Determine the molar mass of an element from the periodic table.*
3. Interconversions among mass, moles, and numbers — *Convert among moles, mass, and number of atoms of any element.*

8-3 The Molar Masses of Compounds

1. Formula weight — *Calculate the formula weight of a compound from its formula and atomic masses of its elements.*
2. One mole of a compound — *Convert among moles, mass, and number of molecules or formula units of a compound.*

SUMMARY OF SECTIONS 8-1 THROUGH 8-3

Questions: How do chemists measure known amounts of atoms without counting? How does the chemist's counting unit relate to a specific mass?

The atom is such a small rascal that we must deal with an almost incomprehensible number of them to manipulate in the laboratory. Still, although we do not need to know the actual number of atoms present in a sample, we often need to known the relative number present. In this endeavor the chemist is not alone. The grocer (vegetables and fruit), the builder (nails), and the farmer (seeds) must often sell, purchase, or plant large numbers of items. Since counting is out of the question we rely on a mass scale to provide a count or a comparative number.

In Chapter 3, we briefly touched on how the mass of one atom compares to that of another. We found that the atomic masses of the elements are found by comparison to the standard (^{12}C = 12.000··· amu). The atomic mass of an element can be thought of as the mass of an "average" atom of that element although an "average" atom does not exist. When the atomic masses of the elements are expressed in some mass unit other than amu (i.e., grams) there are obviously a large number of atoms present. However, we do know that the number of atoms is the *same* for each element. In chemistry, this number is known as "one **mole**" and is referred to as **Avogadro's number.**

$$1.000 \text{ mole} = 6.022 \times 10^{23} \text{ objects or particles}$$

The mass of one mole of the atoms of an element is the atomic mass expressed in grams and is known as the **molar mass of the element**. Thus one mole of an element stands for a certain number of atoms (always 6.022×10^{23}) and a mass in grams (which depends on the element).

	Number of atoms	Mass
1.000 mole of atoms	6.022×10^{23}	32.07 g of S
	6.022×10^{23}	238.0 g of U
	6.022×10^{23}	55.85 g of Fe
	6.022×10^{23}	4.003 g of He

The mathematical conversions among moles, mass, and number turn out to be surprisingly straightforward operations. The confusion often centers on the scientific notation necessary in the definition of the mole. Students using the factor-label method (and a little neatness and organization) have little trouble with these problems. This section is extremely important, however, since most of the following chapters require a familiarity with mole relationships.

The mole unit is easily applied to molecules as well as atoms. Again, one mole of molecules implies both a number and a mass. For molecules, the mass of one mole (known as the **molar mass of the compound**) is the **formula weight** expressed in grams. The formula weight of a specific compound is determined from the atomic mass of each element and the number of atoms of that element in one molecule of the compound. For example, consider the compound, N_2O_3.

Element	Number of Atoms		Atomic Mass		Total Mass of Element
N	2	x	14.01 amu	=	28.02 amu
O	3	x	16.00 amu	=	48.00 amu
			Formula Weight	=	76.02 amu

	Number of Molecules or Formula Units	Mass
1.000 mole of molecules or formula units	6.022×10^{23}	76.02 g of N_2O_3
	6.022×10^{23}	28.01 g of CO
	6.022×10^{23}	138.2 g of K_2CO_3
	6.022×10^{23}	352.0 g of UF_6

Conversions among moles, mass, and number of molecules are carried out in the same manner as with moles of atoms. Two examples of the calculations follow.

Example A-1 Converting Moles to Mass

What is the mass of 0.175 moles of N_2F_4?

PROCEDURE

The molar mass of the compound is used as a conversion factor between moles and mass. The molar mass of N_2F_4 is

N	2 x 14.01 amu =	28.02 amu
F	4 x 19.00 amu =	76.00 amu
	Formula Weight =	104.02 amu
	Molar Mass = 104.0 g/mol	

In this case, the conversion factor has grams in the numerator (requested) and moles in the denominator (given).

SOLUTION

$$0.175 \text{ mol } N_2F_4 \times \frac{104.0 \text{ g } N_2F_4}{\text{mol } N_2F_4} = 18.2 \text{ g } N_2F_4$$

Given — This conversion factor converts moles to grams. — Requested

Example A-2 Converting Mass to Number of Molecules

How many individual molecules are in 651 grams of PCl_3?

PROCEDURE

In this case we must convert from grams of PCl_3 to moles of PCl_3 using the molar mass of PCl_3 as a conversion factor. In a second step we convert moles of PCl_3 to a number of molecules using Avogadro's number as a conversion factor.

SOLUTION

The molar mass of PCl_3 is

P	1 x 30.97 amu =	30.97 amu
Cl	3 x 35.45 amu =	106.35 amu
	Formula weight =	137.32 amu
	Molar mass = 137.3 g/mol	

$$651 \text{ g } PCl_3 \times \frac{1 \text{ mol } PCl_3}{137.3 \text{ g } PCl_3} \times \frac{6.022 \times 10^{23} \text{ molecules}}{\text{mol } PCl_3} = 2.86 \times 10^{24} \text{ molecule}$$

Given — This factor converts grams to moles. — This factor converts moles to molecules. — Requested

NEW TERMS

Avogadro's number
Formula weight
Molar mass (compound)
Molar mass (element)
Mole

SELF-TEST

A-1 Multiple Choice

___ 1. The molar mass of oxygen atoms is

(a) 32.00 g/mol
(b) 16.00 g/mol
(c) 8.00 g/mol
(d) 16.00 amu
(e) 32.00 amu

___ 2. A 0.50-mole quantity of Ne atoms has a mass of

(a) 20 g (b) 5.0 g (c) 10 g (d) 10 amu (e) 40 g

___ 3. The number 6.022×10^{22} is

(a) 1.000 mole
(b) 0.1000 mole
(c) 10.00 mole
(d) 5.000 mole
(e) no such number exists

___ 4. The mass of one atom of carbon is about

(a) 2×10^{-23} g
(b) 1×10^{-23} g
(c) 2×10^{23} g
(d) 12.0 g
(e) 6.0 g

___ 5. How many moles of O atoms are in two moles of $Fe_2(C_2O_4)_3$?

(a) 4 (b) 3 (c) 24 (d) 12 (e) 6

___ 6. The molar mass of SO_2 is

(a) 64.07 g
(b) 32.00 g
(c) 64.07 amu
(d) 44.07 g
(e) 48.00 amu

___ 7. How many moles of P atoms are in 62 g of P_4?

(a) 0.50 (b) 2.0 (c) 4.0 (d) 1.0 (e) 8.0

___ 8. Which of the following statements is false about 1.00 mole of O_2?

(a) It contains 16.0 g of O atoms.
(b) It contains 1.20 x 10^{24} atoms of O.
(c) It contains 6.02 x 10^{23} molecules of O_2.
(d) It contains two moles of O atoms.
(e) It has a mass of 32.0 g.

A-2 Problems

1. Relative numbers of items

(a) A farmer wishes to plant a certain field in corn rather than soybeans. An average soybean seed weighs 25.00 mg and an average corn seed weighs 47.50 mg. In the previous year, the farmer used 148 lb of soybeans. How many pounds of corn are needed to provide the same number of seeds as was present in the soybeans?

(b) What mass of calcium contains the same number of atoms as 54.8 kg of carbon?

(c) The formula of a compound is K_2O. What mass of potassium is present for each 10.0 lb of oxygen?

2. Find the mass of the following:

(a) 0.50 mole of K

(b) 125 moles of Ca

(c) 3.22 x 10^{21} atoms of Br

3. Find the number of moles in the following.

(a) 765 g of Al

(b) 4.42×10^{18} atoms of B

4. Find the formula weight of the following.

(a) $HMnO_4$

(b) $Al_2(SO_3)_3$

(c) $C_6H_{12}O_6$

5. Calculate the following.

(a) the number of moles in 45.5 g of CO

(b) the mass of 15.2 moles of N_2O

(c) the number of moles in 4.75×10^{23} molecules of SO_2

(d) the mass of 7.20×10^{24} formula units of K_2O

(e) the number of molecules in 324 kg of UO_2

Review Section B *The Masses of Compounds and Their Component Elements*

OUTLINE / OBJECTIVES

8-4 The Composition of Compounds

1. The mole composition of compounds — *Calculate the number of moles of each element in a given amount of compound.*

2. The mass composition of compounds. — *Calculate the mass of each element present in a given amount of a compound.*

3. The mass percent of compounds — *Calculate the percent composition of a compound from its formula.*

8-5 Empirical and Molecular Formulas

1. Mass percent and empirical formula — *Calculate the empirical formula of a compound from its percent or mass composition.*

2. Molecular formulas from empirical formulas — *Use the molar mass and empirical formula of a compound to calculate its molecular formula.*

SUMMARY OF SECTIONS 8-4 AND 8-5

Questions: *How does the formula of a compound relate to its component elements? What do the masses of the elements in a compound tell us about its formula?*

To complete the discussion of the quantitative relationships implied by a chemical formula we now examine the composition of compounds. The moles and mass of a certain element present in a given amount of compound can be calculated using the formula as a conversion factor between moles of a component atom and moles of compound as illustrated in the following example.

Example B-1 Calculation of the Mole and Mass Composition of a Compound

(a) How many moles of aluminum and (b) what mass of chromium are present in 382 g of $Al_2(Cr_2O_7)_3$?

PROCEDURE

Calculate the molar mass and from this, the number of moles of $Al_2(Cr_2O_7)_3$ present in 382 g of $Al_2(Cr_2O_7)_3$. For part (a) calculate the number of moles of Al present using the conversion factor $\frac{2 \text{ mol Al}}{1 \text{ mol } Al_2(Cr_2O_7)_3}$. For part (b) calculate the mass of Cr present using $\frac{6 \text{ mol Cr}}{1 \text{ mol } Al_2(Cr_2O_7)_3}$ and $\frac{52.00 \text{ g Cr}}{\text{mol Cr}}$.

SOLUTION

(2 x 26.98 amu) + (6 x 52.00 amu) + (21 x 16.00 amu) = 702.0 amu (702.0 g/mol)

$$382 \text{ g } Al_2(Cr_2O_7)_3 \times \frac{1 \text{ mol } Al_2(Cr_2O_7)_3}{702.0 \text{ g } Al_2(Cr_2O_7)_3} = 0.544 \text{ mol } Al_2(Cr_2O_7)_3$$

$$\text{(a) } 0.544 \text{ mol } Al_2(Cr_2O_7)_3 \times \frac{2 \text{ mol Al}}{1 \text{ mol } Al_2(Cr_2O_7)_3} = \underline{1.09 \text{ mol Al}}$$

$$\text{(b) } 0.544 \text{ mol } Al_2(Cr_2O_7)_3 \times \frac{6 \text{ mol Cr}}{1 \text{ mol } Al_2(Cr_2O_7)_3} \times \frac{52.00 \text{ g Cr}}{\text{mol Cr}} = \underline{170 \text{ g Cr}}$$

The **percent composition** can be calculated from a knowledge of the molecular formula. Given the percent composition, however, only the **empirical formula** can be determined. A knowledge of the empirical formula and the molar mass are needed to determine the **molecular formula**.

The percent composition of a compound is easily found from the formula of a compound.

$$\% \text{ composition} = \frac{\text{total mass of component element}}{\text{molar mass of compound}} \times 100\%$$

The percent composition of the elements in a compound is the same for all compounds with the same empirical formula. The empirical formula of a compound is the simplest whole number ratio of atoms.

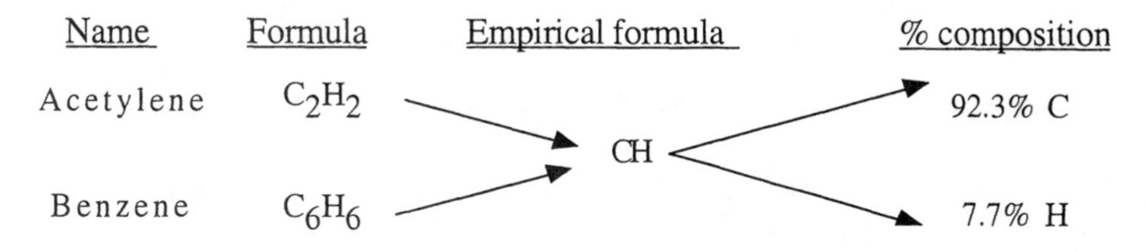

Conversely, if percent composition is given (or the masses of the elements comprising a compound), the empirical formula can be determined. A sample calculation following the procedure outlined in Section 8-4 in the text follows.

Example B-2 Calculation of Empirical and Molecular Formulas

A component is composed of 56.4% P and 43.6% O. (a) What is the empirical formula of the compound? (b) If the molar mass is 220 g/mol, what is the molecular formula of the compound?

PROCEDURE

In a 100-g sample of the compound, there are 56.4 g of P and 43.6 g of O. The element ratio is the same as the mole ratio so the next step is to convert the masses of each element to moles.

SOLUTION

$$\text{(a) P } \quad 56.4 \text{ g P} \times \frac{1 \text{ mol P}}{30.97 \text{ g P}} = 1.82 \text{ mol P}$$

$$O \quad 43.6\ \cancel{g\ O} \times \frac{1\ mol\ O}{16.00\ \cancel{g\ O}} = 2.72\ mol\ O$$

The ratios of the moles of the elements are:

$$P: \frac{1.82}{1.82} = 1.0 \quad O: \frac{2.72}{2.72} = 1.5 \quad PO_{1.5} = P_2O_3$$

(b) The mass of one empirical unit is (2 x 30.97) + (3 x 16.00) = 109.9 g/emp. unit

$$\frac{220\ \cancel{g}/mol}{109.9\ \cancel{g}/emp.\ unit} = 2\ emp.\ unit/mol$$

The molecular formula is therefore P_4O_6 which contains two empirical units.

NEW TERMS

Empirical formula
Molecular formula
Percent composition

SELF-TEST

B-1 Multiple Choice

___ 1. The percent composition of carbon in carbon monoxide is

(a) 50.0% (b) 27.3% (c) 42.9% (d) 60.0%

___ 2. There is 140 g of boron in a 450-g sample of a compound. What is the percent composition of boron in the compound?

(a) 23.7% (b) 31.1% (c) 45.2% (d) 68.9%

___ 3. Which of the following is an empirical formula?

(a) $C_4H_8O_2$
(b) C_9H_{15}
(c) $B_2(OH)_4$
(d) N_2H_4
(e) $C_3H_6O_2$

___ 4. The molar mass of a compound with the empirical formula CH_2 is 56 g. The molecular formula is

(a) $C_2N_2H_4$
(b) C_4H_8
(c) C_3H_6
(d) C_3H_4O
(e) C_7H_{14}

B-2 Problems

1. How many moles of each atom are present in 38.0 g of $Ba(ClO_3)_2$?

2. What mass of sulfur is present in 782 g of $K_2S_2O_3$?
   (Calculate by converting to moles of sulfur then to mass of sulfur.)

3. What is the percent composition of the elements in:

   (a) $NaClO_2$

   (b) Ammonium carbonate

4. What is the mass of nitrogen in 195 lb of ammonium carbonate?
   (Calculate using percent composition.)

5. A 5.67-g quantity of chlorine reacts with 8.94 g of oxygen to form a compound. What is its empirical formula?

6. A compound is 53.3% C, 11.1% H, and 35.6% O. Its molar mass is 90 g/mol. What is its molecular formula?

Review Section C *Mass and Energy Changes in Chemical Reactions*

OUTLINE / OBJECTIVES

8-6 Stoichiometry

1. Mole ratios

   Use the balanced equation to obtain mole relationships among reactants and products.

2. Stoichiometry problems

   Make the following stoichiometric conversions: (a) mole to mole, (b) mole to mass, (c) mass to mass, (d). mass to number.

8-7 Limiting Reactant

1. Stoichiometric mixtures
2. Nonstoichiometric mixtures

   Identify the limiting reactant, the reactant in excess, and the yield from given amounts of reactants.

8-8 Percent Yield

1. Measured and calculated yields

   Calculate the percent yield from the actual yield and the theoretical yield.

2. Incomplete reactions

   Give the two reasons why certain reactions may be incomplete

8-9 Heat Energy in Chemical Reactions

1. Thermochemical equations

   Write thermochemical equations in two ways.

2. Stoichiometry and heat energy

   Use the balanced equation to calculate the amount of heat energy involved in a specified reaction.

SUMMARY OF SECTIONS 8-6 THROUGH 8-9

Questions: *What does a chemical equation tell us about the proportions of substances that react? What are the limits on how much of a given reactant is consumed in a reaction? Are one or more reactants always completely consumed in a reaction? Can heat energy be included in a chemical equation?*

A critical skill to be learned in chemistry is the ability to carry out **stoichiometry** calculations. This concerns conversions between the amount (moles, mass, or number) of one reactant or product and that of another reactant or product. The key conversion factor for this endeavor relates moles of all reactants and products by the coefficients in the balanced equation. Thus in the balanced equation

$$Si_2H_6 + 8H_2O \longrightarrow 2Si(OH)_4 + 7H_2$$

conversion factors (**mole ratios**) between H_2O and H_2 are

$$\frac{8 \text{ mol } H_2O}{7 \text{ mol } H_2} \quad \text{and} \quad \frac{7 \text{ mol } H_2}{8 \text{ mol } H_2O}$$

The first ratio is used to convert moles of H_2 to H_2O and the second (which is simply the reciprocal of the first) is used to convert moles of H_2O to H_2. All other conversions (e.g., mass to moles, moles to number) were studied earlier in this chapter. The conversions are represented in the following general scheme. The origin of the conversion factor for each step is indicated next to the arrow.

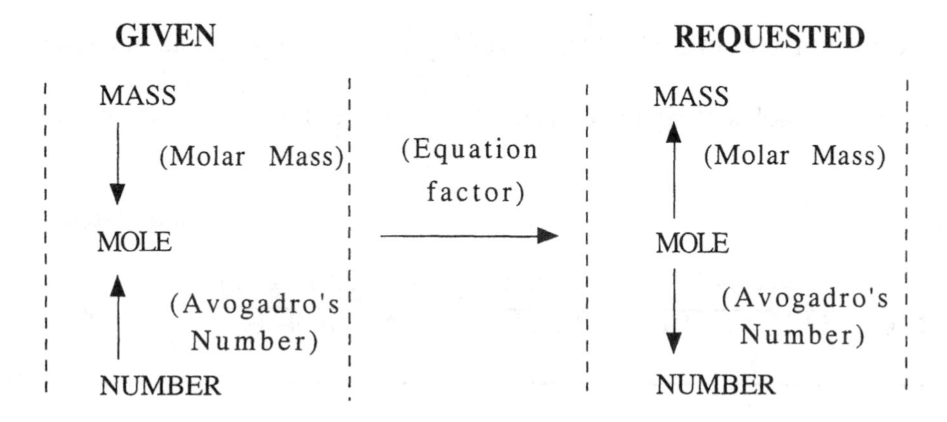

The following two examples illustrate stoichiometry problems.

Example C-1 Mole to Mole Conversions

How many moles of H_2 are produced from 36.5 moles of H_2O according to the following balanced equation?

$$Si_2H_6 + 8H_2O \longrightarrow 2Si(OH)_4 + 7H_2$$

PROCEDURE

From the balanced equation, the mole ratio that converts moles to what's given (H_2O) to moles of what's requested (H_2) is

$$\frac{7 \text{ mol } H_2}{8 \text{ mol } H_2O}$$

SOLUTION

Given x mole ratio = Requested

$$36.5 \cancel{\text{mol } H_2O} \times \frac{7 \text{ mol } H_2}{8 \cancel{\text{mol } H_2O}} = \underline{31.9 \text{ mol } H_2}$$

Example C-2 Mass to Mass Conversions

What mass of H_2 is produced from 165 g of Si_2H_6? (Use the balanced equation from problem C-1.)

PROCEDURE

In this case, the following conversions are necessary:

1. Convert mass of Si_2H_6 (Given) to moles of Si_2H_6 using molar mass of Si_2H_6.
2. Convert moles of Si_2H_6 to moles of H_2 using the appropriate mole ratio.
3. Convert moles of H_2 to grams of H_2 (Requested) using the molar mass of H_2.

Given **1** **2** **3** **Requested**

$$165\ \cancel{g\ Si_2H_6} \times \frac{1\ \cancel{mol\ Si_2H_6}}{62.23\ \cancel{g\ Si_2H_6}} \times \frac{7\ \cancel{mol\ H_2}}{1\ \cancel{mol\ Si_2H_6}} \times \frac{2.016\ g\ H_2}{\cancel{mol\ H_2}} = \underline{37.4\ g\ H_2}$$

Also, we must realize that reactants are not always mixed in exact proportions so that all are completely consumed (this would be a stoichiometric mixture). In that case, the reactant that is completely consumed limits the yield of products and is called the **limiting reactant**. A quantity of the other reactant (or reactants) is left over and is present in excess. This is illustrated by the following problem.

Example C-3 The Limiting Reactant

In the reaction between Si_2H_6 and H_2O, a 50.0-g quantity of Si_2H_6 is mixed with 100 g of H_2O. How many moles of $Si(OH)_4$ are formed?

PROCEDURE

Find the yield in moles from each of the two reactants. The one that produces the smaller yield is the limiting reactant.

$$\text{mass} \begin{bmatrix} H_2O \\ Si_2H_6 \end{bmatrix} \longrightarrow \text{moles} \begin{bmatrix} H_2O \\ Si_2H_6 \end{bmatrix} \longrightarrow \text{moles } Si(OH)_4$$

SOLUTION

$$H_2O: \quad 100\ \cancel{g\ H_2O} \times \frac{1\ \cancel{mol\ H_2O}}{18.02\ \cancel{g\ H_2O}} \times \frac{2\ mol\ Si(OH)_4}{8\ \cancel{mol\ H_2O}} = 1.39\ mol\ Si(OH)_4$$

$$Si_2H_6:\quad 50.0\ \cancel{g\ Si_2H_6} \times \frac{1\ \cancel{mol\ Si_2H_6}}{62.23\ \cancel{g\ Si_2H_6}} \times \frac{2\ mol\ Si(OH)_4}{1\ \cancel{mol\ Si_2H_6}} = 1.61\ mol\ Si(OH)_4$$

The limiting reactant is H_2O and the yield is <u>1.39 mol $Si(OH)_4$</u>.

There are many examples of chemical reactions that do not go to completion (100% to the right). There are two major reasons for this: (1) a reaction may be a reversible reaction where reactants and products reach a point of equilibrium, and (2) other competing reactions may occur between the same reactants. In these cases, it is convenient to express the **percent yield** of a product. The percent yield is obtained by dividing the **actual yield** by the **theoretical yield** (the amount of a product that would be produced if at least one of the reactants was completely converted to products) times 100%.

$$\text{Percent yield} = \frac{\text{Actual yield (measured)}}{\text{Theoretical yield (calculated)}} \times 100\%$$

An example of this type of problem follows.

Example C-4 Percent Yield

In the reaction of HCl with O_2, 46.0 g of Cl_2 is formed. If all of the HCl had reacted, 49.5 g of Cl_2 would form. What is the percent yield for this reaction?

PROCEDURE

Substitute the following into the equation used to calculate percent yield.

actual yield = 46.0 g theoretical yield = 49.5 g

SOLUTION

$$\text{percent yield} = \frac{46.0\ \text{g}}{49.5\ \text{g}} \times 100\% = \underline{92.9\%\ \text{yield}}$$

The study of **chemical thermodynamics** includes the heat energy that is involved in chemical changes. When heat energy is included, the equation is known as a **thermochemical equation**. In endothermic reactions the reactants absorb energy from the surroundings when a reaction occurs. The origin of this energy is in the difference in the potential energy (in the form of chemical energy) of the reactants and products involved.

An exothermic thermochemical reaction can be written in two ways. In the first, the heat energy is represented as a product. In the second, it is shown as a negative value for **ΔH** [the change in heat content (**enthalpy**)].

(1) $2CO\ (g) + O_2\ (g) \rightarrow 2CO_2\ (g) + 568\ kJ$

(2) $2CO\ (g) + O_2\ (g) \rightarrow 2CO_2\ (g)\quad \Delta H = -568\ kJ$

NEW TERMS

Actual yield
Chemical thermodynamics
Enthalpy
Limiting reactant
Mole ratio
Percent yield
Stoichiometry
Thermochemical equation
Theoretical yield

SELF-TEST

C-1 Problems

1. Given the following balanced equation illustrating the combustion of butane (C_4H_{10}).

$$2C_4H_{10}(l) + 13O_2(g) \longrightarrow 8CO_2(g) + 10H_2O(l)$$

(a) Write the mole ratio that converts moles of O_2 to moles of H_2O.

(b) Write the mole ratio that converts moles of CO_2 to moles of H_2O.

(c) Write the two conversion factors that are needed to convert grams of C_4H_{10} to moles of O_2.

(d) Write the two conversion factors that are needed to convert moles of C_4H_{10} to number of molecules of CO_2.

(e) Write the three conversion factors that convert grams of H_2O to grams of O_2.

(f) How many moles of CO_2 are produced from 0.115 mole of O_2?

(g) What mass of CO_2 is produced from 8.67 moles of C_4H_{10}?

(h) What mass of O_2 completely reacts with 155 g of C_4H_{10}?

(i) How many individual H_2O molecules are produced from 6.64×10^{-6} g of O_2?

2. Given the following balanced equation:

$$4NH_3(g) + 5O_2 \longrightarrow 4NO(g) + 6H_2O(g)$$

What mass of H_2O is produced from a mixture of 100 g of NH_3 and 200 g of O_2?

3. When a 100-kg quantity of C_4H_{10} is burned in a closed system, 280 kg of CO_2 is formed. What is the percent yield of the reaction? Use the balanced equation from problem 1 in this section.

4. An impure sample containing solid gold(III) oxide is heated. The gold(III) oxide decomposes to form metallic gold and gaseous oxygen. If the original sample has a mass of 365 g and a total of 146 g of gold is produced, what percent of the original sample was composed of gold(III) oxide?

5. A 235-kJ quantity of heat must be supplied to decompose sodium bicarbonate to one mole each of $H_2O(g)$, $CO_2(g)$, and $Na_2CO_3(s)$. Write the thermochemical equation in both ways discussed. What mass of sodium bicarbonate is decomposed by the input of 175 kJ of heat energy?

Review Section D

CHAPTER SUMMARY SELF-TEST

D-1 Problems

1. Fill in the blanks.

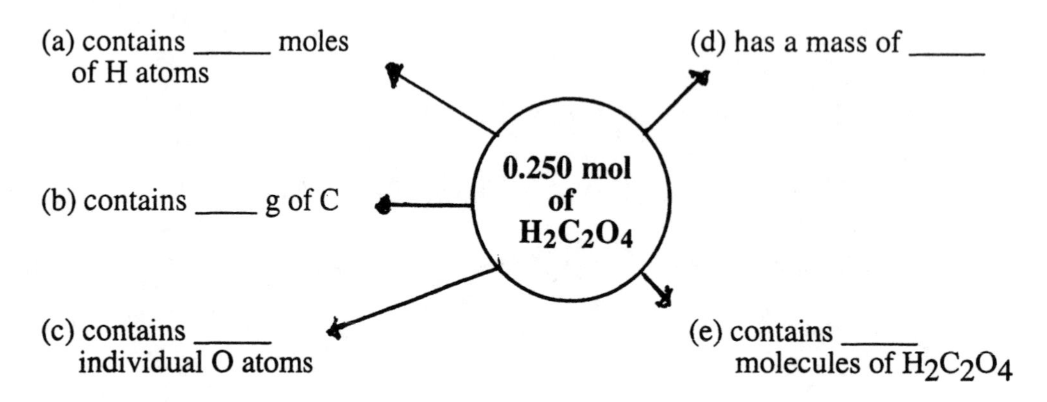

(f) The empirical formula for $H_2C_2O_4$ is __________.

(g) The percent composition of oxygen in $H_2C_2O_4$ _________.

2.

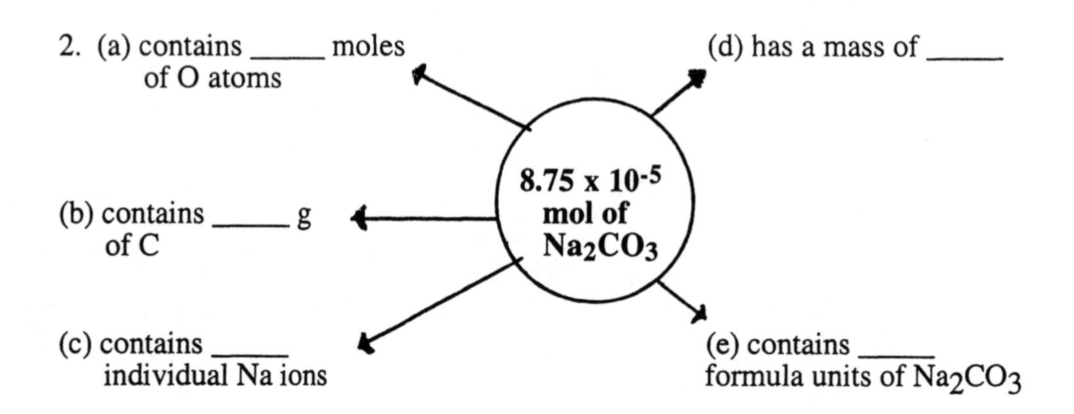

3. Ethylenediaminetetraacetic acid (EDTA) is a compound that is 41.1% C, 5.5% H, 9.6% N, and 43.8% O. There are two empirical units per molecular unit.

    (a) Its empirical formula is ________________________.

    (b) Its molecular formula is ________________________.

    (c) 1.22 moles has a mass of ________________________.

    (d) 5.00 x 10^{24} molecules have a mass of __________.

4. The combustion of liquid methyl alcohol (CH_4O), a possible substitute for gasoline, produces carbon dioxide and water. Write the balanced equation and determine what mass of water would be produced from the complete combustion of 964 g of methyl alcohol.

5. In the preceding problem, it was found that only 1.15 kg of CO_2 was formed. What is the percent yield of CO_2? The remainder of the alcohol that did not form CO_2 burns to form CO and water. Write the balanced equation illustrating this. What mass of CO is formed in the reaction?

6. Solid potassium superoxide (KO_2) is used in space vehicles to remove carbon dioxide gas and to regenerate oxygen gas. In the reaction KO_2 reacts with carbon dioxide to produce solid potassium carbonate and oxygen gas. Write the balanced equation and determine what mass of oxygen is produced from 83.0 g of CO_2.

7. Solid zinc sulfide reacts with oxygen to form solid zinc oxide and gaseous sulfur dioxide. Write the balanced equation and determine what mass of SO_2 is produced from a mixture of 85.0 g of zinc sulfide and 60.0 g of oxygen.

8. In the preceding problem, it is found that 396 kJ of heat energy is released in the reaction. What is the value of ΔH for the reaction?

Answers to Self-Tests

A-1 Multiple Choice

1. **b** 16.00 g/mol

2. **c** 10 g $[0.50\ \cancel{mol} \times \frac{20\ g}{\cancel{mol}} = 10\ g]$

3. **b** 0.100 mol $[6.02 \times 10^{22}\ \cancel{atoms} \times \frac{1\ mol}{6.02 \times 10^{23}\ \cancel{atoms}} = 0.100\ mol]$

4. **a** 2×10^{-23} g $[\frac{12\ g}{\cancel{mol}} \times \frac{1\ \cancel{mol}}{6.0 \times 10^{23}\ atoms} = 2 \times 10^{-23}\ g/atom]$

5. **c** 24 $[2\ \cancel{mol\ Fe_2(C_2O_4)_3} \times \frac{12\ mol\ O}{\cancel{mol\ Fe_2(C_2O_4)_3}} = 24\ mol\ O]$

6. **a** 64.07 g [(S) 32.07 g + (O) (2 x 16.00 g)] = 64.07 g]

7. **b** 2.0 $[62\ \cancel{g\ P_4} \times \frac{1\ \cancel{mol\ P_4}}{123.9\ \cancel{g\ P_4}} \times \frac{4\ mol\ P\ atoms}{1\ \cancel{mol\ P_4}} = 2.0\ mol\ P\ atoms]$

8. **a** It contains 16 g of O atoms. (It contains 32.0 g of O_2 molecules and 32.0 g of O atoms.)

A-2 Problems

1. (a) $148\ \cancel{lb\ soybean} \times \frac{47.50\ lb\ corn}{25.00\ \cancel{lb\ soybean}} = \underline{281\ lb\ corn}$

   (b) $54.8\ \cancel{kg\ carbon} \times \frac{40.08\ kg\ calcium}{12.01\ \cancel{kg\ carbon}} = \underline{183\ lb\ calcium}$

   (c) $10.0\ \cancel{lb\ oxygen} \times \frac{39.10\ lb\ potassium \times 2}{16.00\ \cancel{lb\ oxygen} \times 1} = \underline{48.9\ lb\ oxygen}$

2. (a) $0.50 \cancel{\text{mol K}} \times \frac{39.10 \text{ g K}}{\cancel{\text{mol K}}} = \underline{20 \text{ g K}}$

(b) $125 \cancel{\text{mol Ca}} \times \frac{40.08 \text{ g}}{\cancel{\text{mol Ca}}} = \underline{5010 \text{ g Ca}}$

(c) $3.22 \times 10^{21} \cancel{\text{atoms}} \times \frac{1 \cancel{\text{mol Br}}}{6.022 \times 10^{23} \cancel{\text{atoms}}} \times \frac{79.90 \text{ g}}{\cancel{\text{mol Br}}} = \underline{0.427 \text{ g}}$

3. (a) $765 \cancel{\text{g Al}} \times \frac{1 \text{ mol Al}}{26.98 \cancel{\text{g Al}}} = \underline{28.4 \text{ mol Al}}$

(b) $4.42 \times 10^{18} \cancel{\text{atoms B}} \times \frac{1 \text{ mol B}}{6.022 \times 10^{23} \cancel{\text{atoms B}}} = 7.34 \times 10^{-6} \text{ mol B}$

4. (a) $HMnO_4$ [1.008 (H) + 54.94 (Mn) + (4 x 16.00) (O)] = $\underline{119.9 \text{ amu}}$

(b) $Al_2(SO_3)_3$ [(2 x 26.98)(Al) + (3 x 32.07)(S)
+ (9 x 16.00) (O)] = $\underline{294.2 \text{ amu}}$

(c) $C_6H_{12}O_6$ [(6 x 12.01)(C) + (12 x 1.008)(H) + (6 x 16.00)(O)] =
$\underline{180.2.\text{amu}}$

5. (a) $45.5 \cancel{\text{g CO}} \times \frac{1 \text{ mol}}{28.01 \cancel{\text{g CO}}} = \underline{1.62 \text{ mol}}$

(b) $15.2 \cancel{\text{mol } N_2O} \times \frac{44.02 \text{ g}}{\cancel{\text{mol } N_2O}} = \underline{669 \text{ g}}$

(c) $4.75 \times 10^{23} \cancel{\text{molecules}} \times \frac{1 \text{ mol}}{6.022 \times 10^{23} \cancel{\text{molecules}}} = \underline{0.789 \text{ mol}}$

(d) $7.20 \times 10^{24} \cancel{\text{form. units}} \times \frac{1 \cancel{\text{mol } K_2O}}{6.022 \times 10^{23} \cancel{\text{form. units}}}$
$\times \frac{94.20 \text{ g } K_2O}{\cancel{\text{mol } K_2O}} = \underline{1.13 \times 10^3 \text{ g } K_2O}$

(e) $324 \cancel{\text{kg } UO_2} \times \frac{1 \cancel{\text{g } UO_2}}{10^{-3} \cancel{\text{kg } UO_2}} \times \frac{1 \cancel{\text{mol } UO_2}}{270.0 \cancel{\text{g } UO_2}}$
$\times \frac{6.022 \times 10^{23} \text{ molecules}}{\cancel{\text{mol } UO_2}} = \underline{7.23 \times 10^{26} \text{ molecules}}$

B-1 Multiple Choice

1. **c** 12.0 g/28.0 g x 100% = 42.9%

2. **b** 140 g/450 g x 100% = 31.1%

3. **e** $C_3H_6O_2$

4. **b** C_4H_8 [CH_2 = 14 g/emp. unit; $\frac{56 \cancel{\text{g}}/\text{mol}}{14 \cancel{\text{g}}/\text{emp. unit}} = 4$ $(CH_2) \times 4 = C_4H_8$]

B-2 Problems

1. $Ba(ClO_3)_2$ formula weight = [137.3 + (2 x 35.45) + (6 x 16.00)] = 304.2 g/mol

$$38.0 \text{ g } Ba(ClO_3)_2 \times \frac{1 \text{ mol } Ba(ClO_3)_2}{304.2 \text{ g } Ba(ClO_3)_2} = 0.125 \text{ mol } Ba(ClO_3)_2$$

$$0.125 \text{ mol } Ba(ClO_3)_2 \times \frac{1 \text{ mol Ba}}{\text{mol } Ba(ClO_3)_2} = \underline{0.125 \text{ mol Ba}}$$

$$0.125 \text{ mol } Ba(ClO_3)_2 \times \frac{2 \text{ mol Cl}}{\text{mol } Ba(ClO_3)_2} = \underline{0.250 \text{ mol Cl}}$$

$$0.125 \text{ mol } Ba(ClO_3)_2 \times \frac{6 \text{ mol O}}{\text{mol } Ba(ClO_3)_2} = \underline{0.750 \text{ mol O}}$$

2. $K_2S_2O_3$: formula weight = [(2 x 39.10) + (2 x 32.07) + (3 x 16.00)]
=190.3 g/mol

$$782 \text{ g } K_2S_2O_3 \times \frac{1 \text{ mol } K_2S_2O_3}{190.3 \text{ g } K_2S_2O_3} \times \frac{2 \text{ mol S}}{\text{mol } K_2S_2O_3} \times \frac{32.07 \text{ g S}}{\text{mol S}} = \underline{264 \text{ g S}}$$

3. (a) Na - 25.4%, Cl - 39.2%, O - 35.4%

   (b) $(NH_4)_2CO_3$ N - 29.2%, H - 8.3%, C - 12.5%, O - 50.0%

4. $$195 \text{ lb } (NH_4)_2CO_3 \times \frac{29.2 \text{ lb N}}{100 \text{ lb } (NH_4)_2CO_3} = \underline{56.9 \text{ lb N}}$$

5. Convert mass of each element to moles.

$$\text{Cl} \quad 5.67 \text{ g Cl} \times \frac{1 \text{ mol Cl}}{35.45 \text{ g Cl}} = 0.160 \text{ mol Cl}$$

$$\text{O} \quad 8.94 \text{ g O} \times \frac{1 \text{ mol O}}{16.00 \text{ g O}} = 0.559 \text{ mol O}$$

$$\text{Cl: } \frac{0.160}{0.160} = 1.0 \quad \text{O: } \frac{0.559}{0.160} = 3.5 \quad ClO_{3.5} = \underline{Cl_2O_7}$$

6. Convert percent to moles. In 100 g of compound there are 53.3 g C, 11.1 g of H, and 35.6 g of O.

$$\text{C} \quad 53.3 \text{ g C} \times \frac{1 \text{ mol C}}{12.01 \text{ g C}} = 4.44 \text{ mol C}$$

$$\text{H} \quad 11.1 \text{ g H} \times \frac{1 \text{ mol H}}{1.008 \text{ g H}} = 11.0 \text{ mol H}$$

$$\text{O} \quad 35.6 \text{ g O} \times \frac{1 \text{ mol O}}{16.00 \text{ g O}} = 2.22 \text{ mol O}$$

$$\text{C: } \frac{4.44}{2.22} = 2.0; \quad \text{H: } \frac{11.0}{2.22} = 5.0; \quad \text{O: } \frac{2.22}{2.22} = 1.0$$

Empirical formula = C_2H_5O Emp. wt. = 45.06 g/emp. unit

$$\frac{90 \text{ g/mol}}{45.06 \text{ g/emp. unit}} = 2.0 \text{ emp. unit/mole}$$

Molecular formula = $\underline{C_4H_{10}O_2}$

C-1 Problems

1. $2C_4H_{10} + 13O_2 \longrightarrow 8CO_2 + 10H_2O$

(a) $\dfrac{10 \text{ mol } H_2O}{13 \text{ mol } O_2}$ (b) $\dfrac{10 \text{ mol } H_2O}{8 \text{ mol } CO_2}$

(c) $\dfrac{1 \text{ mol } C_4H_{10}}{58.12 \text{ g } C_4H_{10}}$; $\dfrac{10 \text{ mol } O_2}{2 \text{ mol } C_4H_{10}}$

(d) $\dfrac{8 \text{ mol } CO_2}{2 \text{ mol } C_4H_{10}}$; $\dfrac{6.022 \times 10^{23} \text{ molecules } CO_2}{\text{mol } CO_2}$

(e) $\dfrac{1 \text{ mol } H_2O}{18.02 \text{ g } H_2O}$; $\dfrac{13 \text{ mol } O_2}{10 \text{ mol } H_2O}$; $\dfrac{32.00 \text{ g } O_2}{\text{mol } O_2}$

(f) convert moles of O_2 (Given) to moles of CO_2 (Requested)

using the mole ratio $\dfrac{8 \text{ mol } CO_2}{13 \text{ mol } CO_2}$

$$0.115 \cancel{\text{mol } O_2} \times \frac{8 \text{ mol } CO_2}{13 \cancel{\text{mol } O_2}} = \underline{0.0708 \text{ mol } CO_2}$$

(g) (1) convert moles of C_4H_{10} (Given) to moles CO_2 using

the mole ratio $\dfrac{8 \text{ mol } CO_2}{2 \text{ mol } C_4H_{10}}$

(2) convert moles of CO_2 to mass of CO_2 (Requested)

(1) (2)

$$8.67 \cancel{\text{mol } C_4H_{10}} \times \frac{8 \cancel{\text{mol } CO_2}}{2 \cancel{\text{mol } C_4H_{10}}} \times \frac{44.01 \text{ g } CO_2}{\cancel{\text{mol } CO_2}} = 1.53 \times 10^3 \text{ g } CO_2$$

(h) (1) convert mass of C_4H_{10} (Given) to moles of C_4H_{10}

(2) convert moles of C_4H_{10} to moles of O_2 using $\dfrac{13 \text{ mol } O_2}{2 \text{ mol } C_4H_{10}}$

(3) convert moles of O_2 to mass of O_2 (Requested)

$$155\ \text{g}\ C_4H_{10} \times \overset{(1)}{\frac{1\ \text{mol}\ C_4H_{10}}{58.12\ \text{g}\ C_4H_{10}}} \times \overset{(2)}{\frac{13\ \text{mol}\ O_2}{2\ \text{mol}\ C_4H_{10}}} \times \overset{(3)}{\frac{32.00\ \text{g}\ O_2}{\text{mol}\ O_2}} = \underline{555\ \text{g}\ O_2}$$

(i) (1) convert mass of O_2 (Given) to moles of O_2

(2) convert moles of O_2 to moles of H_2O using $\frac{10\ \text{mol}\ H_2O}{13\ \text{mol}\ O_2}$

(3) convert moles of H_2O to molecules of H_2O

$$6.64 \times 10^{-6}\ \text{g}\ O_2 \times \overset{(1)}{\frac{1\ \text{mol}\ O_2}{32.00\ \text{g}\ O_2}} \times \overset{(2)}{\frac{10\ \text{mol}\ H_2O}{13\ \text{mol}\ O_2}}$$

$$\times \overset{(3)}{\frac{6.022 \times 10^{23}\ \text{molecules}\ H_2O}{\text{mol}\ H_2O}} = \underline{9.61 \times 10^{16}\ \text{molecules}\ H_2O}$$

2. (1) Convert mass of NH_3 to moles of NH_3 then to moles of H_2O.
(2) Convert mass of O_2 to moles of O_2 then to moles of H_2O.
(3) Convert the smaller of the two above to mass of H_2O.

$$(1)\ NH_3:\ 100\ \text{g}\ NH_3 \times \frac{1\ \text{mol}\ NH_3}{17.03\ \text{g}\ NH_3} \times \frac{6\ \text{mol}\ H_2O}{4\ \text{mol}\ NH_3} = 8.81\ \text{mol}\ H_2O$$

$$(2)\ O_2:\ 200\ \text{g}\ O_2 \times \frac{1\ \text{mol}\ O_2}{32.00\ \text{g}\ O_2} \times \frac{6\ \text{mol}\ H_2O}{5\ \text{mol}\ O_2} = 7.50\ \text{mol}\ H_2O$$

(3) O_2 is the limiting reactant so $7.50\ \text{mol}\ H_2O \times \frac{18.02\ \text{g}\ H_2O}{\text{mol}\ H_2O} = \underline{135\ \text{g}\ H_2O}$.

3. (1) Convert kg C_4H_{10} to g C_4H_{10}, then to moles C_4H_{10}.

(2) Convert moles of C_4H_{10} to moles of CO_2.

(3) Convert moles of CO_2 to g of CO_2, then to kg of CO_2.
This is the theoretical yield.

$$100\ \text{kg}\ C_4H_{10} \times \frac{1\ \text{g}\ C_4H_{10}}{10^{-3}\ \text{kg}\ C_4H_{10}} \times \frac{1\ \text{mol}\ C_4H_{10}}{58.12\ \text{g}\ C_4H_{10}}$$

$$\times \frac{8\ \text{mol}\ CO_2}{2\ \text{mol}\ C_4H_{10}} \times \frac{44.01\ \text{g}\ CO_2}{\text{mol}\ CO_2} \times \frac{10^{-3}\ \text{kg}\ CO_2}{\text{g}\ CO_2} = 303\ \text{kg}\ CO_2$$

$$\%\ \text{yield} = \frac{280\ \text{kg}}{303\ \text{kg}} \times 100\% = \underline{92.4\ \%}$$

4. $2Au_2O_3(s) \rightarrow 4\,Au(s) + 3O_2(g)$

(1) Convert mass of Au to moles of Au.

(2) Convert moles of Au to moles of Au_2O_3 using $\frac{2 \text{ mol } Au_2O_3}{4 \text{ mol Au}}$.

(3) Convert moles of Au_2O_3 to mass of Au_2O_3.

(4) Convert mass of Au_2O_3 to percent of original sample.

$$146 \cancel{\text{g Au}} \times \underset{(1)}{\frac{1 \cancel{\text{mol Au}}}{197.0 \cancel{\text{g Au}}}} \times \underset{(2)}{\frac{2 \cancel{\text{mol } Au_2O_3}}{4 \cancel{\text{mol Au}}}} \times \underset{(3)}{\frac{442.0 \text{ g } Au_2O_3}{\cancel{\text{mol } Au_2O_3}}} = 164 \text{ g } Au_2O_3$$

(4)

$$\frac{164 \text{ g}}{365 \text{ g}} \times 100\% = \underline{44.9\% \text{ pure}}$$

5. $2NaHCO_3(s) + 235 \text{ kJ} \rightarrow H_2O(g) + CO_2(g) + Na_2CO_3(s)$

$2NaCO_3(s) \rightarrow H_2O(g) + CO_2(g) + Na_2CO_3(s) \quad \Delta H = 235 \text{ kJ}$

$$\text{kJ} \rightarrow \text{mol } NaHCO_3 \rightarrow \text{g } NaHCO_3$$

$$175 \cancel{\text{kJ}} \times \frac{2 \cancel{\text{mol } NaHCO_3}}{235 \cancel{\text{kJ}}} \times \frac{84.01 \text{ g } NaHCO_3}{\cancel{\text{mol } NaHCO_3}} = \underline{125 \text{ g } NaHCO_3}.$$

D-1 Problems

1. 0.250 mole $H_2C_2O_4$

(a) $0.250 \cancel{\text{mol } H_2C_2O_4} \times \frac{2 \text{ mol H atoms}}{\cancel{\text{mol } H_2C_2O_4}} = \underline{0.500 \text{ mol H atom}}$

(b) $0.250 \cancel{\text{mol } H_2C_2O_4} \times \frac{2 \text{ mol C atoms}}{\cancel{\text{mol } H_2C_2O_4}} \times \frac{12.01 \text{ g C}}{\text{mol C atoms}} = \underline{6.01 \text{ g C}}$

(c) $0.250 \cancel{\text{mol } H_2C_2O_4} \times \frac{4 \cancel{\text{mol O atoms}}}{\cancel{\text{mol } H_2C_2O_4}} \times$

$\frac{6.022 \times 10^{23} \text{ atoms O}}{\cancel{\text{mol O atoms}}} = \underline{6.02 \times 10^{23} \text{ atoms O}}$

(d) $0.250 \cancel{\text{mol } H_2C_2O_4} \times \frac{90.04 \text{ g}}{\cancel{\text{mol } H_2C_2O_4}} = \underline{22.5 \text{ g}}$

(e) $0.250 \cancel{\text{mol } H_2C_2O_4} \times \frac{6.022 \times 10^{23} \text{ molecules}}{\cancel{\text{mol } H_2C_2O_4}} =$

$\underline{1.51 \times 10^{23} \text{ molecules}}$

(f) The empirical formula is HCO_2.

(g) 64.0 g/90.0 g x 100% = 71.1% oxygen

2. 8.75×10^{-5} mole Na_2CO_3

(a) $8.75 \times 10^{-5}\ \text{mol Na}_2\text{CO}_3 \times \dfrac{3\ \text{mol O atoms}}{\text{mol Na}_2\text{CO}_3} = 2.63 \times 10^{-4}\ \text{mol O}$

(b) $8.75 \times 10^{-5}\ \text{mol Na}_2\text{CO}_3 \times \dfrac{1\ \text{mol C atoms}}{1\ \text{mol Na}_2\text{CO}_3} \times \dfrac{12.01\ \text{g C}}{\text{mol C atoms}} = 1.05 \times 10^{-3}\ \text{g}$

(c) $8.75 \times 10^{-5}\ \text{mol Na}_2\text{CO}_3 \times \dfrac{2\ \text{mol Na ions}}{1\ \text{mol Na}_2\text{CO}_3} \times \dfrac{6.022 \times 10^{23}\ \text{Na ions}}{\text{mol Na ions}} = 1.05 \times 10^{20}\ \text{Na ions}$

(d) $8.75 \times 10^{-5}\ \text{mol Na}_2\text{CO}_3 \times \dfrac{106.0\ \text{g Na}_2\text{CO}_3}{\text{mol Na}_2\text{CO}_3} = 9.28 \times 10^{-3}\ \text{g Na}_2\text{CO}_3$

(e) $8.75 \times 10^{-5}\ \text{mol Na}_2\text{CO}_3 \times \dfrac{6.022 \times 10^{23}\ \text{form. units}}{\text{mol Na}_2\text{CO}_3} = 5.27 \times 10^{19}\ \text{form. units}$

3. (a) C: $41.1\ \text{g} \times \dfrac{1\ \text{mol}}{12.01\ \text{g}} = 3.42\ \text{mol}$ H: $5.5\ \text{g} \times \dfrac{1\ \text{mol}}{1.008\ \text{g}} = 5.5\ \text{mol}$

N: $9.6\ \text{g} \times \dfrac{1\ \text{mol}}{14.01\ \text{g}} = 0.69\ \text{mol}$ O: $43.8\ \text{g} \times \dfrac{1\ \text{mol}}{16.00\ \text{g}} = 2.74\ \text{mol}$

C: $\dfrac{3.42}{0.69} = 5.0$ H: $\dfrac{5.5}{0.69} = 8.0$ N: $\dfrac{0.69}{0.69} = 1.0$ O: $\dfrac{2.74}{0.69} = 4.0$

The empirical formula is $C_5H_8NO_4$.

(b) The molecular formula is $C_{10}H_{16}N_2O_8$.

(c) The molar mass of EDTA is (10 x 12.01) + (16 x 1.008) + (2 x 14.01) + (8 x 16.00) = 292.2 g

$1.22\ \text{mol} \times \dfrac{292.2\ \text{g}}{\text{mol}} = 356\ \text{g}$

(d) $5.00 \times 10^{24}\ \text{molecules} \times \dfrac{1\ \text{mol}}{6.022 \times 10^{23}\ \text{molecules}} \times \dfrac{292.2\ \text{g}}{\text{mol}} = 2.43 \times 10^{3}\ \text{g}$

4. $2CH_4O(l) + 3O_2(g) \longrightarrow 2CO_2(g) + 4H_2O(l)$

(1) Convert mass of CH_4O to moles of CH_4O.

(2) Convert moles of CH_4O to moles of H_2O using $\frac{4 \text{ mol } H_2O}{2 \text{ mol } CH_4O}$.

(3) Convert moles of H_2O to mass of H_2O.

$$964 \text{ g } CH_4O \times \overset{(1)}{\frac{1 \text{ mol } CH_4O}{32.04 \text{ g } CH_4O}} \times \overset{(2)}{\frac{4 \text{ mol } H_2O}{2 \text{ mol } CH_4O}} \times \overset{(3)}{\frac{18.02 \text{ g } H_2O}{\text{mol } H_2O}} = \underline{1.08 \times 10^3 \text{ g } H_2O}$$

5. (1) Convert mass of CH_4O to moles of CH_4O.

(2) Convert moles of CH_4O to moles of CO_2 using $\frac{2 \text{ mol } CO_2}{2 \text{ mol } CH_4O}$

(3) Convert moles of CO_2 to mass of CO_2 (the theoretical yield).

$$964 \text{ g } CH_4O \times \overset{(1)}{\frac{1 \text{ mol } CH_4O}{32.04 \text{ g } CH_4O}} \times \overset{(2)}{\frac{2 \text{ mol } CO_2}{2 \text{ mol } CH_4O}} \times \overset{(3)}{\frac{44.01 \text{ g } CO_2}{\text{mol } CO_2}} = 1.32 \times 10^3 \text{ g } CO_2$$

Theoretical yield = 1.32×10^3 g = 1.32 kg

Actual yield = 1.15 kg   Percent yield = $\frac{1.15 \text{ kg}}{1.32 \text{ kg}} \times 100\% = 87.1\%$

$CH_4O(l) + O_2(g) \longrightarrow CO(g) + 2H_2O(l)$

If 86.5% of the CH_4O is converted to CO_2, the remainder or 13.5% is converted to CO. Therefore, 0.135 x 964 g = 130 g CH_4O forms CO

g CH_4O $\longrightarrow$ mol CH_4O $\longrightarrow$ mol CO $\longrightarrow$ g CO

$$130 \text{ g } CH_4O \times \frac{1 \text{ mol } CH_4O}{32.04 \text{ g } CH_4O} \times \frac{1 \text{ mol CO}}{1 \text{ mol } CH_4O} \times \frac{28.01 \text{ g CO}}{\text{mol CO}} = \underline{114 \text{ g CO}}$$

6. $4KO_2(s) + 2CO_2(g) \longrightarrow 2K_2CO_3(s) + 3O_2(g)$

(1) Convert mass of CO_2 to moles of CO_2.

(2) Convert moles of CO_2 to moles of O_2 using $\frac{3 \text{ mol } O_2}{2 \text{ mol } CO_2}$.

(3) Convert moles of O_2 to mass of O_2.

(1) (2) (3)

$$83.0\ \text{g } CO_2 \times \frac{1\ \text{mol } CO_2}{44.01\ \text{g } CO_2} \times \frac{3\ \text{mol } O_2}{2\ \text{mol } CO_2} \times \frac{32.00\ \text{g } O_2}{\text{mol } O_2} = \underline{90.5\ \text{g } O_2}$$

7. $2ZnS(s) + 3O_2(g) \longrightarrow 2ZnO(s) + 2SO_2(g)$

(1) Convert mass of ZnS to moles of ZnS then to moles of SO_2.
(2) Convert mass of O_2 to moles of O_2 then to moles of SO_2.
(3) Convert the smaller of the two above to mass of SO_2.

(1) ZnS: $85.0\ \text{g ZnS} \times \frac{1\ \text{mol ZnS}}{97.46\ \text{g ZnS}} \times \frac{2\ \text{mol } SO_2}{2\ \text{mol ZnS}} = 0.872\ \text{mol } SO_2$

(2) O_2: $60.0\ \text{g } O_2 \times \frac{1\ \text{mol } O_2}{32.00\ \text{g } O_2} \times \frac{2\ \text{mol } SO_2}{3\ \text{mol } O_2} = 1.25\ \text{mol } SO_2$

(3) <u>ZnS is the limiting reactant</u> $0.872\ \text{mol } SO_2 \times \frac{64.07\ \text{g } SO_2}{\text{mol } SO_2}$
$= \underline{55.9\ \text{g } SO_2}$.

8. Since ZnS is the limiting reactant, 55.9 g (0.872 mol) of SO_2 forms.
Therefore, $\frac{396\ \text{kJ}}{0.872\ \text{mol } SO_2} = 454\ \text{kJ/mol } SO_2$
Since the equation calls for two mol of SO_2 and heat is evolved,

$$\Delta H = 2\ \text{mol } SO_2 \times \frac{-454\ \text{kJ}}{\text{mol } SO_2} = \underline{-908\ \text{kJ}}$$

Solutions to Black Text Problems

8-1 $1.24\ \text{lb nickels} \times \frac{2.47\ \text{lb pennies}}{5.03\ \text{lb nickels}} = \underline{0.609\ \text{lb of pennies}}$

8-3 $145\ \text{g Au} \times \frac{108\ \text{g Ag}}{197.0\ \text{g Au}} = \underline{79.5\ \text{g Ag}}$

8-4 $212\ \text{lb Al} \times \frac{12.01\ \text{lb C}}{26.98\ \text{lb Al}} = \underline{94.4\ \text{lb C}}$

8-6 $18.0\ \text{g O} \times \frac{63.55\ \text{g Cu}}{16.00\ \text{g O}} = \underline{71.5\ \text{g Cu}}$

8-8 $25.0\ \text{g C} \times \frac{x\ \text{g}}{12.01\ \text{g C}} = 33.3\ \text{g}$ $x = 16.0\ \text{g (O)}$ The compound is CO.

8-10 $60.0 \text{ lb O} \times \frac{32.07 \text{ lb S}}{16.00 \text{ lb O}} \times \frac{1}{3} = \underline{40.1 \text{ lb S}}$

8-11 $6.022 \times 10^{23} \text{ units} \times \frac{1 \text{ sec}}{2 \text{ units}} \times \frac{1 \text{ min}}{60 \text{ sec}} \times \frac{1 \text{ hr}}{60 \text{ min}} \times \frac{1 \text{ day}}{24 \text{ hr}}$

$\times \frac{1 \text{ year}}{365 \text{ day}} = \underline{9.548 \times 10^{15} \text{ years (9.548 quadrillion)}}$

$\frac{9.548 \times 10^{15} \text{ years}}{5.5 \times 10^{9}} = \underline{1.7 \times 10^{6} \text{ years (1.7 million)}}$

8-13 6.022×10^{26} (if mass in kg); 6.022×10^{20} (if mass in mg)

8-14 (a) P: $14.5 \text{ g P} \times \frac{1 \text{ mol P}}{30.97 \text{ g P}} = \underline{0.468 \text{ mol P}}$

$0.468 \text{ mol P} \times \frac{6.022 \times 10^{23} \text{ atoms}}{\text{mol P}} = \underline{2.82 \times 10^{23} \text{ atoms P}}$

(b) Rb: $1.75 \text{ mol Rb} \times \frac{85.47 \text{ g Rb}}{\text{mol Rb}} = \underline{150 \text{ g Rb}}$

$1.75 \text{ mol Rb} \times \frac{6.022 \times 10^{23} \text{ atoms}}{\text{mol Rb}} = \underline{1.05 \times 10^{24} \text{ atoms}}$

(c) Al: 27.0 g, 1.00 mol, 6.02×10^{23} atoms

(d) $3.01 \times 10^{24} \text{ atoms} \times \frac{1 \text{ mol X}}{6.022 \times 10^{23} \text{ atoms}} = 5.00 \text{ mol X}$

$\frac{363 \text{ g X}}{5.00 \text{ mol X}} = 76.2$ g/mol; From the periodic table, $\underline{X = Ge}$

(e) Ti: $1 \text{ atom} \times \frac{1 \text{ mol Ti}}{6.022 \times 10^{23} \text{ atoms}} = \underline{1.66 \times 10^{-24} \text{ mol}}$

$1.66 \times 10^{-24} \text{ mol Ti} \times \frac{47.88 \text{ g Ti}}{\text{mol Ti}} = \underline{7.95 \times 10^{-23} \text{ g Ti}}$

8-18 (a) $32.1 \text{ mol S} \times \frac{6.022 \times 10^{23} \text{ atoms}}{\text{mol S}} = \underline{1.93 \times 10^{25} \text{ atoms}}$

(b) $32.1 \text{ g S} \times \frac{1 \text{ mol S}}{32.07 \text{ g S}} \times \frac{6.022 \times 10^{23} \text{ atoms}}{\text{mol S}} = \underline{6.03 \times 10^{23} \text{ atoms}}$

(c) $32.0 \text{ g O} \times \frac{1 \text{ mol O}}{16.00 \text{ g O}} \times \frac{6.022 \times 10^{23} \text{ atoms}}{\text{mol O}} = \underline{1.20 \times 10^{24} \text{ atoms}}$

8-20 $50.0 \text{ g Al} \times \frac{1 \text{ mol Al}}{26.98 \text{ g Al}} = 1.85 \text{ mol Al}$

$50.0 \text{ g Fe} \times \frac{1 \text{ mol Fe}}{55.85 \text{ g Fe}} = 0.895 \text{ mol Fe}$

There are more moles of atoms (more atoms) in 50.0 g of Al.

8-21 $20.0\ \text{g Ni} \times \dfrac{1\ \text{mol Ni}}{58.69\ \text{g Ni}} = 0.341\ \text{mol Ni}$

$2.85 \times 10^{23}\ \text{atoms} \times \dfrac{1\ \text{mol Ni}}{6.022 \times 10^{23}\ \text{atoms}} = 0.473\ \text{mol Ni}$

The 2.85×10^{23} atoms of Ni contain more atoms than 20.0 g.

8-23 $1.40 \times 10^{21}\ \text{atoms} \times \dfrac{1\ \text{mol}}{6.022 \times 10^{23}\ \text{atoms}} = 2.32 \times 10^{-3}\ \text{mol}$

$0.251\ \text{g}/2.32 \times 10^{-3}\ \text{mol} = \underline{108\ \text{g/mol (silver)}}$

8-25 (a) $KClO_2$ 39.10 + 35.45 +(2 x 16.00) = 106.6 amu

(b) SO_3 32.07 + (3 x 16.00) = 80.07 amu

(c) N_2O_5 (2 x 14.01) + (5 x 16.00) = 108.0 amu

(d) H_2SO_4 (2 x 1.008) + 32.07 + (4 x 16.00) = 98.09. amu

(e) Na_2CO_3 (2 x 22.99) + 12.01 + (3 x 16.00) = 106.0 amu

(f) CH_3COOH ($C_2H_4O_2$) (2 x 12.01) + (4 x 1.008) + (2 x 16.00) = 60.05 amu

(g) $Fe_2(CrO_4)_3$ (2 x 55.85) + (3 x 52.00) + (12 x 16.00) = 459.7 amu

8-27 $Cr_2(SO_4)_3$ (2 x 52.00) + (3 x 32.07) + (12 x 16.00) = 392.2 amu

8-29 (a) H_2O: $10.5\ \text{mol}\ H_2O \times \dfrac{18.01\ \text{g}\ H_2O}{\text{mol}\ H_2O} = \underline{189\ \text{g}\ H_2O}$

$10.5\ \text{mol}\ H_2O \times \dfrac{6.022 \times 10^{23}\ \text{molecules}}{\text{mol}\ H_2O} = \underline{6.32 \times 10^{24}\ \text{molecules}}$

(b) BF_3: $3.01 \times 10^{21}\ \text{molecules} \times \dfrac{1\ \text{mol}}{6.022 \times 10^{23}\ \text{molecules}} = \underline{5.00 \times 10^{-3}\ \text{mol}\ BF_3}$

$5.00 \times 10^{-3}\ \text{mol}\ BF_3 \times \dfrac{67.81\ \text{g}\ BF_3}{\text{mol}\ BF_3} = \underline{0.339\ \text{g}\ BF_3}$

(c) SO_2: $14.0\ \text{g}\ SO_2 \times \dfrac{1\ \text{mol}\ SO_2}{64.07\ \text{g}\ SO_2} = \underline{0.219\ \text{mol}\ SO_2}$

$0.219\ \text{mol}\ SO_2 \times \dfrac{6.022 \times 10^{23}\ \text{molecules}}{\text{mol}\ SO_2} = \underline{1.32 \times 10^{23}\ \text{molecules}}$

(d) K_2SO_4: $1.20 \times 10^{-4}\ \text{mol}\ K_2SO_4 \times \dfrac{174.3\ \text{g}\ K_2SO_4}{\text{mol}\ K_2SO_4} = \underline{0.209\ \text{g}\ K_2SO_4}$

$1.20 \times 10^{-4}\ \text{mol}\ K_2SO_4 \times \dfrac{6.022 \times 10^{23}\ \text{formula units}}{\text{mol}\ K_2SO_4} = \underline{7.23 \times 10^{19}\ \text{formula units}}$

(e) SO_3: 4.50×10^{24} ~~molecules~~ x $\frac{1 \text{ mol } SO_3}{6.022 \times 10^{23} \text{ molecules}}$ = 7.47 mol SO_3

7.47 ~~mol SO_3~~ x $\frac{80.07 \text{ g } SO_3}{\text{mol } SO_3}$ = 598 g SO_3

(f) $N(CH_3)_3$: 0.450 ~~g $N(CH_3)_3$~~ x $\frac{1 \text{ mol } N(CH_3)_3}{59.11 \text{ g } N(CH_3)_3}$ = 7.61×10^{-3} mol

7.61×10^{-3} ~~mol $N(CH_3)_3$~~ x $\frac{6.022 \times 10^{23} \text{ molecules}}{\text{mol } N(CH_3)_3}$ = 4.58×10^{21} molecules

8-31 21.5 g/0.0684 mol = 314 g/mol

8-33 1.07×10^{24} ~~molecules~~ x $\frac{1 \text{ mol}}{6.022 \times 10^{23} \text{ molecules}}$ = 1.78 mol $\frac{287 \text{ g}}{1.78 \text{ mol}}$ = 161 g/mol

8-35 C: 2.55 ~~mol C_2H_6O~~ x $\frac{2 \text{ mol C}}{\text{mol } C_2H_6O}$ = 5.10 mol C

H: 15.3 mol; O: 2.55 mol Total = 23.0 mol of atoms

C: 5.10 ~~mol C~~ x $\frac{12.01 \text{ g C}}{\text{mol C}}$ = 61.3 g C

H: 15.4 g; O: 40.8 g Total mass = 117.5 g

8-36 28.0 ~~g $Ca(ClO_3)_2$~~ x $\frac{1 \text{ mol}}{207.0 \text{ g } Ca(ClO_3)_2}$ = 0.135 mol $Ca(ClO_3)_2$

Ca: 0.135 ~~mol $Ca(ClO_3)_2$~~ x $\frac{1 \text{ mol Ca}}{\text{mol } Ca(ClO_3)_2}$ = 0.135 mol Ca

Cl: 0.270 mol Cl O: 0.810 mol O Total = 1.215 mol of atoms

8-38 1.50 ~~mol H_2SO_4~~ x $\frac{2 \text{ mol H}}{\text{mol } H_2SO_4}$ x $\frac{1.008 \text{ g H}}{\text{mol H}}$ = 3.02 g H

1.50 ~~mol H_2SO_4~~ x $\frac{1 \text{ mol S}}{\text{mol } H_2SO_4}$ x $\frac{32.07 \text{ g S}}{\text{mol S}}$ = 48.1 g S

1.50 ~~mol H_2SO_4~~ x $\frac{4 \text{ mol O}}{\text{mol } H_2SO_4}$ x $\frac{16.00 \text{ g O}}{\text{mol O}}$ = 72.0 g O

8-40 1.20×10^{22} ~~molecules~~ x $\frac{1 \text{ mol } O_2}{6.022 \times 10^{23} \text{ molecules}}$ = 0.0199 mol O_2

0.0199 ~~mol O_2~~ x $\frac{2 \text{ mol O atoms}}{\text{mol } O_2}$ = 0.0398 mol O atoms

0.0199 ~~mol O_2~~ x $\frac{32.00 \text{ g } O_2}{\text{mol } O_2}$ = 0.637 g O_2 The mass is the same.

8-42 Total mass of compound = 1.375 + 3.935 = 5.310 g

N: $\frac{1.375 \text{ g}}{5.310 \text{ g}}$ x 100% = $\underline{25.89\% \text{ N}}$ O: $\frac{3.935 \text{ g}}{5.310 \text{ g}}$ x 100% = $\underline{74.11\% \text{ O}}$

8-43 Si: $\frac{2.27 \text{ g}}{4.86 \text{ g}}$ x 100% = $\underline{46.7\% \text{ Si}}$ O: 100% - 46.7% = $\underline{53.3\%\text{O}}$

8-45 (a) $\mathbf{C_2H_6O}$ Formula weight = (2 x 12.01) + (6 x 1.008) + 16.00 = 46.07 amu

C: $\frac{24.02 \text{ amu}}{46.07 \text{ amu}}$ x 100% = $\underline{52.14\% \text{ C}}$ H: $\frac{6.048 \text{ amu}}{46.07 \text{ amu}}$ x 100% = $\underline{13.13\% \text{ H}}$

O: 100.00% - (52.14 + 13.13)% = $\underline{34.73\% \text{ O}}$

(b) $\mathbf{C_3H_6}$ Formula weight = (3 x 12.01) + (6 x 1.008) = 42.08 amu

C: $\frac{36.03 \text{ amu}}{42.08 \text{ amu}}$ x 100% = $\underline{85.62\% \text{ C}}$ H: 100.00% - 85.62% = $\underline{14..38\% \text{ H}}$

(c) $\mathbf{C_9H_{18}}$ Formula weight =(9 x 12.01) +(18 x 1.008)= 126.2 amu

C: $\frac{108.1 \text{ amu}}{126.2 \text{ amu}}$ x 100% = $\underline{85.66\% \text{ C}}$ H = 100% - 85.66% = $\underline{14.34\% \text{ H}}$

(b) and (c) are actually the same. The difference comes from rounding off.

(d) $\mathbf{Na_2SO_4}$ Formula weight = (2 x 22.99) + 32.07 + (4 x 16.00) = 142.1 amu

Na: $\frac{45.98 \text{ amu}}{142.1 \text{ amu}}$ x 100% = $\underline{32.36\% \text{ Na}}$ S: $\frac{32.07 \text{ amu}}{142.1 \text{ amu}}$ x 100% = $\underline{22.57\% \text{ S}}$

O: 100.00% - (32.36 + 22.57)% = $\underline{45.07\% \text{ O}}$

(e) $\mathbf{(NH_4)_2CO_3}$ Formula weight = (2 x 14.01)+ (8 x 1.008) + 12.01 + (3 x 16.00)= 96.09 amu

N: $\frac{28.02 \text{ amu}}{96.09 \text{ amu}}$ x 100% = $\underline{29.16\% \text{ N}}$ H: $\frac{8.064 \text{ amu}}{96.09 \text{ amu}}$ x 100% = $\underline{8.392\% \text{ H}}$

C: $\frac{12.01 \text{ amu}}{96.09 \text{ amu}}$ x 100% = $\underline{12.50\%\text{C}}$ O: $\frac{48.00 \text{ amu}}{96.09 \text{ amu}}$ x 100% = $\underline{49.95\% \text{ O}}$

8-47 $Na_2B_4O_7$ $10H_2O$ (or $Na_2B_4O_{17}H_{20}$) Formula weight =

(2 x 22.99) + (4 x 10.81) + (17 x 16.00) + (20 x 1.008) = 381.4 amu

Na: $\frac{45.98 \text{ amu}}{381.4 \text{ amu}}$ x 100% = $\underline{12.06\% \text{ Na}}$ B: $\frac{43.24 \text{ amu}}{381.4 \text{ amu}}$ x 100% = $\underline{11.34\% \text{ B}}$

O: $\frac{272.0 \text{ amu}}{381.4 \text{ amu}}$ x 100% = $\underline{71.31\% \text{ O}}$ H: $\frac{20.16 \text{ amu}}{381.4 \text{ amu}}$ x 100% = $\underline{5.286\% \text{ H}}$

8-49 $C_7H_5SNO_3$ Formula weight = (7 x 12.01) + (5 x 1.008)+ 32.07+ 14.01 + (3 x 16.00)= 183.2 amu

C: $\frac{84.07 \text{ amu}}{183.2 \text{ amu}}$ x 100% = $\underline{45.89\% \text{ C}}$ H: $\frac{5.040 \text{ amu}}{183.2 \text{ amu}}$ x 100% = $\underline{2.751\% \text{ H}}$

S: $\frac{32.07 \text{ amu}}{183.2 \text{ amu}}$ x 100% = $\underline{17.51\% \text{ S}}$ N: $\frac{14.01 \text{ amu}}{183.2 \text{ amu}}$ x 100% = $\underline{7.647\% \text{ N}}$

O: 100% - (45.89 + 2.751 + 17.51 + 7.647) = $\underline{26.20\% \text{ O}}$

8-51 $Na_2C_2O_4$ Formula weight = (2 x 22.99) + (2 x 12.01) + (4 x 16.00) = 134.0 amu

$$125\ \text{g } Na_2C_2O_4 \times \frac{1\ \text{mol } Na_2C_2O_4}{134.0\ \text{g } Na_2C_2O_4} \times \frac{2\ \text{mol C}}{\text{mol } Na_2C_2O_4} \times \frac{12.01\ \text{g C}}{\text{mol C}} = \underline{22.4\ \text{g C}}$$

8-52 Na_3PO_4 Formula weight = (3 x 22.99) + 30.97 + (4 x 16.00) = 163.9 amu

There is 30.97 lb of P in 163.9 lb of compound.

$$25.0\ \text{lb } Na_3PO_4 \times \frac{30.97\ \text{lb P}}{163.9\ \text{lb } Na_3PO_4} = \underline{4.72\ \text{lb P}}$$

8-54 Fe_2O_3 Formula weight = (2 x 55.85) + (3 x 16.00) = 159.7 amu

There is 111.7 lb of Fe (2 x 55.85) in 159.7 lb of Fe_2O_3.

$$2000\ \text{lb } Fe_2O_3 \times \frac{111.7\ \text{lb Fe}}{159.7\ \text{lb } Fe_2O_3} = \underline{1.40 \times 10^3\ \text{lb Fe}}$$

8-56 (a) FeS (b) SrI_2 (c) $KClO_3$ (d) I_2O_5

(e) $Fe_2O_{2.66} = Fe_{6/3}O_{8/3} = Fe_6O_8 = Fe_3O_4$

(f) $C: \frac{4.22}{4.22} = 1.0 \quad H: \frac{7.03}{4.22} = 1.66 \quad Cl: \frac{4.22}{4.22} = 1.0 \quad CH_{1.66}Cl = CH_{5/3}Cl = \underline{C_3H_5Cl_3}$

8-58 Assume 100 g of compound. There are then 63.1 g of O and 36.8 g of N per 100 g.

$$63.1\ \text{g O} \times \frac{1\ \text{mol O}}{16.00\ \text{g O}} = 3.94\ \text{mol O}$$

$$36.8\ \text{g N} \times \frac{1\ \text{mol N}}{14.01\ \text{g N}} = 2.63\ \text{mol N}$$

$$O: \frac{3.94}{2.63} = 1.5 \quad N: \frac{2.63}{2.63} = 1.0 \quad NO_{1.5} = \underline{N_2O_3}$$

8-60 $8.25\ \text{g K} \times \frac{1\ \text{mol K}}{39.10\ \text{gK}} = 0.211\ \text{mol K}$

$$6.75\ \text{g O} \times \frac{1\ \text{mol O}}{16.00\ \text{g O}} = 0.422\ \text{mol O} \quad \frac{0.211}{0.211} = 1.0 \quad \frac{0.422}{0.211} = 2.0 \quad \underline{KO_2}$$

8-62 In 100 g of compound there are 21.6 g of Mg, 21.4 g of C, and 57.0 g of O.

$$21.6\ \text{g Mg} \times \frac{1\ \text{mol Mg}}{24.31\ \text{g Mg}} = 0.889\ \text{mol Mg} \quad 21.4\ \text{g C} \times \frac{1\ \text{mol C}}{12.01\ \text{g C}} = 1.78\ \text{mol C}$$

$$57.0\ \text{g O} \times \frac{1\ \text{mol O}}{16.00\ \text{g O}} = 3.56\ \text{mol O}$$

$$Mg: \frac{0.889}{0.889} = 1.0 \quad C: \frac{1.78}{0.889} = 2.0 \quad O: \frac{3.56}{0.889} = 4.0 \quad \underline{MgC_2O_4}$$

8-63 $C:\ 9.90\ \text{g C} \times \frac{1\ \text{mol C}}{12.01\ \text{g C}} = 0.824\ \text{mol C} \qquad H:\ 1.65\ \text{g H} \times \frac{1\ \text{mol H}}{1.008\ \text{g H}} = 1.64\ \text{mol H}$

$$Cl:\ 29.3\ \text{g Cl} \times \frac{1\ \text{mol Cl}}{35.45\ \text{g Cl}} = 0.827\ \text{mol Cl}$$

$$C: \frac{0.824}{0.824} = 1.0 \quad H: \frac{1.64}{0.824} = 2.0 \quad Cl: \frac{0.827}{0.824} = 1.0 \quad \underline{CH_2Cl}$$

8-65 $\% O = 100\% - (24.1 + 6.90 + 27.6)\% = 41.4\% O$

$$N: 24.1\ \cancel{g\,N} \times \frac{1\ mol\ N}{14.01\ \cancel{g\,N}} = 1.72\ mol\ N \qquad H: 6.90\ \cancel{g\,H} \times \frac{1\ mol\ H}{1.008\ \cancel{g\,H}} = 6.85\ mol\ H$$

$$S: 27.6\ \cancel{g\,S} \times \frac{1\ mol\ S}{32.07\ \cancel{g\,S}} = 0.861\ mol\ S \qquad O: 41.4\ \cancel{g\,O} \times \frac{1\ mol\ O}{16.00\ \cancel{g\,O}} = 2.59\ mol\ O$$

$$N: \frac{1.72}{0.861} = 2.0 \quad H: \frac{6.85}{0.861} = 8.0 \quad S: \frac{0.861}{0.861} = 1.0 \quad O: \frac{2.59}{0.861} = 3.0$$

$\underline{N_2H_8SO_3}$

8-66

$$C: 63.2\ \cancel{g\,C} \times \frac{1\ mol\ C}{12.01\ \cancel{g\,C}} = 5.26\ mol\ C \qquad O: 31.6\ \cancel{g\,O} \times \frac{1\ mol\ O}{16.00\ \cancel{g\,O}} = 1.98\ mol\ O$$

$$H: 5.26\ \cancel{g\,H} \times \frac{1\ mol\ H}{1.008\ \cancel{g\,H}} = 5.22\ mol\ H$$

$$C: \frac{5.26}{1.98} = 2.66\ (\tfrac{8}{3}) \quad O: \frac{1.98}{1.98} = 1.0 \quad H: \frac{5.22}{1.98} = 2.63\ (\tfrac{8}{3})$$

$C_{8/3}H_{8/3}O = \underline{C_8H_8O_3}$

8-68 In 100 g of compound: 20.0 g of C, 2.2 g of H, and 77.8 g of Cl.

$$20.0\ \cancel{g\,C} \times \frac{1\ mol\ C}{12.01\ \cancel{g\,C}} = 1.66\ mol\ C \qquad 2.2\ \cancel{g\,H} \times \frac{1\ mol\ H}{1.008\ \cancel{g\,H}} = 2.18\ mol\ H$$

$$77.8\ \cancel{g\,Cl} \times \frac{1\ mol\ Cl}{35.45\ \cancel{g\,Cl}} = 2.19\ mol\ Cl$$

$$C: \frac{1.66}{1.66} = 1.0 \quad H: \frac{2.18}{1.66} = 1.31\ \left(\frac{4}{3}\right) \quad Cl: \frac{2.19}{1.66} = 1.32\ \left(\frac{4}{3}\right)$$

$CH_{4/3}Cl_{4/3} = \underline{C_3H_4Cl_4}$ (empirical formula)

empirical mass = (3 x 12.01) + (4 x 1.0) + (4 x 35.5) = 182 amu = 182 g/emp. unit

$$\frac{545\ \cancel{g}/mol}{182\ \cancel{g}/emp.\ unit} = 3\ emp.\ unit/mol \qquad C_{(3 \times 3)}H_{(3 \times 4)}Cl_{(3 \times 4)} = \underline{C_9H_{12}Cl_{12}}$$

8-70

$$B: 18.7\ \cancel{g\,B} \times \frac{1\ mol\ B}{10.81\ \cancel{g\,B}} = 1.73\ mol\ B \qquad C: 20.7\ \cancel{g\,C} \times \frac{1\ mol\ C}{12.01\ \cancel{g\,C}} = 1.72\ mol\ C$$

$$H: 5.15\ \cancel{g\,H} \times \frac{1\ mol\ H}{1.008\ \cancel{g\,H}} = 5.11\ mol\ H \qquad O: 55.4\ \cancel{g\,O} \times \frac{1\ mol\ O}{16.00\ \cancel{g\,O}} = 3.46\ mol\ O$$

$$B: \frac{1.73}{1.72} = 1.0 \quad C: \frac{1.72}{1.72} = 1.0 \quad H: \frac{5.11}{1.72} = 3.0 \quad O: \frac{3.46}{1.72} = 2.0$$

Emp. formula = BCH_3O_2 Emp. mass = 57.84 g/emp. unit

$$\frac{115\ \cancel{g}/mol}{57.84\ \cancel{g}/emp.\ unit} = 2\ emp.\ units/mol \qquad \underline{B_2C_2H_6O_4}\ (molecular\ formula)$$

8-71 $34.9\ \text{g K} \times \frac{1\ \text{mol K}}{39.10\ \text{g K}} = 0.893\ \text{mol K}$ $21.4\ \text{g C} \times \frac{1\ \text{mol C}}{12.01\ \text{g C}} = 1.78\ \text{mol C}$

$12.5\ \text{g N} \times \frac{1\ \text{mol N}}{14.01\ \text{g N}} = 0.892\ \text{mol N}$ $2.68\ \text{g H} \times \frac{1\ \text{mol H}}{1.008\ \text{g H}} = 2.66\ \text{mol H}$

$28.6\ \text{g O} \times \frac{1\ \text{mol O}}{16.00\ \text{g O}} = 1.79\ \text{mol O}$

K: $\frac{0.893}{0.892} = 1.0$ C: $\frac{1.78}{0.892} = 2.0$ N: $\frac{0.892}{0.892} = 1.0$ H: $\frac{2.66}{0.892} = 3.0$ O: $\frac{1.79}{0.892} = 2.0$

Empirical formula = $KC_2NH_3O_2$ Empirical mass = 112.2 g/ emp. unit

$\frac{224\ \text{g/mol}}{112.2\ \text{g/emp. unit}} = 2$ emp. units/mol $K_2C_4N_2H_6O_4$ (molecular formula)

8-73 In a 20.0-g sample, there are 18.3 g of I and 1.7 g of C.

$18.3\ \text{g I} \times \frac{1\ \text{mol I}}{126.9\ \text{g I}} = 0.144\ \text{mol I}$ $1.7\ \text{g C} \times \frac{1\ \text{mol C}}{12.01\ \text{g C}} = 0.14\ \text{mol C}$

Emp. formula = IC Emp. mass= 138.9 g/emp. unit

$\frac{834\ \text{g/mol}}{138.9\ \text{g/emp. unit}} = 6$ emp. units/mol I_6C_6 (molecular formula)

8-78 (a)

10.0 mol Al

- $\times \frac{1\ \text{mol}\ Al_2O_3}{3\ \text{mol Al}} = 3.33\ \text{mol}\ Al_2O_3$
- $\times \frac{1\ \text{mol}\ AlCl_3}{3\ \text{mol Al}} = 3.33\ \text{mol}\ AlCl_3$
- $\times \frac{3\ \text{mol NO}}{3\ \text{mol Al}} = 10.0\ \text{mol NO}$
- $\times \frac{6\ \text{mol}\ H_2O}{3\ \text{mol Al}} = 20.0\ \text{mol}\ H_2O$

(b)

3.00 mol NH_4ClO_4

- $\times \frac{1\ \text{mol}\ Al_2O_3}{3\ \text{mol}\ NH_4ClO_4} = 1.00\ \text{mol}\ Al_2O_3$
- $\times \frac{1\ \text{mol}\ AlCl_3}{3\ \text{mol}\ NH_4ClO_4} = 1.00\ \text{mol}\ AlCl_3$
- $\times \frac{3\ \text{mol NO}}{3\ \text{mol}\ NH_4ClO_4} = 3.00\ \text{mol NO}$
- $\times \frac{6\ \text{mol}\ H_2O}{3\ \text{mol}\ NH_4ClO_4} = 6.00\ \text{mol}\ H_2O$

8-80 (a) $2\ \cancel{\text{mol NH}_3} \times \dfrac{3\ \text{mol O}_2}{2\ \cancel{\text{mol NH}_3}} = \underline{15.0\ \text{mol O}_2}$

$2\ \cancel{\text{mol NH}_3} \times \dfrac{2\ \text{mol CH}_4}{2\ \cancel{\text{mol NH}_3}} = \underline{10.0\ \text{mol CH}_4}$

(b) $10.0\ \cancel{\text{mol O}_2} \times \dfrac{2\ \text{mol HCN}}{3\ \cancel{\text{mol O}_2}} = \underline{6.67\ \text{mol HCN}}$

$10.0\ \cancel{\text{mol O}_2} \times \dfrac{6\ \text{mol H}_2\text{O}}{3\ \cancel{\text{mol O}_2}} = \underline{20.0\ \text{mol H}_2\text{O}}$

8-82 (a) $4.86\ \cancel{\text{mol HF}} \times \dfrac{1\ \cancel{\text{mol SiF}_4}}{4\ \cancel{\text{mol HF}}} \times \dfrac{104.1\ \text{g SiF}_4}{\cancel{\text{mol SiF}_4}} = \underline{126\ \text{g SiF}_4}$

$4.86\ \cancel{\text{mol HF}} \times \dfrac{2\ \cancel{\text{mol H}_2\text{O}}}{4\ \cancel{\text{mol HF}}} \times \dfrac{18.02\ \text{g H}_2\text{O}}{\cancel{\text{mol H}_2\text{O}}} = \underline{43.8\ \text{g H}_2\text{O}}$

(b) $4.86\ \cancel{\text{mol HF}} \times \dfrac{1\ \cancel{\text{mol SiO}_2}}{4\ \cancel{\text{mol HF}}} \times \dfrac{60.09\ \text{g SiO}_2}{\cancel{\text{mol SiO}_2}} = \underline{73.0\ \text{g SiO}_2}$

8-83 (a) **mol H_2O → mol H_2**

$0.400\ \cancel{\text{mol H}_2\text{O}} \times \dfrac{2\ \text{mol H}_2}{2\ \cancel{\text{mol H}_2\text{O}}} = \underline{0.400\ \text{mol H}_2}$

(b) **g O_2 → mol O_2 → mol H_2O**

$0.640\ \cancel{\text{g O}_2} \times \dfrac{1\ \cancel{\text{mol O}_2}}{32.00\ \cancel{\text{g O}_2}} \times \dfrac{2\ \text{mol H}_2\text{O}}{1\ \cancel{\text{mol O}_2}} = \underline{0.0400\ \text{mol H}_2\text{O}}$

(c) **g O_2 → mol O_2 → mol H_2**

$0.032\ \cancel{\text{g O}_2} \times \dfrac{1\ \cancel{\text{mol O}_2}}{32.00\ \cancel{\text{g O}_2}} \times \dfrac{2\ \text{mol H}_2}{1\ \cancel{\text{mol O}_2}} = \underline{0.0020\ \text{mol H}_2}$

(d) **g H_2 → mol H_2 → mol H_2O → g H_2O**

$0.400\ \cancel{\text{g H}_2} \times \dfrac{1\ \cancel{\text{mol H}_2}}{2.016\ \cancel{\text{g H}_2}} \times \dfrac{2\ \cancel{\text{mol H}_2\text{O}}}{2\ \cancel{\text{mol H}_2}} \times \dfrac{18.02\ \text{g H}_2\text{O}}{\cancel{\text{mol H}_2\text{O}}} = \underline{3.58\ \text{g H}_2\text{O}}$

8-84 (a)

$$0.450\ \cancel{\text{mol C}_3\text{H}_8} \begin{cases} \times \dfrac{3\ \text{mol CO}_2}{1\ \cancel{\text{mol C}_3\text{H}_8}} = \underline{1.35\ \text{mol CO}_2} \\ \times \dfrac{4\ \text{mol H}_2\text{O}}{1\ \cancel{\text{mol C}_3\text{H}_8}} = \underline{1.80\ \text{mol H}_2\text{O}} \\ \times \dfrac{5\ \text{mol O}_2}{1\ \cancel{\text{mol C}_3\text{H}_8}} = \underline{2.25\ \text{mol O}_2} \end{cases}$$

(b) **mol CO_2 → mol H_2O → g H_2O**

$0.200\ \cancel{\text{mol CO}_2} \times \dfrac{4\ \cancel{\text{mol H}_2\text{O}}}{3\ \cancel{\text{mol CO}_2}} \times \dfrac{18.02\ \text{g H}_2\text{O}}{\cancel{\text{mol H}_2\text{O}}} = \underline{4.81\ \text{g H}_2\text{O}}$

(c) **g H_2O → mol H_2O → mol C_3H_8 → g C_3H_8**

$$1.80\ \text{g } H_2O \times \frac{1\ \text{mol } H_2O}{18.02\ \text{g } H_2O} \times \frac{1\ \text{mol } C_3H_8}{4\ \text{mol } H_2O} \times \frac{44.09\ \text{g } C_3H_8}{\text{mol } C_3H_8} = \underline{1.10\ \text{g } C_3H_8}$$

(d) **g O_2 → mol O_2 → mol C_3H_8 → g C_3H_8**

$$160\ \text{g } O_2 \times \frac{1\ \text{mol } O_2}{32.00\ \text{g } O_2} \times \frac{1\ \text{mol } C_3H_8}{5\ \text{mol } O_2} \times \frac{44.09\ \text{g } C_3H_8}{\text{mol } C_3H_8} = \underline{44.1\ \text{g } C_3H_8}$$

(e) **molecules O_2 → mol O_2 → mol CO_2 → g CO_2**

$$1.20 \times 10^{23}\ \text{molecules} \times \frac{1\ \text{mol } O_2}{6.022 \times 10^{23}\ \text{molecules}} \times \frac{3\ \text{mol } CO_2}{5\ \text{mol } O_2} \times \frac{44.09\ \text{g } CO_2}{\text{mol } CO_2} = \underline{5.27\ \text{g } CO_2}$$

(f) **molecules CO_2 → mol CO_2 → mol H_2O**

$$4.50 \times 10^{22}\ \text{molecules} \times \frac{1\ \text{mol } CO_2}{6.022 \times 10^{23}\ \text{molecules}} \times \frac{4\ \text{mol } H_2O}{3\ \text{mol } CO_2} = \underline{0.0996\ \text{mol } H_2O}$$

8-86 $N_2 + O_2 \rightarrow 2NO$

$2NO + O_2 \rightarrow 2NO_2$

$N_2 + 2O_2 \rightarrow 2NO_2$ (total reaction)

g NO_2 → mol NO_2 → mol N_2 → g N_2

$$155\ \text{g } NO_2 \times \frac{1\ \text{mol } NO_2}{46.01\ \text{g } NO_2} \times \frac{1\ \text{mol } N_2}{2\ \text{mol } NO_2} \times \frac{28.02\ \text{g } N_2}{\text{mol } N_2} = \underline{47.2\ \text{g } N_2}$$

8-88 g $CaCO_3$ → mol $CaCO_3$ → mol HCl → g HCl

$$1.00\ \text{g } CaCO_3 \times \frac{1\ \text{mol } CaCO_3}{100.1\ \text{g } CaCO_3} \times \frac{2\ \text{mol HCl}}{1\ \text{mol } CaCO_3} \times \frac{36.46\ \text{g HCl}}{\text{mol HCl}} = \underline{0.728\ \text{g HCl}}$$

8-90 mol FeS_2 → mol H_2S → molecules H_2S

$$0.520\ \text{mol } FeS_2 \times \frac{1\ \text{mol } H_2S}{1\ \text{mol } FeS_2} \times \frac{6.022 \times 10^{23}\ \text{molecules}}{\text{mol } H_2S}$$

$$= \underline{3.13 \times 10^{23}\ \text{molecules}}$$

8-91 **kg NO_2 → g NO_2 → mol NO_2 → mol HNO_3 → g HNO_3**

$$18.5 \text{ kg } NO_2 \times \frac{10^3 \text{ g } NO_2}{\text{kg } NO_2} \times \frac{1 \text{ mol } NO_2}{46.01 \text{ g } NO_2} \times \frac{2 \text{ mol } HNO_3}{3 \text{ mol } NO_2} \times \frac{63.02 \text{ g } HNO_3}{\text{mol } HNO_3} = \underline{16{,}900 \text{ g } (16.9 \text{ kg}) \; HNO_3}$$

8-93 **mol sugar → mol alcohol → g alcohol**

$$25.0 \text{ mol } C_6H_{12}O_6 \times \frac{2 \text{ mol } C_2H_5OH}{1 \text{ mol } C_6H_{12}O_6} \times \frac{46.07 \text{ g } C_2H_5OH}{\text{mol } C_2H_5OH} = \underline{2.30 \times 10^3 \text{ g } (2.30 \text{ kg}) \; C_2H_5OH}$$

8-94 **molecules CH_4 → mol CH_4 → mol CO → g CO**

$$8.75 \times 10^{25} \text{ molecules } CH_4 \times \frac{1 \text{ mol } CH_4}{6.022 \times 10^{23} \text{ molecules } CH_4} \times \frac{2 \text{ mol CO}}{1 \text{ mol } CH_4} \times \frac{28.01 \text{ g CO}}{\text{mol CO}} = \underline{8140 \text{ g CO}}$$

8-95 (a) $3.00 \text{ mol CuO} \times \frac{1 \text{ mol } N_2}{3 \text{ mol CuO}} = 1.00 \text{ mol } N_2$ - limiting reactant

$3.00 \text{ mol } NH_3 \times \frac{1 \text{ mol } N_2}{2 \text{ mol } NH_3} = 1.50 \text{ mol } N_2$

(b) Stoichiometric mixture producing 1.00 mol N_2

(c) $1.00 \text{ mol } NH_3 \times \frac{1 \text{ mol } N_2}{2 \text{ mol } NH_3} = 0.500 \text{ mol } N_2$ - limiting reactant

(d) $0.628 \text{ mol CuO} \times \frac{1 \text{ mol } N_2}{3 \text{ mol CuO}} = 0.209 \text{ mol } N_2$ - limiting reactant

$0.430 \text{ mol } NH_3 \times \frac{1 \text{ mol } N_2}{2 \text{ mol } NH_3} = 0.215 \text{ mol } N_2$

(e) $5.44 \text{ mol CuO} \times \frac{1 \text{ mol } N_2}{3 \text{ mol CuO}} = 1.81 \text{ mol } N_2$

$3.50 \text{ mol } NH_3 \times \frac{1 \text{ mol } N_2}{2 \text{ mol } NH_3} = 1.75 \text{ mol } N_2$ - limiting reactant

8-96 (a) $3.00 \text{ mol CuO} \times \frac{2 \text{ mol } NH_3}{3 \text{ mol CuO}} = 2.00 \text{ mol } NH_3$ used

$3.00 - 2.00 = 1.00 \text{ mol } NH_3$ in excess

(c) 1.50 mol CuO in excess

8-99 First, find the limiting reactant

$$0.800 \text{ mol Al} \times \frac{3 \text{ mol } H_2}{2 \text{ mol Al}} = 1.20 \text{ mol } H_2$$

$$1.00 \text{ mol } H_2SO_4 \times \frac{3 \text{ mol } H_2}{3 \text{ mol } H_2SO_4} = 1.00 \text{ mol } H_2$$

Therefore, H_2SO_4 is the limiting reactant and the yield of H_2 is 1.00 mole.

Now convert moles of H_2SO_4 to moles of Al.

$$1.00 \text{ mol } H_2SO_4 \times \frac{2 \text{ mol Al}}{3 \text{ mol } H_2SO_4} = 0.667 \text{ mol of Al used}$$

0.800 - 0.667 = 0.133 mol Al remaining

8-100 $$3.44 \text{ mol } C_5H_6 \times \frac{10 \text{ mol } CO_2}{2 \text{ mol } C_5H_6} = 17.2 \text{ mol } CO_2$$

$$20.6 \text{ mol } O_2 \times \frac{10 \text{ mol } CO_2}{13 \text{ mol } O_2} = 15.8 \text{ mol } CO_2$$

Since O_2 is the limiting reactant: $15.8 \text{ mol } CO_2 \times \frac{44.01 \text{ g } CO_2}{\text{mol } CO_2} = 695 \text{ g } CO_2$

8-102 **g O_2 → mol O_2 → mol N_2 ← mol NH_3**

$$40.0 \text{ g } O_2 \times \frac{1 \text{ mol } O_2}{32.00 \text{ g } O_2} \times \frac{2 \text{ mol } N_2}{3 \text{ mol } O_2} = 0.833 \text{ mol } N_2$$

$$1.50 \text{ mol } NH_3 \times \frac{2 \text{ mol } N_2}{4 \text{ mol } NH_3} = 0.750 \text{ mol } N_2$$

Since NH_3 produces the least N_2, it is the limiting reactant and the yield of N_2 is 0.750 mol.

8-103 **g $AgNO_3$ → mol $AgNO_3$ → mol AgCl → g AgCl**

g $CaCl_2$ → mol $CaCl_2$ → mol AgCl

$$20.0 \text{ g } AgNO_3 \times \frac{1 \text{ mol } AgNO_3}{169.9 \text{ g } AgNO_3} \times \frac{2 \text{ mol AgCl}}{2 \text{ mol } AgNO_3} = 0.118 \text{ mol AgCl}$$

$$10.0 \text{ g } CaCl_2 \times \frac{1 \text{ mol } CaCl_2}{111.0 \text{ g } CaCl_2} \times \frac{2 \text{ mol AgCl}}{1 \text{ mol } CaCl_2} = 0.180 \text{ mol AgCl}$$

Since $AgNO_3$ produces the least AgCl, it is the limiting reactant.

$$0.118 \text{ mol AgCl} \times \frac{143.4 \text{ g AgCl}}{\text{mol AgCl}} = 16.9 \text{ g AgCl}$$

Convert moles of AgCl (the limiting reactant) to grams of $CaCl_2$ used.

$$0.118 \text{ mol AgCl} \times \frac{1 \text{ mol } CaCl_2}{2 \text{ mol AgCl}} \times \frac{111.0 \text{ g } CaCl_2}{\text{mol } CaCl_2} = 6.55 \text{ g } CaCl_2 \text{ used}$$

$$10.0 \text{ g} - 6.55 \text{ g} = \underline{3.5 \text{ g } CaCl_2 \text{ remaining}}$$

8-105 $\mathbf{g\ HNO_3 \rightarrow mol\ HNO_3}$ ↘ $\mathbf{mol\ H_2O \rightarrow g\ H_2O}$

$\mathbf{g\ H_2S \rightarrow mol\ H_2S}$ ↗

$$10.0 \text{ g } HNO_3 \times \frac{1 \text{ mol } HNO_3}{63.02 \text{ g } HNO_3} \times \frac{4 \text{ mol } H_2O}{2 \text{ mol } HNO_3} = 0.317 \text{ mol } H_2O$$

$$5.00 \text{ g } H_2S \times \frac{1 \text{ mol } H_2S}{34.09 \text{ g } H_2S} \times \frac{4 \text{ mol } H_2O}{3 \text{ mol } H_2S} = 0.196 \text{ mol } H_2O \text{ (limiting reactant)}$$

$$0.196 \text{ mol } H_2O \times \frac{18.02 \text{ g } H_2O}{\text{mol } H_2O} = \underline{3.53 \text{ g } H_2O}$$

Convert moles H_2O to grams of other two products.

$$0.196 \text{ mol } H_2O \times \frac{3 \text{ mol S}}{4 \text{ mol } H_2O} \times \frac{32.07 \text{ g S}}{\text{mol S}} = \underline{4.71 \text{ g S}}$$

$$0.196 \text{ mol } H_2O \times \frac{2 \text{ mol NO}}{4 \text{ mol } H_2O} \times \frac{30.01 \text{ g NO}}{\text{mol NO}} = \underline{2.94 \text{ g NO}}$$

Convert mol H_2O to grams of excess reactant used.

$$0.196 \text{ mol } H_2O \times \frac{2 \text{ mol } HNO_3}{4 \text{ mol } H_2O} \times \frac{63.02 \text{ g } HNO_3}{\text{mol } HNO_3} = 6.18 \text{ g } HNO_3$$

$$10.0 \text{ g} - 6.18 \text{ g} = \underline{3.8 \text{ g } HNO_3 \text{ remaining}}$$

8-106 $\mathbf{g\ SO_2 \rightarrow mol\ SO_2 \rightarrow mol\ SO_3 \rightarrow g\ SO_3}$

$$24.0 \text{ g } SO_2 \times \frac{1 \text{ mol } SO_2}{64.07 \text{ g } SO_2} \times \frac{2 \text{ mol } SO_3}{2 \text{ mol } SO_2} \times \frac{80.07 \text{ g } SO_3}{\text{mol } SO_3} = 30.0 \text{ g } SO_3 \text{ (theoretical yield)}$$

$$\text{percent yield} = \frac{21.2 \text{ g}}{30.0 \text{ g}} \times 100\% = \underline{70.7\%}$$

8-109 $g\ C_8H_{18} \rightarrow mol\ C_8H_{18} \rightarrow mol\ CO_2 \rightarrow g\ CO_2$

$$57.0\ g\ C_8H_{18} \times \frac{1\ mol\ C_8H_{18}}{114.2\ g\ C_8H_{18}} \times \frac{16\ mol\ CO_2}{2\ mol\ C_8H_{18}} \times \frac{44.01\ g\ CO_2}{mol\ CO_2} = 176\ g\ CO_2$$

$$\frac{152\ g}{176\ g} \times 100\% = \underline{86.4\%}$$

8-110 If 86.4% is converted to CO_2, the remainder (13.6%) is converted to CO. Thus, 0.136 x 57.0 g = 7.75 g of C_8H_{18} is converted to CO. Notice that 1 mole of C_8H_{18} forms 8 moles of CO (because of the eight carbons in C_8H_{18}). Thus

$g\ C_8H_{18} \rightarrow mol\ C_8H_{18} \rightarrow mol\ CO \rightarrow g\ CO$

$$7.75\ g\ C_8H_{18} \times \frac{1\ mol\ C_8H_{18}}{114.2\ g\ C_8H_{18}} \times \frac{8\ mol\ CO}{1\ mol\ C_8H_{18}} \times \frac{28.01\ g\ CO}{mol\ CO} = \underline{15.2\ g\ CO}$$

8-111 Theoretical yield x 0.700 = 250 g (actual yield)
Theoretical yield = 250 g/0.700 = 357 g N_2

$g\ N_2 \rightarrow mol\ N_2 \rightarrow mol\ H_2 \rightarrow g\ H_2$

$$357\ g\ N_2 \times \frac{1\ mol\ N_2}{28.02\ g\ N_2} \times \frac{4\ mol\ H_2}{1\ mol\ N_2} \times \frac{2.016\ g\ H_2}{mol\ H_2} = \underline{103\ g\ H_2}$$

8-114 $2Mg(s) + O_2 \rightarrow 2MgO(s) + 1204\ kJ$

$$2Mg(s) + O_2(g) \rightarrow 2MgO(s) \quad \Delta H = -1204\ kJ$$

8-116 $CaCO_3(s) + 176\ kJ \rightarrow CaO(s) + CO_2(g)$

$$CaCO_3(s) \rightarrow CaO(s) + CO_2(g) \quad \Delta H = 176\ kJ$$

8-117 $g\ C_8H_{18} \rightarrow mol\ C_8H_{18} \rightarrow kJ$

$$1.00\ g\ C_8H_{18} \times \frac{1\ mol\ C_8H_{18}}{114.2\ g\ C_8H_{18}} \times \frac{5480\ kJ}{mol\ C_8H_{18}} = \underline{48.0\ kJ}$$

$$1.00\ g\ CH_4 \times \frac{1\ mol\ CH_4}{16.04\ g\ CH_4} \times \frac{890\ kJ}{mol\ CH_4} = \underline{55.6\ kJ}.$$

8-119 $2Al(s) + Fe_2O_3(s) \rightarrow Al_2O_3(s) + Fe(l) \quad \Delta H = -850\ kJ$

$kJ \rightarrow mol\ Al \rightarrow g\ Al$

$$35.8\ kJ \times \frac{2\ mol\ Al}{850\ kJ} \times \frac{26.99\ g\ Al}{mol\ Al} = \underline{2.27\ g\ Al}$$

8-120 kJ → mol $C_6H_{12}O_6$ → g $C_6H_{12}O_6$

$$975 \text{ kJ} \times \frac{1 \text{ mol } C_6H_{12}O_6}{2519 \text{ kJ}} \times \frac{180.2 \text{ g } C_6H_{12}O_6}{\text{mol } C_6H_{12}O_6} = \underline{69.7 \text{ g } C_6H_{12}O_6}$$

8-122 $\$4.5 \times 10^{12} \times \frac{100 \text{ pennies}}{\$} \times \frac{1 \text{ mol pennies}}{6.022 \times 10^{23} \text{ pennies}} \times = \underline{7.5 \times 10^{-10} \text{ mol pennies}}$

8-123 $0.443 \text{ g N} \times \frac{1 \text{ mol N}}{14.01 \text{ g N}} = 0.0316 \text{ mol N}$

Thus 1.420 g of M also equals 0.0316 mol M since M and N are present in equimolar amounts.

1.420 g /0.0316 mol = 44.9 g/mol [scandium (Sc)]

8-126 $0.344 \text{ g } P_4 \times \frac{1 \text{ mol } P_4}{123.9 \text{ g } P_4} = 2.78 \times 10^{-3} \text{ mol } P_4;$

$$2.78 \times 10^{-3} \text{ mol } P_4 \times \frac{4 \text{ mol P}}{\text{mol } P_4} \times \frac{6.022 \times 10^{23} \text{ atoms P}}{\text{mol P}} = \underline{6.70 \times 10^{21} \text{ atoms P}}$$

8-127 100 mol H_2 = 202 g H_2 therefore

100 H atoms < 100 H_2 molecules < 100 g H_2 < 100 mol H_2

8-129 $2.84 \times 10^{23} \text{ form units} \times \frac{1 \text{ mol}}{6.022 \times 10^{23} \text{ form units}} = 0.472 \text{ mol}$

$\frac{56.6 \text{ g}}{0.472 \text{ mol}} = 120 \text{ g/mol}$ 120 - 55.8 = 64 g of S $\frac{64 \text{ g S}}{32.07 \text{ g S/mol}} = 2 \text{ mol S}$

Formula = $\underline{FeS_2}$

8-130 (a) $2Na^+$ and $S_4O_6^{2-}$

(b) $10.0 \text{ g Na} \times \frac{1 \text{ mol Na}}{22.99 \text{ g Na}} \times \frac{4 \text{ mol S}}{2 \text{ mol Na}} \times \frac{32.07 \text{ g S}}{\text{mol S}} = \underline{27.9 \text{ g S}}$

(c) NaS_2O_3 (d) 270.3 g/mol

(e) $25.0 \text{ g } Na_2S_4O_6 \times \frac{1 \text{ mol } Na_2S_4O_6}{270.3 \text{ g } Na_2S_4O_6} = 0.0925 \text{ mol } Na_2S_4O_6$

$$0.0925 \text{ mol } Na_2S_4O_6 \times \frac{6.022 \times 10^{23} \text{ form units}}{\text{mol } Na_2S_4O_6} = \underline{5.57 \times 10^{22} \text{ formula units}}$$

(f) $\frac{(6 \times 16.00) \text{ g O}}{270.3 \text{ g compound}} \times 100\% = \underline{35.5 \text{ \% oxygen}}$

8-132 $\frac{2N}{2N + x\,O} = 0.368$ $\frac{28.02}{28.02 + 16.00\,x} = 0.368$ $x = 3$ N_2O_3 dinitrogen trioxide

8-134 4.55×10^{22} molecules $\times \frac{1 \text{ mol dioxin}}{6.022 \times 10^{23} \text{ molecules}} = 0.0756$ mol dioxin

$\frac{24.3 \text{ g}}{0.0756 \text{ mol}} = 321$ g/mol C: $\frac{0.456}{0.076} = 6.0$ H & Cl: $\frac{0.152}{0.076} = 2.0$ O: $\frac{0.076}{0.076} = 1.0$

empirical formula = $C_6H_2Cl_2O$

emp. mass= 161 g/emp unit $\frac{321 \text{ g/mol}}{161.0 \text{ g/emp unit}} = 2$ $\underline{C_{12}H_4Cl_4O_2}$ (molecular formula)

8-136 Cr: $14.9 \, \cancel{\text{g Cr}} \times \frac{1 \text{ mol Cr}}{52.00 \, \cancel{\text{g Cr}}} = 0.287$ mol Cr

Cl: $30.4 \, \cancel{\text{g Cl}} \times \frac{1 \text{ mol Cl}}{35.45 \, \cancel{\text{g Cl}}} = 0.858$ mol Cl O: $54.7 \, \cancel{\text{g O}} \times \frac{1 \text{ mol}}{16.00 \, \cancel{\text{g O}}} = 3.42$ mol O

Cr: $\frac{0.287}{0.287} = 1.0$ Cl: $\frac{0.858}{0.287} = 3.0$ O: $\frac{3.42}{0.287} = 12.0$

Empirical formula $CrCl_3O_{12}$ Actual formula = $\underline{Cr(ClO_4)_3}$ <u>chromium(III) perchlorate</u>

8-137 Assume exactly 100 g of compound. There are then 51.1 g H_2O and 48.9 g $MgSO_4$.

H_2O: $51.1 \, \cancel{\text{g } H_2O} \times \frac{1 \text{ mol } H_2O}{18.02 \, \cancel{\text{g } H_2O}} = 2.84$ mol H_2O

$MgSO_4$: $48.9 \, \cancel{\text{g } MgSO_4} \times \frac{1 \text{ mol}}{120.4 \, \cancel{\text{g } MgSO_4}} = 0.406$ mol $MgSO_4$

2.94 mol H_2O/0.406 mol $MgSO_4$ = 7.0 mol H_2O/mol $MgSO_4$

The formula is $\underline{MgSO_4 \cdot 7H_2O}$

8-140 $1.20 \, \cancel{\text{g } CO_2} \times \frac{1 \, \cancel{\text{mol } CO_2}}{44.01 \, \cancel{\text{g } CO_2}} \times \frac{1 \text{ mol C}}{\cancel{\text{mol } CO_2}} = 0.0273$ mol C

$0.489 \, \cancel{\text{g } H_2O} \times \frac{1 \, \cancel{\text{mol } H_2O}}{18.02 \, \cancel{\text{g } H_2O}} \times \frac{2 \text{ mol H}}{\cancel{\text{mol } H_2O}} = 0.0543$ mol H

C: $\frac{0.0273}{0.0273} = 1.0$ H: $\frac{0.0543}{0.0273} = 2.0$ $\underline{CH_2}$

8-143 **g Fe_2O_3 → mol Fe_2O_3 → mol Fe_3O_4 →**

mol FeO → mol Fe → g Fe

$125 \, \cancel{\text{g } Fe_2O_3} \times \frac{1 \, \cancel{\text{mol } Fe_2O_3}}{159.7 \, \cancel{\text{g } Fe_2O_3}} \times \frac{2 \, \cancel{\text{mol } Fe_3O_4}}{3 \, \cancel{\text{mol } Fe_2O_3}} \times \frac{3 \, \cancel{\text{mol FeO}}}{1 \, \cancel{\text{mol } Fe_3O_4}}$

$\times \frac{1 \, \cancel{\text{mol Fe}}}{1 \, \cancel{\text{mol FeO}}} \times \frac{55.85 \text{ g Fe}}{\cancel{\text{mol Fe}}} = \underline{87.4 \text{ g Fe}}$

8-144 $2KClO_3 \rightarrow 2KCl + 3O_2$

Find the mass of $KClO_3$ needed to produce 12.0 g O_2.

g O_2 → mol O_2 → mol $KClO_3$ → g $KClO_3$

$$12.0\ \text{g } O_2 \times \frac{1\ \text{mol } O_2}{32.00\ \text{g } O_2} \times \frac{2\ \text{mol } KClO_3}{3\ \text{mol } O_2} \times \frac{122.6\ \text{g } KClO_3}{\text{mol } KClO_3} = 30.7\ \text{g } KClO_3$$

$$\text{percent purity} = \frac{30.7\ \text{g}}{50.0\ \text{g}} \times 100\% = \underline{61.4\%}$$

8-145 Convert g of SO_2 to g of FeS_2

g SO_2 → mol SO_2 → mol FeS_2 → g FeS_2

$$312\ \text{g } SO_2 \times \frac{1\ \text{mol } SO_2}{64.07\ \text{g } SO_2} \times \frac{4\ \text{mol } FeS_2}{8\ \text{mol } SO_2} \times \frac{120.0\ \text{g } FeS_2}{\text{mol } FeS_2} = 292\ \text{g } FeS_2$$

$$\frac{292\ \text{g}}{6500\ \text{g}} \times 100\% = \underline{4.49\%\ FeS_2}$$

8-147 (1) Find the limiting reactant.

g NH_3 → mol NH_3 ↘

mol NO → g NO

g O_2 → mol O_2 ↗

$$80.0\ \text{g } NH_3 \times \frac{1\ \text{mol } NH_3}{17.03\ \text{g } NH_3} \times \frac{4\ \text{mol NO}}{4\ \text{mol } NH_3} = 4.70\ \text{mol NO (limiting reactant)}$$

$$200\ \text{g } O_2 \times \frac{1\ \text{mol } O_2}{32.00\ \text{g } O_2} \times \frac{4\ \text{mol NO}}{5\ \text{mol } O_2} = 5.00\ \text{mol NO}$$

(2) Find the theoretical yield based on NH_3.

$$4.70\ \text{mol NO} \times \frac{30.01\ \text{g NO}}{\text{mol NO}} = 141\ \text{g NO (theoretical yield)} \qquad \frac{40.0\ \text{g}}{141\ \text{g}} \times 100\% = \underline{28.4\%\ \text{yield}}$$

8-149 $$0.250\ \text{mol } H_2O \times \frac{1\ \text{mol } CaCl_2 \cdot 6H_2O}{5\ \text{mol } H_2O} = 0.0500\ \text{mol } CaCl_2 \cdot 6H_2O\ \text{-limiting reactant}$$

$$9.50 \times 10^{22}\ \text{molecules HCl} \times \frac{1\ \text{mol HCl}}{6.022 \times 10^{23}\ \text{molecules HCl}} \times \frac{1\ \text{mol } CaCl_2 \cdot 6H_2O}{2\ \text{mol HCl}} = 0.0789\ \text{mol } CaCl_2 \cdot 6H_2O$$

$$15.0\ \text{g } CaCO_3 \times \frac{1\ \text{mol } CaCO_3}{100.1\ \text{g } CaCO_3} \times \frac{1\ \text{mol } CaCl_2 \cdot 6H_2O}{1\ \text{mol } CaCO_3} = 0.150\ \text{mol } CaCl_2 \cdot 6H_2O$$

$$0.050\ \text{mol } CaCl_2 \cdot 6H_2O \times \frac{219.1\ \text{g } CaCl_2 \cdot 6H_2O}{\text{mol } CaCl_2 \cdot 6H_2O} = \underline{11.0\ \text{g } CaCl_2 \cdot 6H_2O}$$

8-150 C: $14.1\ \text{g C} \times \dfrac{1\ \text{mol C}}{12.01\ \text{g C}} = 1.17\ \text{mol C}$

H: $2.35\ \text{g H} \times \dfrac{1\ \text{mol H}}{1.008\ \text{g H}} = 2.33\ \text{mol H}$

Cl: $83.5\ \text{g Cl} \times \dfrac{1\ \text{mol Cl}}{35.45\ \text{g Cl}} = 2.36\ \text{mol Cl}$

C: $\dfrac{1.17}{1.17} = 1.0$ H: $\dfrac{2.36}{1.17} = 2.0$ Cl: $\dfrac{2.33}{1.17} = 2.$ Molecular formula = CH_2Cl_2

$CH_4(g) + 2Cl_2(g) \longrightarrow CH_2Cl_2(l) + 2HCl(g)$

$$2.85\ \text{g } CH_4 \times \frac{1\ \text{mol } CH_4}{16.04\ \text{g } CH_4} \times \frac{1\ \text{mol } CH_2Cl_2}{1\ \text{mol } CH_4} = 0.178\ \text{mol } CH_2Cl_2$$

$$15.0\ \text{g } Cl_2 \times \frac{1\ \text{mol } Cl_2}{70.90\ \text{g } Cl_2} \times \frac{1\ \text{mol } CH_2Cl_2}{2\ \text{mol } Cl_2} = 0.106\ \text{mol } CH_2Cl_2$$ - limiting reactant

$$0.106\ \text{mol } CH_2Cl_2 \times \frac{84.93\ \text{g } CH_2Cl_2}{\text{mol } CH_2Cl_2} = \underline{9.00\ \text{g } CH_2Cl_2}$$

8-153 Assume 100 g of compound. There are then 27.4 g Na, 14.3 g C, 57.1 g O, and 1.19 g H per 100 g.

$27.4\ \text{g Na} \times \dfrac{1\ \text{mol Na}}{22.99\ \text{g Na}} = 1.19\ \text{mol Na}$ $14.3\ \text{g C} \times \dfrac{1\ \text{mol C}}{12.01\ \text{g C}} = 1.19\ \text{mol C}$

$57.1\ \text{g O} \times \dfrac{1\ \text{mol O}}{16.00\ \text{g O}} = 3.57\ \text{mol O}$ $1.19\ \text{g H} \times \dfrac{1\ \text{mol H}}{1.008\ \text{g H}} = 1.18\ \text{mol H}$

O: $\dfrac{3.57}{1.18} = 3.0$ Formula = $NaHCO_3$ = $Na^+\ HCO_3^-$

H — O — C(—O⁻)(=O)

The geometry around the C is trigonal planar with the approximate H-O-C angle of 120°.

8-155 In 100 g of compound there are 19.8 g Ca, 1.00 g H, 31.7 g S, and 47.5 g O.

$$19.8 \text{ g Ca} \times \frac{1 \text{mol Ca}}{40.08 \text{ g Ca}} = 0.494 \text{ mol Ca} \qquad 1.00 \text{ g H} \times \frac{1 \text{ mol H}}{1.008 \text{ g H}} = 0.992 \text{ mol H}$$

$$31.7 \text{ g S} \times \frac{1 \text{mol S}}{32.07 \text{ g S}} = 0.988 \text{ mol S} \qquad 47.5 \text{ g O} \times \frac{1 \text{mol O}}{16.00 \text{ g O}} = 2.97 \text{ mol O}$$

$$\text{H: } \frac{0.992}{0.494} = 2.0 \qquad \text{S: } \frac{0.988}{0.494} = 2.0 \qquad \text{O: } \frac{2.97}{0.494} = 6.0$$

Formula = $CaH_2S_2O_6$ = $Ca(HSO_3)_2$ calcium bisulfite or calcium hydrogen sulfite

H — Ö — S̈ — Ö:⁻
|
:O:

The geometry around the S is trigonal pyramid. The H-O-S angle is about 109°.

9

The Gaseous State

Review Section A *The Nature of the Gaseous State and the Effects of Changing Conditions*

OUTLINE	OBJECTIVES
9-1 The Nature of Gases and the Kinetic Molecular Theory	
1. Five common properties of gases	*Describe the qualitative properties common to all gases.*
2. The kinetic molecular theory of gases	*Apply the kinetic molecular theory of gases to explain their common properties.*
3. Graham's law	*Apply Graham's law to the relationship between the average velocity of a gas and its molar mass.*
9-2 The Pressure of a Gas	
1. Atmospheric pressure and the barometer	
2. The definition of pressure	*Distinguish between force and pressure.*
3. Units used to describe pressure	*Convert among the various units used to describe pressure.*
9-3 Boyle's Law	
1. Pressure and the volume of a gas	*Apply Boyle's law to calculate the effect of pressure changes on the volume.*
2. Boyle's law and kinetic theory	*Describe the basis for Boyle's law from kinetic theory.*
9-4 Charles's Law and Gay-Lussac's Law	
1. Temperature and the volume of a gas	
2. Gas volumes and the Kelvin scale	*Use Charles's law to calculate the effect of temperature changes on the volume.*
3. Temperature and the pressure of a gas	*Calculate the effect of a change of temperature on the pressure of a gas by application of Gay-Lussac's law.*
4. Charles's law, Gay-Lussac's law, and kinetic theory	*Apply kinetic theory to describe the basis of Charles's law and Gay-Lussac's law.*
5. The combined gas law	
6. Standard temperature and pressure	*Calculate a missing variable (P, V, or T) for a gas under two sets of conditions.*
9-5 Avogadro's Law	
1. The amount of gas and the volume	*Describe the effect on the volume of a gas of adding a specified amount of gas.*

SUMMARY OF SECTIONS 9-1 THROUGH 9-5

Questions: *What are the common properties of gases? Why do gases have these common properties? How does a gas exert pressure? How are gases affected by changes of conditions such as pressure, temperature, and the amount of gas present?*

The gaseous state is unique compared to the other two states in many respects. Unlike solids and liquids, gases are compressible, have low densities, fill containers uniformly, mix rapidly and thoroughly, and exert pressure evenly in all directions. These and other properties, including what are known as "the gas laws," become obvious from an understanding of the **kinetic molecular theory** applied to gases. The major assumptions of this theory are:

(1) Gas molecules have negligible volume.

(2) Gas molecules undergo random collisions with each other and with the walls of the container thus exerting pressure.

(3) The total energy of all of the collisions is conserved.

(4) Gas molecules have negligible interactions with each other.

(5) The average kinetic energy of the molecules is proportional to the temperature.

An immediate consequence of the last point of kinetic theory is that the average velocity of a gas is related to its formula weight (or molar mass) since K.E. = $1/2mv^2$. The extension of this principle to the **effusion** and **diffusion** of gases is known as **Graham's law.** An example of the application of Graham's law follows.

Example A-1 Graham's Law

Consider two gases, one with a formula weight twice the other. How fast, on the average, do the light molecules travel, effuse, or diffuse relative to the heavy molecules?

PROCEDURE

$$\frac{v(\text{heavy})}{v(\text{light})} = \sqrt{\frac{m_{light}}{m_{heavy}}}$$

Since $m_{heavy} = 2m_{light}$, substitute for m_{heavy} in the equation.

SOLUTION

$$\frac{v(\text{heavy})}{v(\text{light})} = \sqrt{\frac{m_{light}}{2m_{light}}}$$

$v(\text{light}) = \sqrt{2} \times v(\text{heavy}) = \underline{1.41\ v(\text{heavy})}$

It seems obvious now, but the understanding of the nature of gases began with the demonstration that the atmosphere exerts **pressure** that can be measured with a **barometer.** Pressure is the weight of the gas (the **force**) applied per unit area. The pressure of a gas is measured in many different units but all can be compared to the standard unit which is the average pressure of the atmosphere at sea level (**one atmosphere**). The most common unit of pressure in chemistry calculations besides atmosphere is torr, which is the same as mm of mercury. One atmosphere is the pressure that supports a column of Hg 760 mm high, which is thus 760 **torr**.

The gas laws discussed in this section of the review relate the pressure, temperature, and number of moles to the volume of a gas. The laws can be used to calculate how one parameter changes as the others are varied. Examples of these laws are as follows.

Example A-2 Boyle's Law (V and P)

If the volume of a gas is 32.5 L at a pressure of 4.25 atm, what is the volume if the pressure is increased to 5.70 atm?

PROCEDURE

The appropriate expression of Boyle's law under two conditions is

$$P_1V_1 = P_2V_2 \text{ or solving for } V_2, \quad V_2 = V_1 \times \frac{P_1}{P_2}$$

The problem can be solved by substitution or, better yet, by reason. (That way you don't have to remember the formula.)

$$V_{final} = V_{initial} \times P_{corr}$$

P_{corr} symbolizes the factor that converts the given volume to the final volume. In this problem, the pressure increases. According to Boyle's law, an ***increase*** in pressure results in a ***decrease*** in volume. Therefore, we reason that P_{corr} is a factor less than one in order to convert $V_{initial}$ to a smaller value.

In summary, P goes ***up***; therefore, V goes ***down*** , and the P_{corr} is less than one.

SOLUTION

$$V = 32.5 \text{ L} \times \frac{4.25 \text{ atm}}{5.70 \text{ atm}} = \underline{24.2 \text{ L}}$$

Example A-3 Charles's Law (V and T)

If the volume of a gas is 32.5 L at a temperature of 28°C, what is the volume at 112°C?

PROCEDURE

$$V_{final} = V_{initial} \times T_{corr}$$

T goes ***up***; therefore V goes ***up*** and the T_{corr} is greater than one. In these calculations, the *Kelvin* temperature scale, which begins at *absolute zero*.

SOLUTION

$$V = 32.5 \text{ L} \times \frac{(112 + 273)\text{K}}{(28 + 273)\text{K}} = \underline{41.6 \text{ L}}$$

Example A-4 Gay-Lussac's Law (P and T)

If the pressure on a gas is 485 torr and the temperature is 162°C, what is the pressure if the temperature is increased to 215°C?

PROCEDURE

$$P_{final} = P_{initial} \times T_{corr}$$

T goes ***up***; therefore P goes ***up*** and the T_{corr} is greater than one.

SOLUTION

$$P = 485 \text{ torr} \times \frac{(215 + 273)\cancel{K}}{(162 + 273)\cancel{K}} = \underline{544 \text{ torr}}$$

One additional law follows from the previous three laws. This law relates a quantity of gas under two sets of conditions and is known as the **combined gas law.**

Example A-5 Combined Gas Law (P, V, and T)

If the volume of a gas is 285 mL at a temperature of 16°C and a pressure of 685 torr, what is the volume at a temperature of 116°C and a pressure of 842 torr?

PROCEDURE

$$V_{final} = V_{initial} \times P_{corr} \times T_{corr}$$

P goes ***up***; therefore V goes ***down*** and P_{corr} is less than one.
T goes ***up*** ; therefore V goes ***up*** and T_{corr} is greater than one.

SOLUTION

$$V = 285 \text{ mL} \times \frac{685 \cancel{\text{torr}}}{842 \cancel{\text{torr}}} \times \frac{(116 + 273)\cancel{K}}{(16 + 273)\cancel{K}} = \underline{312 \text{ mL}}$$

Example A-6 Avogadro's Law (V and n)

If 1.51 moles of N_2 has a volume of 67.7 L, what is the volume of 1.96 moles of N_2 under the same conditions?

PROCEDURE

$$V_{final} = V_{initial} \times n_{corr}$$

n goes ***up***; therefore V goes ***up*** and n_{corr} is greater than one.

SOLUTION

$$V = 67.7 \text{ L} \times \frac{1.96 \cancel{\text{mol}}}{1.51 \cancel{\text{mol}}} = \underline{87.9 \text{ L}}$$

NEW TERMS

Absolute zero
Atmospheric pressure
Avogadro's law
Barometer
Boyle's law
Charles's law
Combined gas law
Diffusion
Effusion
Force
Gay-Lussac's law
Graham's law
Kelvin temperature
Kinetic molecular theory
Pressure
STP
Torr

SELF-TEST

A-1 Multiple Choice

___ 1. Which of the following is <u>not</u> characteristic of gases?

(a) They fill a container uniformly.
(b) Different gases mix rapidly.
(c) Gases have a low density.
(d) Gases are virtually incompressible.

___ 2. Which of the following is false about the kinetic theory of gases?

(a) Gas molecules are in random motion.
(b) The volume of a gas is composed mostly of matter.
(c) Gas molecules have no attraction for each other.
(d) Collisions between molecules are elastic.
(e) Gas molecules collide with each other and with the sides of the container.

___ 3. Which of the following is false?

(a) At the same temperature light molecules move faster on the average than heavy molecules.
(b) At the same temperature all molecules have the same average kinetic energy.
(c) Gases diffuse faster at higher temperatures.
(d) Gases are compressible.
(e) All gas molecules have the same average speed at the same temperature.

___ 4. Which of the following is not a unit of pressure?

(a) in. of Hg
(b) pascals
(c) torr
(d) kelvins
(e) bars

___ 5. One atmosphere at sea level supports a column of mercury

(a) 76.0 cm high
(b) 14.7 in. high
(c) 760 cm high
(d) 76.0 mm high
(e) 29.9 cm high

___ 6. When the pressure on a gas is doubled at constant temperature, the volume

(a) is doubled (b) stays the same
(c) is halved (d) is quartered

___ 7. The temperature of a volume of gas is increased from 0 ^{o}C to 20°C. The volume

(a) increases by a factor of 20
(b) decreases by a factor of 20
(c) decreases by a factor of 20/273
(d) increases by a factor of 293/273
(e) decreases by a factor of 273/293

___ 8. What is -50°C on the Kelvin scale?

(a) -323 (b) 323 (c) -223 (d) 223 (e) 273

___ 9. Which of the following is a representation of Gay-Lussac's law?

(a) PV = k (b) P α T (c) t(C) α P (d) V/T = k

___ 10. Standard temperature and pressure (STP) is:

(a) one atm pressure and 25°C
(b) 760 torr and 273 K
(c) 760 atm and 0°C
(d) one atm and 273°C
(e) one atm and 760°C

___ 11. When the temperature of a volume of gas increases,

(a) the pressure decreases
(b) the molecules collide more frequently with the sides of the container
(c) the molecules move more slowly
(d) the molecules become heavier
(e) the volume decreases

___ 12. If one mole of a gas occupies 25 L at a certain temperature, what would be the volume of 6.02 x 10^{21} molecules of the gas under the same conditions?

(a) 2.5 L (b) 250 L (c) 0.25 L (d) 75 L (e) 6.25 L

A-2 Problem

1. At the same temperature, how does the average velocity of SF_6 molecules compare to that of Ne atoms?

2. A 575-mL quantity of gas is contained at a pressure of 855 torr. What is the volume if the pressure is decreased to 725 torr?

3. A quantity of gas has a volume of 12.5 L at a temperature of -16°C. What is the temperature if the gas expands to 14.6 L at a constant pressure?

4. A certain quantity of gas in a pressurized can (constant volume) has a pressure of 1.22 atm at a temperature of 20°C. What temperature is required to raise the pressure to 3.75 atm (which causes the can to explode)?

5. A quantity of gas has a volume of 4350 mL at a temperature of 298 K and a pressure of 0.862 atm. What is the volume if the temperature is changed to 0°C and the pressure to 887 torr?

6. A 525-mL quantity of gas under pressure at 25°C expands to 3.62 L when the pressure is released to one atmosphere. If the temperature of the expanded gas is 19 ^{o}C, what was the original pressure in the container in atmosphere?

7. The volume of 50.0 g of O_2 is 37.1 L at a certain temperature and pressure. How many molecules must be *added* to this amount of oxygen to cause the volume to increase to 43.8 L at the same temperature and pressure?

Review Section B *Relationships Among Quantities of Gases, Conditions, and Chemical Reactions*

OUTLINE / OBJECTIVES

9-6 The Ideal Gas Law

1. A general gas law
2. The gas constant

Use the ideal gas law to calculate an unknown property of a sample of gas (P, V, T, or n) when the other quantities are known.

9-7 Dalton's Law of Partial Pressures

1. The pressure exerted by mixtures

Carry out gas law calculations for a mixture of gases.

9-8 The Molar Volume and Density of a Gas

1. The volume of one mole of gas at STP

Convert between moles of gas and volume at STP using the molar volume as a conversion factor.

2. The densities of gases at STP
3. The densities of gases at other conditions

Calculate the density of a gas at STP or other conditions.

9-9 Stoichiometry Involving Gases

1. Calculations with volumes of gases

Apply the ideal gas law or the molar volume relationship to stoichiometry problems involving gases.

SUMMARY OF SECTIONS 9-6 THROUGH 9-9

Questions: *Can all of the gas laws that relate to the volume of a gas be summarized in one relationship? How does the mass of a particular gas relate to its volume? How are the volumes of gases included in stoichiometric calculations?*

The four gas laws discussed above can be joined into one general law. This law, known as the *ideal gas law*, relates Boyle's, Charles's and Avogadro's laws. With this law, one property (temperature, pressure, volume, or moles) can be calculated if the other three are known. An example follows.

Example B-1 Ideal Gas Law

If 0.743 mole of a gas occupies 18.6 L at a pressure of 0.831 atm, what is the temperature (in $^{\circ}C$)?

PROCEDURE

$$PV = nRT \qquad T = \frac{PV}{nR}$$

SOLUTION

$$T = \frac{0.831\ \cancel{atm} \times 18.6\ \cancel{L}}{0.743\ \cancel{mol} \times 0.0821 \dfrac{\cancel{L} \cdot \cancel{atm}}{K \cdot \cancel{mol}}} = 253\ K \qquad (253 - 273) = -20^{\circ}C$$

According to kinetic molecular theory, gas molecules do not interact. Thus properties such as pressure and volume depend only on the amount of gas present and not on its identity. **Dalton's law** is a statement of this observation. It states that the pressure of a mixture of gases is the sum of the partial pressures of the component gases. A sample calculation involving this law follows.

Example B-2 Dalton's Law

If a mixture of gases exerts a pressure of 1250 torr and is composed of 2.35 moles of O_2, 1.76 moles of N_2, and 0.85 moles of CO_2, what is the partial pressure due to N_2?

PROCEDURE

Calculate the decimal fraction of N_2 then multiply that by the total pressure.

SOLUTION

$$\frac{1.76}{1.76 + 2.35 + 0.85} = 0.355$$

$$P(N_2) = 0.355 \times 1250\ \text{torr} = \underline{444\ \text{torr}}$$

Since all gases act the same (physically), one mole of either a pure gas or a mixture has the same volume. The **molar volume** of a gas is 22.4 L, which is the volume occupied by 1.00 mole of a gas at standard temperature (0 $^{\circ}C$) and standard pressure (1 atm). The density of a gas is usually expressed in grams per liter at STP. It is obtained by dividing the molar mass by the molar volume.

$$\frac{g/\cancel{mol}}{L/\cancel{mol}} = g/L$$

The molar volume relationship or the ideal gas law allows us a straightforward way to incorporate gases into the general scheme of stoichiometry introduced in Chapter 8 because this law converts volume at a specified temperature and pressure to moles.

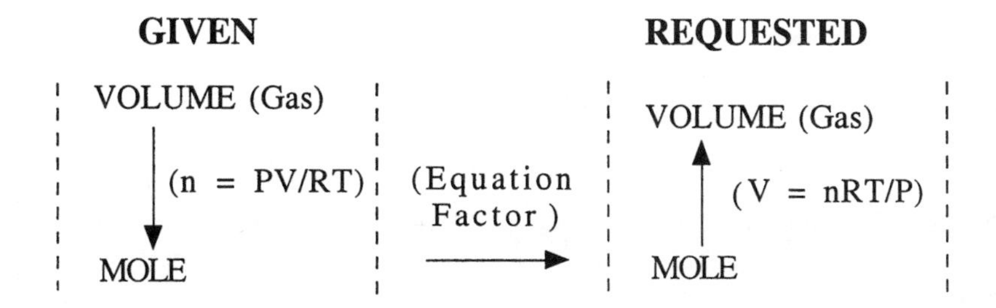

A sample problem follows.

Example B-3 Stoichiometry and Gases

What volume of O_2 is produced from 188 g of KO_2 if the O_2 is measured at 25°C and a pressure of 745 torr?

$$4KO_2(s) + 2CO_2(g) \rightarrow 2K_2CO_3(s) + 3O_2(g)$$

PROCEDURE

1. Convert mass of KO_2 (Given) to moles of KO_2.
2. Convert moles of KO_2 to moles of O_2.
3. Convert moles of O_2 to volume of O_2 (Requested) in a separate calculation.

SOLUTION

$$188\ \cancel{g\ KO_2} \times \overset{1}{\frac{1\ \cancel{mol\ KO_2}}{71.10\ \cancel{g\ KO_2}}} \times \overset{2}{\frac{3\ mol\ O_2}{4\ \cancel{mol\ KO_2}}} = 1.98\ mol\ O_2$$

$$3 \qquad V = \frac{nRT}{P} = \frac{1.98\ \cancel{mol} \times 0.0821\ \frac{L \cdot \cancel{atm}}{\cancel{K} \cdot \cancel{mol}} \times 298\ \cancel{K}}{\frac{745\ \cancel{torr}}{760\ \cancel{torr/atm}}} = \underline{49.4\ L}$$

NEW TERMS

Dalton's law
Gas constant
Ideal gas law
Molar volume

SELF-TEST

B-1 Multiple Choice

___ 1. Which of the following is a correct value of the gas constant?

(a) $62.4 \frac{L \cdot atm}{mol \cdot {}^oC}$ (b) $82.1 \frac{mL \cdot atm}{K \cdot mol}$

(c) $0.0821 \frac{L \cdot atm}{{}^oC \cdot mol}$ (d) $62.4 \frac{mL \cdot torr}{K \cdot mol}$

___ 2. Under what conditions can there be serious deviation from the ideal gas law?

(a) high pressure and temperature
(b) low pressure and temperature
(c) high pressure and low temperature
(d) low pressure and high temperature

___ 3. What is the total pressure of a mixture of gases if 60.0% of the gas is N_2 and the partial pressure of N_2 is 300 torr?

(a) 240 torr (b) 500 torr
(c) 1000 torr (d) 180 torr
(e) 360 torr

___ 4. A quantity of oxygen is collected over water. The pressure of the gas above the water is 755 torr. The pressure that the oxygen alone would exert is 728 torr. The vapor pressure of water at this temperature is

(a) 728 torr (b) 755 torr
(c) 782 torr (d) 27 torr

___ 5. A 224-L quantity of N_2 measured at STP

(a) contains one mole of N_2
(b) has a mass of 28.0 g
(c) contains 6.02×10^{24} N_2 molecules
(d) is 0.10 mole
(e) contains 1.20×10^{24} N atoms

___ 6. The density of a gas is 4.1 g/L at STP. The gas is

(a) NO_2 (b) CO_2 (c) N_2O_4 (d) SO_3 (e) CO

B-2 Problems

1. What is the volume of a gas if 0.334 moles of the gas has a temperature of 11°C and a pressure of 0.900 atm?

2. What mass of SO_2 occupies 47.6 L at a pressure of 745 torr and a temperature of -6°C?

3. How many molecules are in a volume of 1.00 mL at a temperature of 117°C and a pressure of 1.65×10^{-4} torr?

4. A mixture of gases is composed of 46.2% N_2, 31.4% SO_2, and the rest CO_2. If the total pressure is 1.08 atm, what is the partial pressure of each gas?

5. A mixture of gases is composed of 132 g of O_2 and 2.72×10^{24} molecules of N_2. If the total pressure is 685 torr, what is the partial pressure of each gas?

6. What mass of CO_2 occupies 54.2 mL at STP?

7. Given the balanced equation:

$$2HNO_3(aq) + 3H_2S(aq) \rightarrow 2NO(g) + 3S(s) + 4H_2O(l)$$

What mass of H_2S is required to produce 8.62 L of NO measured at STP?

8. What mass of H_2O is produced if 16.3 L of NO measured at 22^oC and 0.954 atm is also formed? (Use the balanced equation in problem 7 above.)

Review Section C

CHAPTER SUMMARY SELF-TEST

C-1 Matching

___ Combined gas law

___ Ideal gas law

___ Charles's law

___ Dalton's law

___ Boyle's law

___ Gay-Lussac's law

___ Avogadro's law

___ Graham's law

(a) $P_{tot} = P_1 + P_2 +$

(b) $V_1T_1 = V_2T_2$

(c) $\frac{P_1}{T_1} = \frac{P_2}{T_2}$

(d) $P = \frac{k}{V}$

(e) $PT = nVT$

(f) $\frac{m_1}{m_2} = \sqrt{\frac{v_1}{v_2}}$

(g) $V \alpha n$

(h) $\frac{T_1}{V_1} = \frac{T_2}{V_2}$

(i) $P = \frac{nRT}{V}$

(j) $V \alpha P$

(k) $v_1 = v_2\sqrt{\frac{m_2}{m_1}}$

(l) $P = X_1P_1$

(m) $V \alpha \frac{T}{P}$

Answers to Self-Tests

A-1 Multiple Choice

1. **d** Gases are virtually incompressible.

2. **b** The volume of a gas is composed mostly of matter. (It is, in fact, almost entirely empty space.)

3. **e** All gas molecules have the same average speed at the same temperature. (They have the same kinetic energy.)

4. **d** kelvins

5. **a** 76.0 cm

6. **c** The volume is halved.

7. **d** increases by a factor of 293/273

8. **d** 223 (-50 + 273 = 223)

9. **b** $P \alpha T$

10. **b** 760 torr and 273 K

11. **b** The molecules collide more frequently with the sides of the container. (This is because they are moving faster.)

12. **c** 0.25 L (6.02×10^{21} = 0.01 mol) $25 \text{ L} \times \frac{0.01 \text{ mol}}{1.00 \text{ mol}} = 0.25 \text{ L}$

A-2 Problem

1. $$\frac{v(SF_6)}{v(Ne)} = \sqrt{\frac{m_{Ne}}{m_{SF_6}}} = \sqrt{\frac{20.2 \text{ amu}}{146 \text{ amu}}} = \sqrt{0.138} = 0.37$$

   $v(SF_6) = 0.37\ v(Ne)$ (An SF_6 molecule has about one-third of the average velocity of a Ne atom.)

2. $V_{initial} = 575$ mL, $P_{in} = 855$ torr $V_{final} = ?$, $P_F = 725$ torr

   Since P goes ***down*** V goes ***up*** and P_{corr} is greater than one.

   $$V_f = V_{in} \times P_{corr} = 575 \text{ mL} \times \frac{855 \text{ torr}}{725 \text{ torr}} = \underline{678 \text{ mL}}$$

3. $V_{Initail} = 12.5$ L, $T_{In} = (-16 + 273) = 257$ K $V_{Final} = 14.6$ L, $T_F = ?$

Since V goes ***up*** T goes ***up*** and V_{corr} is greater than one.

$$T_{F\backslash\backslash f} = T_{in} \times V_{corr} = 257 \text{ K} \times \frac{14.6 \text{ L}}{12.5 \text{ L}} = 300 \text{ K} \qquad t(C) = \underline{27^oC}$$

4. $P_{in} = 1.22$ atm, $T_{in} = (20 + 273) = 293$ K, $P_v = 3.75$ atm, $T_v = ?$

Since P goes ***up*** T goes ***up*** and P_{corr} is greater than one.

$$T = 293 \text{ K} \times \frac{3.75 \text{ atm}}{1.22 \text{ atm}} = 901 \text{ K} \qquad t(C) = (901 - 273) = \underline{628^oC}$$

5. $V_{in} = 4350$ mL, $T_{in} = 298$ K, $P_{in} = 0.862$ atm, $V_f = ?$

$$T_F = 0^oC = 273 \text{ K}, P_F = 887 \text{ torr} \times \frac{1 \text{ atm}}{760 \text{ torr}} = 1.17 \text{ atm}$$

Since T goes ***down*** V goes ***down*** and T_{Corr} is less than one.

Since P goes ***up*** V goes ***down*** and P_{corr} is less than one.

$$V = 4350 \text{ mL} \times \frac{273 \text{ K}}{298 \text{ K}} \times \frac{0.862 \text{ atm}}{1.17 \text{ atm}} = \underline{2940 \text{ mL}}$$

6. Don't let the wording confuse you. The initial situation should be considered after expansion and the compressed gas regarded as the final condition.

$V_{in} = 3.62 \text{ L} = 3.62 \times 10^3$ mL, $T_{in} = (19 + 273) = 292$ K,

$P_{in} = 1.00$ atm, $V_f = 525$ mL, $T_f = (25 + 273) = 298$ K, $P_f = ?$

Since T goes ***up*** P goes ***up*** and T_{corr} is greater than one.

Since V goes ***down*** P goes ***up*** and V_{corr} is greater than one.

$$P_F = 1.00 \text{ atm} \times \frac{298 \text{ K}}{292 \text{ K}} \times \frac{3.62 \times 10^3 \text{ mL}}{525 \text{ mL}} = \underline{7.04 \text{ atm}}$$

7. $V_{in} = 37.1 \text{ L}, n_{in} = 50.0 \text{ g } O_2 \times \dfrac{1 \text{ mol}}{32.00 \text{ g } O_2} = 1.56 \text{ mol}$

$V_f = 43.8$ L, $n_f = 1.56 + X$ (X = number of moles of gas added.)

Since V goes ***up*** n goes ***up*** and n_{corr} is greater than one.

$$V_F = V_{In} \times n_{corr} \qquad 43.8 \text{ L} = 37.1 \text{ L} \times \frac{(1.56 + X) \text{ mol}}{1.56 \text{ mol}}$$

$438 \times 1.56 = 37.1\,(1.56 + X) \quad X = 0.28 \text{ mol added}$

$$0.28 \cancel{\text{mol}} \times \frac{6.022 \times 10^{23} \text{ molecules}}{\cancel{\text{mol}}} = \underline{1.7 \times 10^{23} \text{ molecules added}}$$

B-1 Multiple Choice

1. **b** $82.1 \dfrac{\text{mL} \cdot \text{atm}}{\text{K} \cdot \text{mol}}$

2. **c** high pressure and low temperature

3. **b** 500 torr ($0.600 \times P_T = 300$ torr; $P_T = 500$ torr)

4. **d** 27 torr

5. **c** contains 6.02×10^{24} N_2 molecules (224 L equals 10 moles at STP)

6. **c** N_2O_4 ($4.1 \frac{\text{g}}{\cancel{\text{L}}} \times 22.4 \frac{\cancel{\text{L}}}{\text{mol}} = 92$ g/mol)

B-2 Problems

1. n = 0.334 mol, T = (11 + 273) = 284 K, P = 0.900 atm, V = ?

$$PV = nRT \qquad V = \frac{nRT}{P}$$

$$V = \frac{0.334 \cancel{\text{mol}} \times 0.0821 \dfrac{\text{L} \cdot \cancel{\text{atm}}}{\cancel{\text{K}} \cdot \cancel{\text{mol}}} \times 284 \cancel{\text{K}}}{0.900 \cancel{\text{atm}}} = \underline{8.65 \text{ L}}$$

2. n = ?, T = (-6 + 273) = 267 K, P = 745 torr, V = 47.6 L

$$PV = nRT \qquad n = \frac{PV}{RT} = \frac{\dfrac{745 \cancel{\text{torr}}}{760 \cancel{\text{torr/atm}}} \times 47.6 \cancel{\text{L}}}{0.0821 \dfrac{\cancel{\text{L}} \cdot \cancel{\text{atm}}}{\cancel{\text{K}} \cdot \text{mol}} \times 267 \cancel{\text{K}}} = 2.13 \text{ mol}$$

$$2.13 \cancel{\text{mol SO}_2} \times \frac{64.07 \text{ g } SO_2}{\cancel{\text{mol SO}_2}} = \underline{136 \text{ g } SO_2}$$

3. $n = ?, T = (117 + 273) = 390\ K, P = 1.65 \times 10^{-4}$ torr,
$V = 1.00\ mL = 1.00 \times 10^{-3}\ L$

$$n = \frac{PV}{RT} = \frac{\frac{1.65 \times 10^{-4}\ \cancel{torr}}{760\ \cancel{tor/atm}} \times (1.00 \times 10^{-3}\ \cancel{L})}{390\ \cancel{K} \times 0.0821 \frac{\cancel{L} \cdot \cancel{atm}}{\cancel{K} \cdot mol}} = 6.78 \times 10^{-12}\ mol$$

$$6.78 \times 10^{-12}\ \cancel{mol} \times \frac{6.022 \times 10^{23}\ molecules}{\cancel{mol}} = \underline{4.08 \times 10^{12}\ molecules}$$

4. $P_{Component} = [\frac{\%}{100\%}] \times P_{tot}$

$P(N_2) = 0.462 \times 1.08\ atm = 0.50\ atm$

$P(SO_2) = 0.314 \times 1.08\ atm = 0.34\ atm$

$P(CO_2) = 1.08\ atm - (0.50\ atm + 0.34\ atm) = 0.24\ atm$

5. $P(O_2) = [\frac{\%\ O_2}{100\%}] \times P_{tot}$ $\quad \%\ O_2 = [\frac{moles\ (O_2)}{total\ moles}] \times 100\%$

Calculate the total moles of gas present.

$$n(O_2) = 132\ \cancel{g\ O_2} \times \frac{1\ mol}{32.00\ \cancel{g\ O_2}} = 4.12\ mol$$

$$n(N_2) = 2.72 \times 10^{24}\ \cancel{molecules} \times \frac{1\ mol}{6.022 \times 10^{23}\ \cancel{molecules}} = 4.52\ mol$$

$$P(O_2) = [\frac{4.12}{4.12 + 4.52}] \times 685\ torr = \underline{327\ torr}$$

$$P(N_2) = P_T - P(O_2) = 685\ torr - 327\ torr = \underline{358\ torr}$$

6. $$54.2\ \cancel{mL} \times \frac{10^{-3}\ \cancel{L}}{\cancel{mL}} \times \frac{1\ mol}{22.4\ \cancel{L}\ (STP)} = 2.42 \times 10^{-3}\ mol$$

$$2.42 \times 10^{-3}\ \cancel{mol\ CO_2} \times \frac{44.01\ g}{\cancel{mol\ CO_2}} = \underline{0.107\ g}$$

7. (1) Convert volume of NO to moles of NO using the molar volume relationship since conditions are at STP.
(2) Convert moles of NO to moles of H_2S.
(3) Convert moles of H_2S to mass of H_2S.

$$\begin{array}{ccccc} & (1) & (2) & (3) & \\ 8.62\ \cancel{L\ (STP)} \times & \dfrac{1\ \cancel{mol\ NO}}{22.4\ \cancel{L\ (STP)}} \times & \dfrac{3\ \cancel{mol\ H_2S}}{2\ \cancel{mol\ NO}} \times & \dfrac{34.09\ g\ H_2S}{\cancel{mol\ H_2S}} & = \underline{19.7\ g\ H_2S} \end{array}$$

8. (1) Convert volume of NO to moles of NO using the ideal gas law.
(2) Convert moles of NO to moles of H_2O.
(3) Convert moles of H_2O to mass of H_2O.

$$(1)\quad n = \frac{PV}{RT} = \frac{0.954\ \cancel{atm} \times 16.3\ \cancel{L}}{0.0821\ \dfrac{\cancel{L}\cdot \cancel{atm}}{\cancel{K}\cdot mol} \times 295\ \cancel{K}} = 0.642\ mol\ NO$$

$$\begin{array}{cccc} & (2) & (3) & \\ 0.642\ \cancel{mol\ NO} \times & \dfrac{4\ \cancel{mol\ H_2O}}{2\ \cancel{mol\ NO}} \times & \dfrac{18.02\ g\ H_2O}{\cancel{mol\ H_2O}} & = \underline{23.1\ g\ H_2O} \end{array}$$

C-1 Matching

m Combined gas law is $V \alpha \frac{T}{P}$ or $\frac{PV}{T} = k$

i Ideal gas law $P = \frac{nRT}{V}$ or $PV = nRT$

h Charles's law $\frac{T_1}{V_1} = \frac{T_2}{V_2}$ or $\frac{V_1}{T_1} = \frac{V_2}{T_2}$

a Dalton's law $P_T = P_1 + P_2 + \cdots$

d Boyle's Law $P = \frac{k}{V}$ or $PV = k$

c Gay-Lussac's law $\frac{P_1}{T_1} = \frac{P_2}{T_2}$

g Avogadro's law $V \alpha n$

k Graham's law $v_1 = v_2\sqrt{\frac{m_2}{m_1}}$

Solutions to Black Text Problems

9-6 $\frac{v_1}{v_2} = \sqrt{\frac{m_2}{m_1}}$ $v_2 = 20.0$ mi/hr, $m_2 = 6.00$ kg $= 6000$ g

$v_1 = ?$ $m_1 = 1.50$ g

$v_1 = 20.0 \text{ miles/hr} \times \sqrt{\frac{6000 \text{ g}}{150 \text{ g}}} = \underline{1260 \text{ mi/hr}}$

9-7 The molecule with the largest formula weight travels the slowest. SF_6(146.1 amu) < SO_2(64.07 amu) < N_2O(44.02 amu)< CO_2(44.01 amu) < N_2 (28.02 amu) < H_2(2.016 amu)

9-8 $\frac{r_{N_2}}{r_{Ar}} = \sqrt{\frac{39.9 \text{ amu}}{28.0 \text{ amu}}}$ $\underline{r_{N_2} = 1.19\, r_{Ar}}$

9-11 $\frac{r_{CO}}{r_X} = \sqrt{\frac{MM_X}{MM_{CO}}}$ $\underline{r_{CO} = 2.13\, r_X}$

$\frac{2.13\, r_X}{r_X} = \sqrt{\frac{MM_X}{28.0}}$ $MM_X = \underline{127 \text{ g/mol}}$

9-12 $\frac{r_{(235)}}{r_{(238)}} = \sqrt{\frac{352 \text{ g/mol}}{349 \text{ g/mol}}}$

$\underline{r_{(235)} = 1.004\, r_{(238)}}$ (about 0.4% faster)

9-14 (a) $1650 \text{ torr} \times \frac{1 \text{ atm}}{760 \text{ torr}} = \underline{2.17 \text{ atm}}$

(b) $3.50 \times 10^{-5} \text{ atm} \times \frac{760 \text{ torr}}{\text{atm}} = \underline{0.0266 \text{ torr}}$

(c) $18.5 \text{ lb/in.}^2 \times \frac{1 \text{ atm}}{14.7 \text{ lb/in.}^2} \times \frac{766 \text{ torr}}{\text{atm}} = \underline{9560 \text{ torr}}$

(d) $5.65 \text{ kPa} \times \frac{1 \text{ atm}}{101.3 \text{ kPa}} = \underline{0.0558 \text{ atm}}$

(e) $190 \text{ torr} \times \frac{1 \text{ atm}}{760 \text{ torr}} \times \frac{14.7 \text{ lb/in.}^2}{\text{atm}} = \underline{3.68 \text{ lb/in.}^2}$

(f) $85 \text{ torr} \times \frac{1 \text{ atm}}{760 \text{ torr}} \times \frac{101.3 \text{ kPa}}{\text{atm}} = \underline{11 \text{ kPa}}$

9-15 (a) $30.2 \text{ in. Hg} \times \frac{1 \text{ atm}}{29.9 \text{ in. Hg}} \times \frac{760 \text{ torr}}{\text{atm}} = \underline{768 \text{ torr}}$

(b) $25.7 \text{ kbar} \times \frac{10^3 \text{ bar}}{\text{kbar}} \times \frac{1 \text{ atm}}{1.013 \text{ bar}} = \underline{2.54 \times 10^4 \text{ atm}}$

(c) $57.9 \text{ kPa} \times \frac{1 \text{ atm}}{101.3 \text{ kPa}} \times \frac{14.7 \text{ lb/in.}^2}{\text{atm}} = \underline{8.40 \text{ lb/in.}^2}$

(d) $0.025 \text{ atm} \times \frac{760 \text{ torr}}{\text{atm}} = \underline{19 \text{ torr}}$

9-17 $10.3 \text{ mbar} \times \frac{10^{-3} \text{ bar}}{\text{mbar}} \times \frac{1 \text{ atm}}{1.013 \text{ bar}} = \underline{0.0102 \text{ atm}}$

9-19 Assume a column of Hg has a 1 cm^2 cross section and is 76.0 cm high.
Weight of Hg = $76.0 \text{ cm} \times 1 \text{ cm}^2 \times 13.6 \text{ g/cm}^3 = 1030 \text{ g}$.
If water is substituted, 1030 g of water in the column is required.
$\text{height} \times 1 \text{ cm}^2 \times 1.00 \text{ g/cm}^3 = 1030 \text{ g}$ height = 1030 cm

$1030 \text{ cm} \times \frac{1 \text{ in.}}{2.54 \text{ cm}} \times \frac{1 \text{ ft}}{12 \text{ in.}} = \underline{33.8 \text{ ft}}$

If a well is 40 ft deep, the water cannot be raised in one stage by suction since 33.8 ft is the theoretical maximum height that is supported by the atmosphere.

9-20 $V_2 = 6.85 \text{ L} \times \frac{0.650 \text{ atm}}{0.435 \text{ atm}} = \underline{10.2 \text{ L}}$

9-22 $V_2 = 785 \text{ mL} \times \frac{760 \text{ torr}}{610 \text{ torr}} = \underline{978 \text{ mL}}$

9-23 $P_2 = 62.5 \text{ torr} \times \frac{125 \text{ mL}}{115 \text{ mL}} = \underline{67.9 \text{ torr}}$

9-26 $V_{final} (V_f) = 15 V_{initial} (V_i)$ $\frac{V_f}{V_i} = \frac{P_i}{P_f}$; $\frac{15 V_i}{V_i} = \frac{0.950 \text{ atm}}{P_f}$

$P_f = 0.950 \text{ atm} \times 15 = \underline{14.3 \text{ atm}}$

9-28 $V_2 = 1.55 \text{ L} \times \frac{373 \text{ K}}{298 \text{ K}} = \underline{1.94 \text{ L}}$

9-30 $T_2 = 290 \text{ K} \times \frac{392 \text{ mL}}{325 \text{ mL}} = 350 \text{ K}$ $350 \text{ K} - 273 = \underline{77^\circ\text{C}}$

9-32 $V_2 = 3.66 \times 10^4 \text{ L} \times \frac{323 \text{ K}}{455 \text{ K}} = \underline{2.60 \times 10^4 \text{ L}}$

9-33 $V_2 = 1.25 V_1$ $T_2 = 273 \text{ K} \times \frac{1.25 V_I}{V_I} = 341 \text{ K}$

$341 \text{ K} - 273 = \underline{68^\circ\text{C}}$

9-35 $P_2 = 2.50 \text{ atm} \times \frac{295 \text{ K}}{251 \text{ K}} = \underline{2.94 \text{ atm}}$

9-36 $P_1 = 850\ \cancel{torr} \times \frac{1\ atm}{760\ \cancel{torr}} = 1.12\ atm$

$T_2 = 328\ K \times \frac{0.652\ \cancel{atm}}{1.12\ \cancel{atm}} = 191\ K \quad 191\ K - 273 = \underline{-82^oC}$

9-38 $T_2 = 298\ K \times \frac{2.50\ \cancel{atm}}{1.25\ \cancel{atm}} = 596\ K \quad 596\ K - 273 = \underline{323^oC}$

9-39 $P_2 = 28.0\ lb/in.^2 \times \frac{313\ \cancel{K}}{290\ \cancel{K}} = \underline{30.2\ lb/in.^2}$

9-42 (a) $PV = kT$ and (d) $\frac{P}{T} \ \alpha\ \frac{1}{V}$

9-46 $P_2 = 0.950\ atm \times \frac{5.50\ \cancel{L}}{4.75\ \cancel{L}} \times \frac{308\ \cancel{K}}{273\ \cancel{K}} = \underline{1.24\ atm}$

9-47 $V_2 = 17.5\ L \times \frac{6.00\ \cancel{atm}}{1.00\ \cancel{atm}} \times \frac{273\ \cancel{K}}{373\ \cancel{K}} = \underline{76.8\ L}$

9-49 $T_2 = 223\ K \times \frac{155\ \cancel{torr}}{78.0\ \cancel{torr}} \times \frac{9.55 \times 10^{-5}\ \cancel{mL}}{4.78 \times 10^{-4}\ \cancel{mL}} = 88.5\ K = \underline{-185^oC}$

9-52 $T_2 = 298\ K \times \frac{1.38\ \cancel{L}}{1.55\ \cancel{L}} \times \frac{1.02\ \cancel{atm}}{1.05\ \cancel{atm}} = \underline{258\ K\ (-15^oC)}$

9-53 $V_2 = 35.0\ mL \times \frac{295\ \cancel{K}}{290\ \cancel{K}} \times \frac{11.5\ \cancel{atm}}{1.00\ \cancel{atm}} = \underline{409\ mL}$

9-54 $V_2 = 2.54\ L \times \frac{0.0750\ \cancel{mol}}{0.112\ \cancel{mol}} = \underline{1.70\ L}$

9-55 Let X = total moles needed in the expanded balloon.

$188\ L \times \frac{X}{8.40\ mol} = 275\ L \quad X = 12.3\ mol$

12.3 - 8.4 = <u>3.9 moles must be added</u>.

9-57 $n_2 = 2.50 \times 10^{-3}\ mol \times \frac{164\ \cancel{mL}}{75.0\ \cancel{mL}} = 5.47 \times 10^{-3}\ mol$

$(5.47 \times 10^{-3}) - (2.50 \times 10^{-3}) = 2.97 \times 10^{-3}$ mol added

$2.97 \times 10^{-3}\ \cancel{mol\ N_2} \times \frac{28.02\ g\ N_2}{\cancel{mol\ N_2}} = \underline{0.0832\ g\ N_2}$

9-59 $T = \frac{PV}{nR} = \frac{2.25\ \cancel{atm} \times 4.50\ \cancel{L}}{0.332\ \cancel{mol} \times 0.0821 \frac{\cancel{L} \cdot \cancel{atm}}{K \cdot \cancel{mol}}} = 371\ K \quad 371\ K - 273 = \underline{98^oC}$

9-61 $n = \frac{PV}{RT} = \frac{0.955 \text{ atm} \times 16.4 \text{ L}}{250 \text{ K} \times 0.0821 \frac{\text{L} \cdot \text{atm}}{\text{K} \cdot \text{mol}}} = 0.763 \text{ mol } NH_3$

$0.763 \text{ mol } NH_3 \times \frac{17.03 \text{ g } NH_3}{\text{mol } NH_3} = \underline{13.0 \text{ g } NH_3}$

9-62 $n = 0.250 \text{ g } O_2 \times \frac{1 \text{ mol } O_2}{32.00 \text{ g } O_2} = 7.81 \times 10^{-3} \text{ mol } O_2$

$P = \frac{nRT}{V} = \frac{7.81 \times 10^{-3} \text{ mol} \times 0.0821 \frac{\text{L} \cdot \text{atm}}{\text{K} \cdot \text{mol}} \times 302 \text{ K}}{0.250 \text{ L}} = 0.775 \text{ atm}$

$0.775 \text{ atm} \times 760 \text{ torr/atm} = \underline{589 \text{ torr}}$

9-64 $n = \frac{PV}{RT} = \frac{1.15 \text{ atm} \times 3.50 \text{ L}}{0.0821 \frac{\text{L} \cdot \text{atm}}{\text{K} \cdot \text{mol}} \times 296 \text{ K}} = 0.166 \text{ mol}$

$0.166 \text{ mol Ne} \times \frac{20.18 \text{ g Ne}}{\text{mol Ne}} = \underline{3.35 \text{ g Ne}}$

9-66 $P = \frac{780 \text{ torr}}{760 \text{ torr/atm}} = 1.03 \text{ atm}$

$n = \frac{PV}{RT} = \frac{1.03 \text{ atm} \times 2.5 \times 10^{7} \text{ L}}{0.0821 \frac{\text{L} \cdot \text{atm}}{\text{K} \cdot \text{mol}} \times 300 \text{ K}} = 1.0 \times 10^{6} \text{ mol of gas}$

He: $1.0 \times 10^{6} \text{ mol He} \times \frac{4.003 \text{ g He}}{\text{mol He}} \times \frac{1 \text{ lb}}{453.6 \text{ g}} = \underline{8800 \text{ lb He}}$

Air: $1.0 \times 10^{6} \text{ mol air} \times \frac{29.0 \text{ g air}}{\text{mol air}} \times \frac{1 \text{ lb}}{453.6 \text{ g}} = \underline{64{,}000 \text{ lb air}}$

Lifting power with He = 64,000 - 8800 = $\underline{55{,}000 \text{ lb}}$

H_2: $1.0 \times 10^{6} \text{ mol } H_2 \times \frac{2.016 \text{ g } H_2}{\text{mol } H_2} \times \frac{1 \text{ lb}}{453.6 \text{ g}} = \underline{4400 \text{ lb } H_2}$

Lifting power with H_2 = 64,000 - 4000 = $\underline{60{,}000 \text{ lb}}$

Helium is a noncombustible gas whereas hydrogen forms an explosive mixture with O_2.

9-68 $P_{tot} = P(CO_2) + P(N_2) + P(He) = 250 \text{ torr} + 375 \text{ torr} + 137 \text{ torr} = \underline{762 \text{ torr}}$

9-70 $P(Ar) = \left(\frac{Ar\%}{100\%}\right) \times P_T = 0.0090 \times 756 \text{ torr} = \underline{6.8 \text{ torr}}$

9-73 $P(N_2) = 1050 \text{ torr} \times 0.720 = 756 \text{ torr}$ $P(O_2) = 1050 \text{ torr} \times 0.0800 = 84.0 \text{ torr}$

$P(SO_2) = P_T - [P(N_2) + P(O_2)] = 1050 - (756 + 84) = \underline{210 \text{ torr}}$

9-74 $P(O_2) = (\frac{O_2\%}{100\%}) \times P_T$ 256 torr = (0.35) x P_T P_T = 730 torr

9-76 The partial pressure of each gas must be determined when confined in a 2.00-L volume.

$P(N_2)$ = 300 torr $P(O_2) = 85 \text{ torr} \times \frac{4.00 \text{ L}}{2.00 \text{ L}} = 170 \text{ torr}$

$P(CO_2) = 450 \text{ torr} \times \frac{1.00 \text{ L}}{2.00 \text{ L}} = 225 \text{ torr}$

P_T = 300 + 170 + 225 = 695 torr

9-77 First, find the partial pressure of gas A in the 4.00-L container.

$P_A = 0.880 \text{ atm} \times \frac{2.50 \text{ L}}{4.00 \text{ L}} = 0.550 \text{ atm}$

P_B = 0.850 - 0.550 = 0.300 atm

9-79 $15.0 \text{ g } CO_2 \times \frac{1 \text{ mol } CO_2}{44.01 \text{ g } CO_2} \times \frac{22.4 \text{ L}}{\text{mol}} = 7.63 \text{ L}$

9-81 $3.01 \times 10^{24} \text{ molecules} \times \frac{1 \text{ mol } N_2}{6.022 \times 10^{23} \text{ molecules}} \times \frac{22.4 \text{ L } N_2}{\text{mol } N_2} = 112 \text{ L } N_2$

9-83 $6.78 \times 10^{-4} \text{ L} \times \frac{1 \text{ mol}}{22.4 \text{ L}} \times \frac{46.01 \text{ g}}{\text{mol}} = 1.39 \times 10^{-3} \text{ g}$

9-84 Molar mass of B_2H_6 = 27.67 g/mol $\frac{27.67 \text{ g}}{\text{mol}} \times \frac{1 \text{ mol}}{22.4 \text{ L}} = 1.24 \text{ g/L}$

9-86 $\frac{1.52 \text{ g}}{\text{L}} \times \frac{22.4 \text{ L}}{\text{mol}} = 34.0 \text{ g/mol}$

9-88 Find the volume at STP using the combined gas law.

$1.00 \text{ L} \times \frac{273 \text{ K}}{298 \text{ K}} \times \frac{1.20 \text{ atm}}{1.00 \text{ atm}} = 1.10 \text{ L (STP)}$

$\frac{3.60 \text{ g}}{1.10 \text{ L}} = 3.27 \text{ g/L (STP)}$

9-89 Find moles of N_2 in 1 L at 500 torr and 22°C using the ideal gas law.

$P = \frac{500 \text{ torr}}{760 \text{ torr/atm}} = 0.658 \text{ atm}$ $n = \frac{PV}{RT} = \frac{0.658 \text{ atm} \times 1.00 \text{ L}}{0.0821 \frac{\text{L} \cdot \text{atm}}{\text{K} \cdot \text{mol}} \times 295 \text{ K}} = 0.272 \text{ mol } N_2$

$0.272 \text{ mol } N_2 \times \frac{28.02 \text{ g } N_2}{\text{mol } N_2} = 0.762 \text{ g } N_2$

Density = 0.762 g/L (500 torr and 22°C)

9-91 **g $CaCO_3$ → mol $CaCO_3$ → mol CO_2 → vol. CO_2**

$$115 \text{ g } CaCO_3 \times \frac{1 \text{ mol } CaCO_3}{100.1 \text{ g } CaCO_3} \times \frac{1 \text{ mol } CO_2}{1 \text{ mol } CaCO_3} \times \frac{22.4 \text{ L}}{\text{mol } CO_2} = \underline{25.7 \text{ L } CO_2} \text{ (STP)}$$

9-92 **vol. O_2 → mol O_2 → mol Mg → g Mg**

$$5.80 \text{ L } O_2 \times \frac{1 \text{ mol } O_2}{22.4 \text{ L } O_2} \times \frac{2 \text{ mol Mg}}{1 \text{ mol } O_2} \times \frac{24.31 \text{ g Mg}}{\text{mol Mg}} = \underline{12.6 \text{ g Mg}}$$

9-94 **g H_2O → mol H_2O → mol C_2H_2 → vol. C_2H_2**

$$5.00 \text{ g } H_2O \times \frac{1 \text{ mol } H_2O}{18.02 \text{ g } H_2O} \times \frac{1 \text{ mol } C_2H_2}{2 \text{ mol } H_2O} = 0.139 \text{ mol } C_2H_2$$

$$P = \frac{745 \text{ torr}}{760 \text{ torr/atm}} = 0.980 \text{ atm}$$

$$V = \frac{nRT}{P} = \frac{0.139 \text{ mol} \times 0.0821 \frac{\text{L} \cdot \text{atm}}{\text{K} \cdot \text{mol}} \times 298 \text{ K}}{0.980 \text{ atm}} = \underline{3.47 \text{ L}}$$

9-96 (a) **g C_4H_{10} → mol C_4H_{10} → mol CO_2 → vol. CO_2**

$$85.0 \text{ g } C_4H_{10} \times \frac{1 \text{ mol } C_4H_{10}}{58.12 \text{ g } C_4H_{10}} \times \frac{8 \text{ mol } CO_2}{2 \text{ mol } C_4H_{10}} \times \frac{22.4 \text{ L}}{\text{mol } CO_2} = \underline{131 \text{ L}}$$

(b) **g C_4H_{10} → mol C_4H_{10} → mol O_2 → vol. O_2**

$$85.0 \text{ g } C_4H_{10} \times \frac{1 \text{ mol } C_4H_{10}}{58.12 \text{ g } C_4H_{10}} \times \frac{13 \text{ mol } O_2}{2 \text{ mol } C_4H_{10}} = 9.51 \text{ mol } O_2$$

$$V = \frac{nRT}{P} = \frac{9.51 \text{ mol} \times 0.0821 \frac{\text{L} \cdot \text{atm}}{\text{K} \cdot \text{mol}} \times 400 \text{ K}}{3.25 \text{ atm}} = \underline{96.1 \text{ L } O_2}$$

(c) **vol. C_4H_{10} → mol C_4H_{10} → mol CO_2 → vol. CO_2**

$$C_4H_{10}\text{: } n = \frac{PV}{RT} = \frac{0.750 \text{ atm} \times 45.0 \text{ L}}{0.0821 \frac{\text{L} \cdot \text{atm}}{\text{K} \cdot \text{mol}} \times 298 \text{ K}} = 1.38 \text{ mol}$$

$$1.38\ \text{mol}\ C_4H_{10} \times \frac{8\ \text{mol}\ CO_2}{2\ \text{mol}\ C_4H_{10}} \times \frac{22.4\ \text{L}}{\text{mol}\ CO_2} = \underline{124\ \text{L}\ CO_2}$$

9-97 $n(H_2) = \frac{PV}{RT} = \frac{2.80 \times 10^4\ \text{L} \times 70.0\ \text{atm}}{0.0821\ \frac{\text{L}\cdot\text{atm}}{\text{K}\cdot\text{mol}} \times 523\ \text{K}} = 4.56 \times 10^4\ \text{mol}\ H_2$

mol H_2 ⟶ mol Zr ⟶ g Zr ⟶ kg Zr

$$4.56 \times 10^4\ \text{mol}\ H_2 \times \frac{1\ \text{mol Zr}}{2\ \text{mol}\ H_2} \times \frac{91.22\ \text{g Zr}}{\text{mol Zr}} \times \frac{1\ \text{kg Zr}}{10^3\ \text{g Zr}} = 2080\ \text{kg Zr}$$

$$2080\ \text{kg} \times \frac{2.205\ \text{lb}}{\text{kg}} \times \frac{1\ \text{ton}}{2000\ \text{lb}} = \underline{2.29\ \text{tons Zr}}$$

9-99 **vol. O_2 ⟶ mol O_2 ⟶ mol CO_2 ⟶ vol. CO_2**

$P = \frac{825\ \text{torr}}{760\ \text{torr/atm}} = 1.09\ \text{atm}$ $n(O_2) = \frac{PV}{RT} = \frac{27.5\ \text{L} \times 1.09\ \text{atm}}{0.0821\ \frac{\text{L}\cdot\text{atm}}{\text{K}\cdot\text{mol}} \times 250\ \text{K}} = 1.46\ \text{mol}\ O_2$

$$1.46\ \text{mol}\ O_2 \times \frac{1\ \text{mol}\ CO_2}{2\ \text{mol}\ O_2} = 0.730\ \text{mol}\ CO_2$$

$$V = \frac{nRT}{P} = \frac{0.730\ \text{mol} \times 0.0821\ \frac{\text{L}\cdot\text{atm}}{\text{K}\cdot\text{mol}} \times 300\ \text{K}}{1.50\ \text{atm}} = \underline{12.0\ \text{L}\ CO_2}$$

9-100 Force $= 12.0\ \text{cm}^2 \times 15.0\ \text{cm} \times \frac{13.6\ \text{g}}{\text{cm}^3} = 2450\ \text{g}$

$$P = \frac{2450\ \text{g}}{12.0\ \text{cm}^2} = 204\ \text{g/cm}^2$$

$$1\ \text{atm} = 76.0\ \text{cm} \times \frac{13.6\ \text{g}}{\text{cm}^3} = 1030\ \text{g/cm}^2$$

$$204\ \text{g/cm}^2 \times \frac{1\ \text{atm}}{1030\ \text{g/cm}^2} = \underline{0.198\ \text{atm}}$$

9-102 $n = \frac{1.00\ \text{L} \times 1.45\ \text{atm}}{0.0821\ \frac{\text{L}\cdot\text{atm}}{\text{K}\cdot\text{mol}} \times 308\ \text{K}} = 0.0573\ \text{mol}$ $\frac{8.37\ \text{g}}{0.0573\ \text{mol}} = \underline{146\ \text{g/mol}}$

9-103 (1) Using the ideal gas law, find the molar mass of the compound.

$$n = \frac{PV}{RT} = \frac{4.50\ \text{L} \times 1.00\ \text{atm}}{350\ \text{K} \times 0.0821\ \frac{\text{L}\cdot\text{atm}}{\text{K}\cdot\text{mol}}} = 0.157\ \text{mol}$$

$$\text{Molar mass} = \frac{6.58\ \text{g}}{0.157\ \text{mol}} = \underline{41.9\ \text{g/mol}}$$

(2) Using the percent composition, find the empirical formula.
In 100 g of compound there are 85.7 g of C and 14.3 g of H.

$$85.7\ \cancel{g\ C} \times \frac{1\ mol\ C}{12.01\ \cancel{g\ C}} = 7.14\ mol\ C \qquad 14.3\ \cancel{g\ H} \times \frac{1\ mol\ H}{1.008\ \cancel{g\ H}} = 14.2\ mol\ H$$

$$C: \frac{7.14}{7.14} = 1.0 \quad H: \frac{14.2}{7.14} = 2.0 \qquad \text{Emp. formula} = CH_2$$

(3) Using the empirical formula, molar mass, and empirical mass, find the molecular formula.
Emp. mass = 12.01 + (2 x 1.008) = 14.03 g/emp. unit

$$\frac{41.9\ \cancel{g}/mol}{14.03\ \cancel{g}/emp.\ unit} = 3\ emp.\ units/mol \qquad \text{Molecular formula} = \underline{C_3H_6}$$

9-105 n_T = 0.265 mol O_2 + 0.353 mol N_2 + 0.160 mol CO_2 = 0.778 mol of gas V = 6.92 L

$$P(O_2) = \frac{0.265}{0.778} \times 2.86\ atm = 0.974\ atm;\ P(N_2) = 1.30\ atm;\ P(CO_2) = 0.59\ atm$$

9-106 $$3 \times 10^4\ \cancel{molecules} \times \frac{1 mol}{6.022 \times 10^{23}\ \cancel{molecules}} = 5 \times 10^{-20}\ mol$$

$$P = \frac{nRT}{V} = \frac{5 \times 10^{-20}\ \cancel{mol} \times 0.0821 \frac{\cancel{L} \cdot atm}{\cancel{K} \cdot \cancel{mol}} \times 10\ \cancel{K}}{10^{-3}\ \cancel{L}} = \underline{4 \times 10^{-17}\ atm}$$

9-108 $$\frac{V}{n} = \frac{RT}{P} = \frac{0.0821 \frac{L \cdot \cancel{atm}}{\cancel{K} \cdot mol} \times 298\ \cancel{K}}{1.25\ \cancel{atm}} = 19.6\ L/mol$$

$$\frac{44.01\ g/\cancel{mol}}{19.6\ L/\cancel{mol}} = = \underline{2.25\ g/L\ (density)}$$

9-109 $$\text{Density} = \frac{mass}{V} = \frac{P \times MM}{RT} = \frac{1.00\ \cancel{atm} \times 29.0\ g/\cancel{mol}}{0.0821 \frac{L \cdot \cancel{atm}}{\cancel{K} \cdot \cancel{mol}} \times 673K} = \underline{0.525\ g/L\ (hot)}$$

Density at STP = 1.29 g/L

0.525/1.29 = 0.41 (Hot air is less than half as dense as air at STP.)

9-111 $H_3BCO + 3H_2O(l) \longrightarrow B(OH)_3(aq) + CO(g) + 3H_2(g)$

$$P = \frac{565 \text{ torr}}{760 \text{ torr/atm}} = 0.743 \text{ atm}$$

$$n(H_3BCO) = \frac{PV}{RT} = \frac{0.743 \text{ atm} \times 0.425 \text{ L}}{0.0821 \frac{\text{L} \cdot \text{atm}}{\text{K} \cdot \text{mol}} \times 373 \text{ K}} = 0.0103 \text{ mol of } H_3BCO$$

$$0.0103 \text{ mol } H_3BCO \times \frac{4 \text{ mol gas}}{1 \text{ mol } H_3BCO} = 0.0412 \text{ mol gas}$$

$$V = \frac{nRT}{P} = \frac{0.0412 \text{ mol} \times 0.0821 \frac{\text{L} \cdot \text{atm}}{\text{K} \cdot \text{mol}} \times 298 \text{ K}}{0.900 \text{ atm}} = \underline{1.12 \text{ L}}$$

9-113 $2Al(s) + 3F_2(g) \longrightarrow 2AlF_3(s)$ $P = \frac{725 \text{ torr}}{760 \text{ torr/atm}} = 0.954 \text{ atm}$

$$\text{original } F_2 = \frac{0.954 \text{ atm} \times 8.23 \text{ L}}{0.0821 \frac{\text{L} \cdot \text{atm}}{\text{K} \cdot \text{mol}} \times 308 \text{ K}} = 0.310 \text{ mol } F_2$$

$$\text{leftover } F_2 = \frac{3.50 \text{ g}}{38.00 \text{ g/mol}} = 0.0921 \text{ mol} \qquad 0.310 - 0.092 = 0.218 \text{ mol } F_2 \text{ reacts}$$

$$0.218 \text{ mol } F_2 \times \frac{2 \text{ mol } AlF_3}{3 \text{ mol } F_2} \times \frac{83.98 \text{ g } AlF_3}{\text{mol } AlF_3} = \underline{12.2 \text{ g } AlF_3}$$

9-115 $2.54 \times 10^{24} \text{ molecules} \times \frac{1 \text{ mol } N_2O_3}{6.022 \times 10^{23} \text{ molecules}} \times \frac{2 \text{ mol gas}}{\text{mol } N_2O_3} = 8.44 \text{ mol gas}$

$$V = \frac{nRT}{PV} = \frac{8.44 \text{ mol} \times 0.0821 \frac{\text{L} \cdot \text{atm}}{\text{K} \cdot \text{mol}} \times 308 \text{ K}}{1.58 \text{ atm}} = \underline{135 \text{ L}}$$

9-117 $15.0\ \text{mL} \times \frac{0.917\ \text{g}}{\text{mL}} \times \frac{1\ \text{mol}}{18.02\ \text{g}\ H_2O} = 0.763\ \text{mol}\ H_2O$

$$V = \frac{nRT}{P} = \frac{0.763\ \text{mol} \times 0.0821 \frac{\text{L} \cdot \text{atm}}{\text{K} \cdot \text{mol}} \times 298\ \text{K}}{\frac{22\ \text{torr}}{760\ \text{torr/atm}}} = \underline{645\ \text{L}}$$

9-118 $n = \frac{PV}{RT} = \frac{1.20\ \text{atm} \times 0.0200\ \text{L}}{0.0821 \frac{\text{L} \cdot \text{atm}}{\text{K} \cdot \text{mol}} \times 298\ \text{K}} = 9.81 \times 10^{-4}\ \text{mol}\ NH_3$

$$9.81 \times 10^{-4}\ \text{mol}\ NH_3 \times \frac{1\ \text{mol}\ N_2H_4}{2\ \text{mol}\ NH_3} \times \frac{22.4\ \text{L}\ N_2H_4}{\text{mol}\ N_2H_4} = 0.0110\ \text{L} = \underline{11.0\ \text{mL}\ N_2H_4}$$

9-120 (1) Find the empirical formula of reactant compound.

$$\text{N: } 30.4\ \text{g N} \times \frac{1\ \text{mol N}}{14.01\ \text{g N}} = 2.17\ \text{mol N}$$

$$\text{O: } 69.6\ \text{g O} \times \frac{1\ \text{mol O}}{16.00\ \text{g O}} = 4.35\ \text{mol O}$$

$$\text{N: } \frac{2.17}{2.17} = 1.0 \quad \text{O: } \frac{4.35}{2.17} = 2.0 \quad \text{Empirical formula} = NO_2$$

(2) Find the molar mass of product compound.

$$n = \frac{PV}{RT} = \frac{\frac{715\ \text{torr}}{760\ \text{torr/atm}} \times 1.05\ \text{L}}{0.0821 \frac{\text{L} \cdot \text{atm}}{\text{K} \cdot \text{mol}} \times 273\ \text{K}} = 0.0441\ \text{mol}$$

$$\text{Molar mass} = \frac{2.03\ \text{g}}{0.0441\ \text{mol}} = 46.0\ \text{g/mol}$$

(3) Since one compound decomposes to one other compound, the reactant compound must have the same empirical formula as the product compound. Since the empirical mass of NO_2 = 46.01 g/emp unit then the product must be NO_2(MM = 46.0 g/mol). Since 0.0220 mol of reactant form 0.0441 mol of product (1:2 ratio) the reaction must be

$$N_2O_4(l) \longrightarrow 2NO_2(g)$$

9-122 $6Li(s) + N_2(g) \longrightarrow 2Li_3N(s)$ $3Mg(s) + N_2(g) \longrightarrow Mg_3N_2(s)$

$$n = \frac{PV}{RT} = \frac{\dfrac{985\ \text{torr}}{760\ \text{torr/atm}} \times 256\ \text{L}}{0.0821\ \dfrac{\text{L}\cdot\text{atm}}{\text{K}\cdot\text{mol}} \times 373\ \text{K}} = 10.8\ \text{mol } N_2$$

$$10.8\ \text{mol } N_2 \times \frac{6\ \text{mol Li}}{\text{mol } N_2} \times \frac{6.941\ \text{g Li}}{\text{mol Li}} = \underline{450\ \text{g Li}}$$

$$10.8\ \text{mol } N_2 \times \frac{3\ \text{mol Mg}}{\text{mol } N_2} \times \frac{24.31\ \text{g Mg}}{\text{mol Mg}} = \underline{788\ \text{g Mg}}$$

10

The Solid and Liquid States

Review Section A *The Properties of Condensed States and the Forces Involved*

OUTLINE	OBJECTIVES
10-1 Properties of the Solid and Liquid States	
1. Common properties of liquids and solids	*List the general properties of the liquid and solid states.*
2. The kinetic molecular theory applied to condensed states	*Describe how kinetic molecular theory explains common properties of the condensed states.*
3. The motion of water molecules in the three physical states	*Compare the different motions of water molecules in its three physical states.*
10-2 Intermolecular Forces	
1. Three forces of attraction between molecules	*Describe the three forces of attraction between molecules.*
2. Hydrogen bonding and the unique properties of water	*Explain how and why water molecules interact by hydrogen bonding.*
10-3 The Solid State: Melting Point	
1. Four types of solids and their melting points	*Describe how the forces of interaction between the basic particles of a compound affect its melting point.*

SUMMARY OF SECTIONS 10-1 THROUGH 10-3

Questions: *How do the solid and liquid states compare to the gaseous state? What makes the basic particles in solids and liquids stick together? Are all solids the same?*

The solid and liquid states are known as condensed states primarily because the ions or molecules are close together. This fact accounts for four common properties. Specifically, they have high densities, are incompressible, expand little when heated, and have a definite volume.

In solids and liquids, the basic particles obviously "stick together." It is therefore of interest to describe the **intermolecular forces** that cause them to coalesce. All molecules have inherent forces of attraction between them called **London forces.** London forces are weakest for small, low molar mass compounds but become stronger for larger molecules. For nonpolar molecules, London forces are the only forces of attraction.

Polar molecules can align themselves so that the negative end of one molecule is attracted to the positive end of another. Thus, in addition to London forces, polar molecules have a second attractive force known as a **dipole-dipole force.** Given two compounds of similar molar mass, one polar and one nonpolar, the polar compound will have the stronger intermolecular forces of attraction and is more likely to exist in a condensed state at a specific temperature. It is more difficult to compare the forces of a polar, low-molar mass compound with those of a nonpolar, high-molar mass compound.

A third interaction is known as **hydrogen bonding.** Hydrogen bonding is a considerably stronger interaction than normal dipole-dipole interactions. It accounts for the fact that water is a liquid at room temperature rather than a gas. When one looks more closely at the three-dimensional structure of a water molecule, it becomes clearer how hydrogen bonding interactions occur.

We first look at the solid state and then the liquid state. Solids may be either **amorphous** or **crystalline.** In crystalline solids, the basic particles (atoms, molecules, or ions) exist in an orderly arrangement known as a **crystal lattice.** The temperature at which the lattice breaks down (the melting point) reflects the degree of attraction between the basic particles. **Ionic solids** have high melting points; those of **molecular solids** are generally lower because of weaker forces between basic particles. *Network solids*, such as the two main allotropes of carbon (**diamond and graphite**), generally have high melting points. **Metallic solids** have a wide range of melting points.

NEW TERMS

Amorphous solid
Buckminsterfullerene
Crystal lattice
Crystalline solid
Diamond
Dipole-dipole attraction
Graphite
Hydrogen bonding
Intermolecular forces
Ionic solid
London forces
Metallic solid
Network solid

SELF-TEST

A-1 Matching

Which state or states of matter (liquid, solid, gas) are described by the following statements?

(a) The most compressible __________

(b) The state where molecules mix most slowly __________

(c) The state where molecules have the most motion __________

(d) The condensed state or states of matter __________

(e) The state where molecules move the farthest between collisions __________

(f) The state or states where molecules are free to move past one another __________

(g) The state where molecules are in fixed positions __________

(h) The state or states where water molecules do not form hydrogen bonds __________

(i) The most likely state for ionic compounds at room temperature ____________

(j) The most likely state for nonpolar compounds with a low molar mass at room temperature ____________

A-2 Multiple Choice

____ 1. Which of the following bonds is the most polar?

(a) N-N (b) B-F (c) H-Cl (d) N-Cl

____ 2. Which of the following molecules is predicted to be nonpolar?

(a) S-C-O

(b) C-O

(c) H-Be-H

(d) O-S-O (bent)

(e) H-Se-H (bent)

____ 3. Which of the following compounds could form hydrogen bonds?

(a) CH_3OH (b) PH_3 (c) KCN (d) CH_3CN (e) H_2S

____ 4. How many hydrogen bonds can form to the oxygen in a H_2O molecule?

(a) 4 (b) 3 (c) 2 (d) 1 (e) 0

____ 5. How many hydrogen bonds can form to the nitrogen in a NH_3 molecule?

(a) 4 (b) 3 (c) 2 (d) 1 (e) 0

____ 6. A compound is ionic. Which of the following temperatures is the most likely melting point of the compound?

(a) 658°C (b) 325 K (c) -112°C (d) 15°C

____ 7. Diamond is an example of which kind of solid?

(a) ionic (b) molecular (c) network (d) metallic

____ 8. In what type of solid do only cations occupy lattice positions?

(a) ionic (b) molecular (c) network (d) metallic

____ 9. The bonds of a certain molecule with a low molar mass are polar but the molecule itself is nonpolar. Which of the following temperatures would most likely be its boiling point?

(a) 378 K (b) -20°C (c) 343°C (d) 1200°C

____ 10. Which of the following molecular compounds should have the highest melting point?

(a) BF_3 (polar bonds - nonpolar molecule)
(b) NH_2OH (polar bonds - polar molecule)
(c) SCl_2 (polar bonds - polar molecule)
(d) HI (essentially nonpolar bond)

____ 11. The compound $PbCl_2$ melts at 501°C and $PbCl_4$ melts at -15°C. Which of the following statements is most likely true?

(a) $PbCl_2$ is ionic and $PbCl_4$ is molecular.
(b) $PbCl_2$ is molecular and $PbCl_4$ is ionic.
(c) Both compounds are probably ionic.
(d) Both compounds are probably molecular.
(e) No conclusions can be drawn.

____ 12. Although SF_4 is a polar molecule and SF_6 is nonpolar, SF_6 melts at a higher temperature than SF_4. Which of the following conclusions can be made?

(a) Nonpolar compounds melt at higher temperatures than polar compounds.
(b) SF_6 is actually ionic.
(c) London forces (which are proportional to the molar mass) can be more important than dipole-dipole forces.
(d) The S-F bond is nonpolar.

Review Section B *The Liquid State and Changes in State*

OUTLINE	OBJECTIVES
10-4 The Liquid State: Surface Tension and Viscosity	
1. Intermolecular attractions in the liquid state.	*Correlate intermolecular forces with surface tension and viscosity.*
10-5 Vapor Pressure and Boiling Point	
1. The distribution of molecular energies at a specified temperature	*Explain how a liquid is cooled by evaporation.*
2. Vapor pressure and equilibrium	*Describe how vapor and liquid reach a point of equilibrium.*

3. Vapor pressure curves and boiling points

Describe the difference between a boiling point and a normal boiling point.

10-6 Energy and Changes in State

1. Melting points and the heat of fusion
2. Boiling points and the heat of vaporization

Explain how the melting and boiling points and the heats of fusion and vaporization reflect the degree of attraction between basic particles.

10-7 Heating Curve of Water

1. The heating of ice from -10°C to vapor above 100°C

Describe the changes that occur as ice below 0 °C is heated to vapor above 100°C.

SUMMARY OF SECTIONS 10-4 THROUGH 10-7

Questions: *How does the liquid state differ from the solid state? What is involved when a liquid changes to the gaseous state? How is energy concerned with changes in state?*

In the liquid state, the basic particles are still close together but are not in fixed positions, so they can move past one another. This attraction creates a **surface tension** in the liquid and produces **viscosity.**

Molecules of a liquid can escape to the gaseous phase in a process known as **vaporization**. In a closed container an equilibrium is established between the vapor and the liquid. As the temperature increases, the **vapor pressure** of a liquid increases since a higher fraction of molecules have enough kinetic energy to escape to the gaseous phase. The temperature at which the vapor pressure of a liquid equals the restraining pressure is the **boiling point.** The **normal boiling point** is the temperature at which the vapor pressure is exactly one atmosphere. If a liquid is allowed to **evaporate**, the average kinetic energy of the molecules in the remaining liquid decreases; hence the temperature of the liquid decreases. Other relevant phase changes are known as **condensation** and **sublimation.**

In the last section, we discussed how intermolecular forces affected the temperature at which compounds undergo a change of state (melt or boil). In addition to the temperature at which changes of state occur, the amount of heat energy required to effect these changes varies for different compounds. The amount of heat energy required to cause melting of a specified mass of solid is known as the **heat of fusion.** The amount of heat energy required to cause vaporization of a specified mass of liquid is the **heat of vaporization.** The magnitude of these quantities, like the melting and boiling points, depends once again on the strength of the interactions between the basic particles of the compound. For a small, covalent molecule, water has unusually large values for all of these because of the extent and strength of its hydrogen bonding.

During the processes of melting or boiling a pure substance, heat supplied to a substance is transferred into potential energy so that the temperature remains constant. Water is used as a model compound in a **heating curve** to illustrate what happens on the molecular level as it is heated from the solid state at -10°C to the vapor state above 100°C.

NEW TERMS

Boiling point
Condensation
Evaporation
Fusion
Heat of fusion
Heat of vaporization
Heating curve
Normal boiling point
Sublimation
Surface tension
Vapor pressure
Vaporization
Viscosity

SELF-TEST

B-1 Multiple Choice

____ 1. Which of the following types of compounds has the highest heat of fusion?

(a) nonpolar molecular
(b) polar molecular
(c) hydrogen-bonded molecular
(d) ionic

____ 2. A compound boils at -78°C. It is likely that the compound

(a) has a high melting point.
(b) has a low heat of vaporization.
(c) has a high heat of fusion.
(d) is ionic.

____ 3. What is the vaporization of a solid called?

(a) condensation
(b) evaporization
(c) fusion
(d) sublimation

____ 4. Which of the following would cool the fastest if allowed to evaporate under the same conditions? (Refer to Fig. 10-15 in the text.)

(a) water at 95°C
(b) ethyl ether at 0°C
(c) water at 25°C
(d) water ice at 0°C

____ 5. Of the liquids discussed in Figure 10-15 in the text, which would exist as a gas at 25°C and 300 torr?

(a) all three liquids
(b) alcohol and ether
(c) ether
(d) alcohol and water
(e) alcohol

____ 6. The heat of vaporization of PCl_3 is 217 J/g. For BCl_3 it is 160 J/g. Which of the following statements is a possible explanation for the order of these values?

(a) Both compounds are ionic.
(b) PCl_3 is ionic and BCl_3 is molecular.
(c) BCl_3 is polar and PCl_3 is nonpolar.
(d) PCl_3 has a greater molar mass than BCl_3.

____ 7. Carbon tetrachloride has a vapor pressure of 680 torr at 70°C. Which temperature would most likely be its normal boiling point?

(a) 50°C (b) 70°C (c) 135°C (d) 76°C

B-2 Problems

(For the following problems, use these values: heat of fusion, 105 J/g at the melting point of ethyl alcohol of -114°C; heat of vaporization, 854 J/g at the boiling point of 78°C. The specific heat of liquid ethyl alcohol is 2.26 J/(g · C).

1. How many joules are required to melt 285 g of ethyl alcohol?

2. What mass of ethyl alcohol can be vaporized at the boiling point by 10.7 kJ?

3. If 3.70 kJ of heat energy is added to 150 g of liquid ethyl alcohol, how many Celsius degrees does the temperature rise?

4. How many kilojoules are required to change 15.0 g of solid ethyl alcohol at -114°C to vapor at 78°C?

Review Section C

CHAPTER SUMMARY SELF-TEST

C-1 Problems

1. Fill in the following values from those listed below.

	Melting point	Heat of fusion	Boiling point	Heat of vaporization
CH_4	________	________	________	________
NH_3	________	________	________	________
KF	________	________	________	________

melting points: -78°C, 880°C, -183°C

boiling points: 1500°C, -3 °C, -156°C

heats of fusion: 452 J/g, 350 J/g, 61 J/g

heats of vaporization: 577 J/g, 15.1 kJ/g, 1.36 kJ/g

2. The normal boiling point of carbon disulfide is 46°C. Circle the correct answer in the following statements;

(a) The heat of vaporization of CS_2 is (higher, lower) than H_2O.

(b) The attractions between CS_2 molecules are (stronger, weaker) than between H_2O molecules in the liquid state.

(c) At room temperature CS_2 is (more, less) volatile than H_2O.

(d) The freezing point of CS_2 is probably (higher, lower) than H_2O.

3. In Chapter 6 we learned that metals and nonmetals usually combine to form ionic compounds. A compound formed between titanium and chlorine ($TiCl_4$) has a heat of fusion of 49.4 J/g and a melting point of -25°C.

(a) What does this information tell you about the nature of the Ti-Cl bond?

(b) When a 500 g quantity of $TiCl_4$ melts, how many kJ are released? How does this compare to the kJ released when the same mass of water is allowed to melt?

Answers to Self-Tests

A-1 Matching

(a) gas

(b) solid

(c) all the same at the same temperature

(d) solid and liquid

(e) gas

(f) liquid and gas

(g) solid

(h) gas

(i) solid

(j) gas

A-2 Multiple Choice

1. **b** B-F (These two atoms have the largest difference in electronegativity.)
2. **c** H-Be-H (A linear molecule.)
3. **a** CH_3OH (The C-H bond in CH_3CN is nearly nonpolar.)
4. **c** (There are two unshared pairs of electrons on an oxygen.)
5. **d** 1 (There is one lone pair of electrons on the nitrogen.)
6. **a** 658°C
7. **c** network
8. **d** metallic
9. **b** -20°C
10. **b** NH_2OH (Has hydrogen bonds.)
11. **a** $PbCl_2$ is ionic and $PbCl_4$ is molecular.
12. **c** London forces can be more important than dipole-dipole forces.

B-1 Multiple Choice

1. **d** ionic
2. **b** The compound likely has a low heat of vaporization.
3. **d** sublimation
4. **a** Water at 95°C has the highest vapor pressure so it cools the fastest.
5. **c** ether
6. **d** PCl_3 has a greater molar mass than BCl_3. (PCl_3 is also polar and BCl_3 is nonpolar.)
7. **d** 76°C

B-2 Problems

1. $285\ \cancel{g} \times 105\ J/\cancel{g} = 2.99 \times 10^4\ J$

2. $10.7\ \cancel{kJ} \times \frac{10^3\ \cancel{J}}{\cancel{kJ}} \times \frac{1\ g}{854\ \cancel{J}} = 12.5\ g$

3. $3.70\ \cancel{kJ} \times \frac{10^3\ \cancel{J}}{\cancel{kJ}} = 150\ \cancel{g} \times \frac{2.26\ \cancel{J}}{\cancel{g} \cdot {}^{o}C}\quad {}^{o}C = 11^{o}C$ rise

4. melt the solid: $15.0\ \cancel{g} \times 10^5\ J/\cancel{g} = 1580\ J = 1.58\ kJ$

   heat the liquid: $15.0\ \cancel{g} \times 192\ \cancel{{}^{o}C} \times \frac{2.26\ J}{\cancel{g} \cdot \cancel{{}^{o}C}} = 6{,}510\ J = 6.51\ kJ$

   vaporize the liquid: $15.0\ \cancel{g} \times 854\ J/\cancel{g} = 12{,}800\ J = 12.8\ kJ$

   Total heat: 1.58 kJ + 6.51 kJ + 12.8 kJ = <u>20.9 kJ</u>

C-1 Problems

1. CH_4 is a nonpolar molecule with a small molar mass. Therefore, it has the smallest values of the given properties. NH_3 is polar covalent with hydrogen bonding. It has intermediate properties. KF is ionic and has large values for these properties.

   CH_4: melting point, -183°C; heat of fusion, 61 J/g; boiling point, -156°C; heat of vaporization, 577 J/g.

   NH_3: melting point, -78°C; heat of fusion, 350 J/g; boiling point, -33°C; heat of vaporization, 1.36 kJ/g.

   KF: melting point, 880°C; heat of fusion, 452 J/g; boiling point, 1500°C; heat of vaporization, 15.1 kJ/g.

2. (a) lower (b) weaker (c) more (d) lower

3. (a) Since the melting point and the heat of fusion of $TiCl_4$ are typical of molecular compounds, we can conclude that the Ti-Cl bond is primarily covalent rather than ionic.

   (b) $500\ \cancel{g} \times 49.4\ J/\cancel{g} = 24{,}700\ J = \underline{24.7\ kJ}$

   For H_2O: $500\ \cancel{g} \times 334\ J/\cancel{g} = 167{,}000\ J = \underline{167\ kJ}$

Solutions to Black Text Problems

10-53 $18.0\ \cancel{g} \times \frac{393\ J}{\cancel{g}} = \underline{7.07 \times 10^3\ J}$

10-55 $850\ \cancel{J} \times \frac{1.00\ g}{334\ \cancel{J}} = \underline{2.54\ g\ H_2O}$ $850\ \cancel{J} \times \frac{1.00\ g}{519\ \cancel{J}} = \underline{1.64\ g\ NaCl}$

$850\ \cancel{J} \times \frac{1.00\ g}{127\ \cancel{J}} = \underline{6.69\ g\ benzene}$

10-57 $25.0\ \cancel{g} \times \frac{22.2\ cal}{\cancel{g}} = \underline{555\ cal\ (ether)}$ $25.0\ \cancel{g} \times \frac{79.8\ cal}{\cancel{g}} = 2000\ cal\ (H_2O)$

Water would be more effective.

$125\ \cancel{g} \times \frac{104\ J}{\cancel{g}} = \underline{1.30 \times 10^4\ J\ (13.0\ kJ)}$

10-59 NH_3: $450 \text{ g} \times \frac{1.36 \text{ kJ}}{\text{g}} \times \frac{10^3 \text{ J}}{\text{kJ}} = 6.12 \times 10^5 \text{ J}$

Freon: $450 \text{ g} \times \frac{161 \text{ J}}{\text{g}} = 7.25 \times 10^4 \text{ J}$ Ammonia is more effective.

10-61 Conden.: $275 \text{ g} \times \frac{2260 \text{ J}}{\text{g}} = 62.2 \times 10^4 \text{ J}$

Cooling: $275 \text{ g} \times 75 ^\circ\text{C} \times \frac{4.184 \text{ J}}{\text{g} \cdot ^\circ\text{C}} = 8.6 \times 10^4 \text{ J}$

Total = $(62.2 \times 10^4) + (8.6 \times 10^4) = 70.8 \times 10^4 \text{ J} = \underline{7.08 \times 10^5 \text{ J } (708 \text{ kJ})}$

10-63 $120 \text{ g} \times 53 ^\circ\text{C} \times \frac{0.590 \text{ cal}}{\text{g} \cdot ^\circ\text{C}} = 3750 \text{ cal}$

$120 \text{ g} \times \frac{204 \text{ cal}}{\text{g}} = 24{,}500 \text{ cal}$ Total = 28,300 cal = $\underline{2.83 \times 10^4 \text{ cal}}$

10-65 heat ice: $132 \text{ g} \times 20 ^\circ\text{C} \times \frac{0.492 \text{ cal}}{\text{g} \cdot ^\circ\text{C}} = 1300 \text{ cal}$

melt ice: $132 \text{ g} \times \frac{79.8 \text{ cal}}{\text{g}} = 10{,}500 \text{ cal}$

heat H_2O: $132 \text{ g} \times 100 ^\circ\text{C} \times \frac{1.00 \text{ cal}}{\text{g} \cdot ^\circ\text{C}} = 13{,}200 \text{ cal}$

vap. H_2O: $132 \text{ g} \times \frac{540 \text{ cal}}{\text{g}} = 71{,}300 \text{ cal}$ Total = $\underline{96{,}300 \text{ cal } (96.3 \text{ kcal})}$

10-67 Let Y = the mass of the sample in grams. Then

$$\left(\frac{2260 \text{ J}}{\text{g}} \times Y\right) + \left(25.0 ^\circ\text{C} \times Y \times \frac{4.184 \text{ J}}{\text{g} \cdot ^\circ\text{C}}\right) = 28{,}400 \text{ J} \qquad Y = 12.0 \text{ g}$$

10-69 Find the heat released by condensation: $10.0 \text{ g} \times \frac{393 \text{ J}}{\text{g}} = 3930 \text{ J}$

The remainder of the 5000 J is released by the benzene as the liquid cools.
5000 - 3930 = 1070 J Let Y = temperature change in $^\circ$C, then

$$10.0 \text{ g} \times Y ^\circ\text{C} \times \frac{1.72 \text{ J}}{\text{g} \cdot ^\circ\text{C}} = 1070 \text{ J}$$

Y = 62 $^\circ$C change Final temp. = 80 - 62 = $\underline{18 ^\circ\text{C}}$

10-87 $2000 \text{ lb} \times \frac{453.6 \text{ g}}{\text{lb}} \times \frac{266 \text{ J}}{\text{g}} = 2.41 \times 10^8 \text{ J}$

Heating 1 g of H_2O from 25.0°C to 100.0°C and the vaporizing the water requires

$(75.0 ^\circ\text{C} \times 4.184 \text{ J}/^\circ\text{C}) + 2{,}260 \text{ J} = 2.57 \times 10^3 \text{ J/g } H_2O$

$$\frac{2.41 \times 10^8 \text{ J}}{2.57 \times 10^3 \text{ J/g } H_2O} = 9.38 \times 10^4 \text{ g } H_2O = \underline{93.8 \text{ kg } H_2O}$$

10-88 V = 100 L, T = 34 + 273 = 307 K, $P(H_2O) = 0.700 \times 39.0$ torr = 27.3 torr

Use the ideal gas law to find moles of water.

$$n = \frac{PV}{RT} = \frac{\frac{27.3\ \text{torr}}{760\ \text{torr/atm}} \times 100\ \text{L}}{0.0821 \frac{\text{L}\cdot\text{atm}}{\text{K}\cdot\text{mol}} \times 307\ \text{K}} = 0.143\ \text{mol}$$

$$0.143\ \text{mol}\ H_2O \times \frac{18.02\ \text{g}\ H_2O}{\text{mol}\ H_2O} = \underline{2.58\ \text{g}\ H_2O}$$

10-90 First calculate the heat required to heat the water from 25°C to 100°C then to vaporize the water.

$$1.00\ \text{kg} \times \frac{10^3\ \text{g}}{\text{kg}} \times (100 - 25)^\circ\text{C} \times \frac{4.184\ \text{J}}{\text{g}\cdot{}^\circ\text{C}} = 3.10 \times 10^5\ \text{J (to heat water)}$$

$$1.00\ \text{kg} \times \frac{10^3\ \text{g}}{\text{kg}} \times 2{,}260 \frac{\text{J}}{\text{g}} = 2.260 \times 10^6\ \text{J (to vaporize water)}$$

$$[3.10 \times 10^5\ \text{J} \times \frac{1\ \text{kJ}}{10^3\ \text{J}}] + [2.260 \times 10^6\ \text{J} \times \frac{1\ \text{kJ}}{10^3\ \text{J}}] = 2570\ \text{kJ}$$

$$\text{Cr:} \quad \frac{2570\ \text{kJ}}{21.0\ \text{kJ/mol}} \times \frac{52.00\ \text{g Cr}}{\text{mol Cr}} \times \frac{1\ \text{kg}}{10^3\ \text{g}} = \underline{6.36\ \text{kg Cr}}$$

$$\text{Mo:} \quad \frac{2570\ \text{kJ}}{28.0\ \text{kJ/mol}} \times \frac{95.94\ \text{g Mo}}{\text{mol Mo}} \times \frac{1\ \text{kg}}{10^3\ \text{g}} = \underline{8.81\ \text{kg Mo}}$$

$$\text{W:} \quad \frac{2570\ \text{kJ}}{35.0\ \text{kJ/mol}} \times \frac{183.9\ \text{g W}}{\text{mol W}} \times \frac{1\ \text{kg}}{10^3\ \text{g}} = \underline{13.5\ \text{kg W}}$$

10-91 The element must be mercury (Hg) since it is a liquid at room temperature and must be a metal since it forms a +2 ion.

The 15.0 kJ represents the heat required to melt X g of Hg, heat X g from -39°C to 357°C (i.e., 396°C), and vaporize X g of Hg.

$$(11.5\ \text{J/g} \times X) + [396^\circ\text{C} \times 0.139 \frac{\text{J}}{\text{g}\ {}^\circ\text{C}} \times X] + (29.5\ \text{J/g} \times X) = 15.00\ \text{kJ} \times 10^3\ \text{J/kJ}$$

$$11.5\ \text{J/g}\ X + 55.0\ \text{J/g}\ X + 29.5\ \text{J/g}\ X = 15{,}000\ \text{J} \qquad X = \underline{156\ \text{g Hg}}$$

10-93 Calculate the mass of water in the vapor state in the room at -5°C (268 K) using the ideal gas law.

$$n = \frac{PV}{RT} = \frac{\frac{2.50\ \text{torr}}{760\ \text{torr/atm}} \times 20000\ \text{L}}{0.0821 \frac{\text{L}\cdot\text{atm}}{\text{K}\cdot\text{mol}} \times 268\ \text{K}} = 2.99\ \text{mol}\ H_2O\ \text{(in vapor)}$$

$$2.99\ \text{mol}\ H_2O \times \frac{18.02\ \text{g}\ H_2O}{\text{mol}\ H_2O} = 53.9\ \text{g}\ H_2O\ \text{(capacity of the room)}$$

Eventually, all of the 50.0 g should sublime since it is less than the room could contain.

11

Aqueous Solutions

Review Section A *Solutions and the Quantities Involved*

OUTLINE	OBJECTIVES
11-1 The Nature of Aqueous Solutions	
1. A solution - solvent plus solute	*Define the terms miscible and immiscible.*
2. The hydration of ions by water	
3. The forces between ions	*Describe the forces involved in the solution of an ionic compound in water.*

4. The solution of ionic compounds in water — *Write equations illustrating the solution of soluble ionic compounds in water.*

5. The solution of molecular compounds in water — *Describe how a polar molecular compound dissolves in water.*

11-2 Solubility and Temperature

1. Definition of solubility terms — *Define the following terms - solubility, saturated, unsaturated, and supersaturated.*

2. The effect of temperature on solubility — *Determine the concentration of a given ionic compound at a specified temperature using a solubility graph.*

11-3 Concentration: Percent by Mass

1. Relative concentration: concentrated and dilute
2. Percent by mass of solute — *Convert among the masses of solute, solvent, and solution using percent by mass.*

11-4 Concentration: Molarity

1. The amount of a solute in a solution — *Apply the definition of molarity to convert among concentration, quantity, and volume of a specified solution.*

11-5 Dilution of Concentrated Solutions

1. Changing the molarity by adding more solvent — *Calculate the volume or concentration of a specified solution when it is diluted.*

11-6 Stoichiometry Involving Solutions

1. Molarity and stoichiometry — *Include solutions of known molarity in stoichiometry calculations.*

SUMMARY OF SECTIONS 11-1 THROUGH 11-6

***Questions:** How does water disperse an ionic compound to form a solution? How does temperature affect solubility of compounds? How do we indicate "how much" of a compound is dissolved in a solvent? What happens when we add water to a solution? How are solutions involved in stoichiometry?*

If chemists were to design the ideal **solvent** in which to study chemical reactions, the solvent would probably have the following properties: (a) It would be inexpensive and plentiful. (b) It would be nontoxic. (c) It would be a liquid at room temperature with a long temperature range in the liquid state. And (d), it would dissolve a large number of **solutes** to form **solutions**. Fortunately such a solvent exists, and it is just ordinary water. Chemists use many other solvents, but none is quite so useful and versatile as H_2O.

When two liquids dissolve in each other to form a solution, the liquids are said to be **miscible**. If they do not mix, they are said to be **immiscible**.

As we learned in Chapter 10, water molecules are attracted to each other in the liquid state by electrostatic interactions called hydrogen bonds. An ionic crystal is also held together by electrostatic forces that are known as ion-ion forces. When an ionic compound dissolves in water, there are electrostatic interactions between solvent and solute called **ion-dipole forces**. Ionic compounds dissolve in water when the ion-dipole forces between water and ion can overcome the ion-ion forces between oppositely charged ions in the crystal. The solution process of an ionic compound in water can be represented by an equation. The ions present in the original compound are now present in solution as hydrated ions [indicated by the (aq)].

$$Na_2SO_4(s) \xrightarrow{H_2O} 2Na^+(aq) + SO_4^{2-}(aq)$$

Certain polar covalent compounds also dissolve in water. In some cases (e.g., HCl), the compound undergoes **ionization** in solution. In other cases (e.g., CH_3OH), the compound is dissolved without ion formation. Nonpolar compounds do not generally dissolve in polar solvents such as water, however. Nonpolar solutes do dissolve in nonpolar solvents.

We now turn our attention to the quantitative aspects of solutions. In laboratory situations it is often necessary to know the amount of solute in a certain mass or volume of solution, which is known as the concentration of the solute. Concentration can be expressed in several ways with each way being useful for a certain purpose. The **percent by mass** of solute is simply

$$\frac{\text{mass of solute}}{\text{mass of solution}} \times 100\% = \text{percent by mass}$$

Even smaller units of concentration are obtained by using **parts per million (ppm)** or even **parts per billion (ppb)**. In ppm, the ratio of mass of solute to solution above is multiplied by 10^6 and in ppb, the ratio is multiplied by 10^9.

Molarity (M) is the most commonly used unit of concentration and is defined as:

$$M = \frac{\text{number of moles of solute (n)}}{\text{liters of solution (V)}}$$

Knowledge of two of the three variables of molarity allows calculation of the third, as illustrated by the following example.

Example A-1 Conversion of Molarity to Mass

What mass of K_2CO_3 is dissolved in 350 mL of a 0.455 M solution?

PROCEDURE

From V and M we can calculate the moles of K_2CO_3. With the molar mass, we can convert moles to mass.

SOLUTION

$$M = \frac{n}{V} \qquad n = M \times V = 0.455 \text{ mol/L} \times 0.350 \text{ L} = 0.159 \text{ mol } K_2CO_3$$

$$0.159 \text{ \sout{mol K}}_2\text{\sout{CO}}_3 \times \frac{138.2 \text{ g } K_2CO_3}{\text{\sout{mol K}}_2\text{\sout{CO}}_3} = \underline{22.0 \text{ g } K_2CO_3}$$

A common laboratory procedure is the **dilution** of a concentrated solution. Addition of solvent lowers the molarity of a solution according to the following equation:

$$M_{con} \times V_{con} = M_{dil} \times V_{dil}$$

An example of a dilution problem follows.

Example A-2 Molarity of a Diluted Solution

If 625 mL of a 0.837 M solution of HCl is diluted to 2.75 L, what is the molarity of the dilute solution?

PROCEDURE

Solve the dilution equation for the molarity of the dilute solution.

$$M_{dil} = \frac{M_{con} \times V_{con}}{V_{dil}}$$

Convert 625 mL to 0.625 L

SOLUTION

$$M_{dil} = \frac{0.837 \text{ mol/L} \times 0.625 \cancel{\text{L}}}{2.75 \cancel{\text{L}}} = \underline{0.190 \text{ mol/L}}$$

Since molarity and volume relate to moles, the general procedure for stoichiometry problems discussed in Chapter 8 and 9 can be expanded to include solutions.

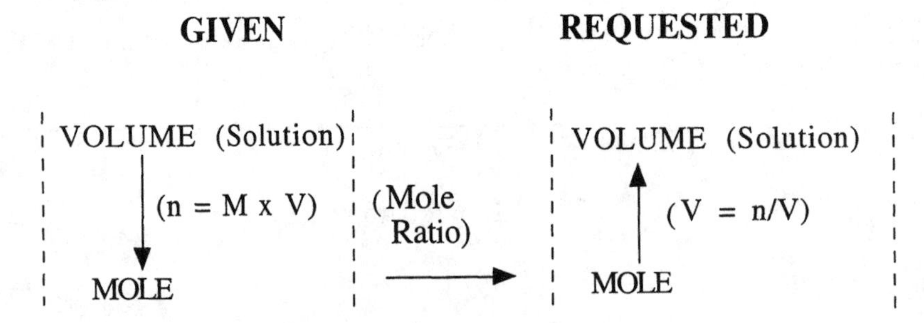

A sample problem involving stoichiometry and solutions is as follows.

Example A-3 Stoichiometry and Molarity

Given the following balanced equation:

$$2AgNO_3(aq) + Na_2CrO_4(aq) \rightarrow Ag_2CrO_4(s) + 2NaNO_3(aq)$$

If 214 mL of a 0.182 M $AgNO_3$ solution is added to a solution containing excess sodium chromate, what mass of silver chromate precipitates?

PROCEDURE

1. Find the moles of $AgNO_3$ from M and V.
2. Convert moles of $AgNO_3$ to moles of Ag_2CrO_4.
3. Convert moles of Ag_2CrO_4 to mass of Ag_2CrO_4.

SOLUTION

$$0.214\,\cancel{L} \times \overset{1}{\frac{0.182\ \cancel{mol\ AgNO_3}}{\cancel{L}}} \times \overset{2}{\frac{1\ \cancel{mol\ Ag_2CrO_4}}{2\ \cancel{mol\ AgNO_3}}} \times \overset{3}{\frac{331.8\ g\ Ag_2CrO_4}{\cancel{mol\ Ag_2CrO_4}}}$$

$$= \underline{6.46\ g\ Ag_2CrO_4}$$

NEW TERMS

Concentration
Dilution
Immiscible
Ion-dipole force
Miscible
Molarity
Parts per billion (ppb)
Parts per million (ppm)
Percent by mass
Recrystallization
Saturated
Supersaturated
Unsaturated

SELF-TEST

A-1 Multiple Choice

____ 1. What are the electrostatic forces between solute and solvent when an ionic compound dissolves in water?

(a) ion-ion
(b) ion-dipole
(c) dipole-dipole
(d) none of these

____ 2. Which of the following statements is false?

(a) Ion-ion forces refer to the forces holding an ionic compound together.
(b) Only ionic compounds dissolve in water.
(c) Most compounds are more soluble in water at a higher temperature.
(d) An ion in aqueous solution is hydrated.

____ 3. When one mole of potassium sulfate dissolves in water, which of the following are present?

(a) 2 mol K^+
(b) 1 mol K^{2+}
(c) 2 mol SO_4^{2-}
(d) 1 mol K^+
(e) 2 mol K^{2+}

____ 4. When one liquid dissolves in another, we say that the two liquids

(a) form a supersaturated solution.
(b) are immiscible.
(c) are both solvents.
(d) form an unsaturated solution.
(e) are miscible.

____ 5. An ionic compound has a solubility of 20 g/100 g H_2O. If 12 g of solute is present in 67 g of water, the solution is

(a) saturated
(b) supersaturated
(c) unsaturated
(d) heterogeneous

A-2 Problems

1. A solution of K_2SO_3 contains 0.0441 g of K_2SO_3 dissolved in 1.00 g of of H_2O. What is the percent by mass of K_2SO_3?

2. A solution is 12.0% by mass alcohol (C_2H_6O). What mass of alcohol is dissolved in each gram of water?

3. Calculate the molarity of the following:

(a) 0.117 mol of HNO_3 in 0.644 L of solution.

(b) 126 g of H_3PO_4 in 1.45 L of solution.

(c) 8.79 x 10^{22} molecules of H_2SO_3 in 450 mL of solution.

4. What volume of 0.335 M HNO_3 is needed to provide 18.0 g of HNO_3?

5. A solution of NaOH has a density of 1.23 g/mL and is 10.0% by mass NaOH. What is the molarity of the solution?

6. If 500 mL of a 0.330 M solution of NaOH is required, what volume of 6.00 M NaOH is needed to dilute with water?

7. When 10.0 mL of 2.00 M HCl is diluted to 115 mL, what is the molarity of the dilute solution?

8. Given the following balanced equation:

$$2Na_3PO_4(aq) + 3Ca(NO_3)_2(aq) \rightarrow Ca_3(PO_4)_2(s) + 6NaNO_3(aq)$$

(a) What volume of 0.662 M Na_3PO_4 completely reacts with 450 mL of 0.752 M $Ca(NO_3)_2$?

(b) What is the molarity of a $Ca(NO_3)_2$ solution if 8500 mL of the solution produced 230 g of $Ca_3(PO_4)_2$?

9. Given the balanced equation:

$$CaCO_3(s) + 2HCl(aq) \rightarrow CO_2(g) + CaCl_2(aq) + H_2O$$

What volume of CO_2 gas measured at 25 °C and 0.945 atm is produced from the complete reaction of 1.27 L of 0.125 M HCl?

Review Section B *The Effects of the Presence of Solutes on Water*

OUTLINE	OBJECTIVES
11-7 The Physical Properties of Solutions	
1. Conductors and nonconductors of electricity	
2. The nature of nonelectrolytes, strong electrolytes, and weak electrolytes	*Explain how specified compounds affect the conductivity of water.*
3. Vapor pressure lowering by a nonvolatile solute	
4. Boiling point elevation	

5. Freezing point lowering — *Describe how the vapor pressure, freezing point, and boiling point are affected by the presence of a solute.*

6. Calculation of molality, boiling point elevation, and freezing point lowering — *Calculate the actual freezing point or boiling point of a solution using molality.*

7. Osmotic pressure — *Describe how solutes affect the movement of solvent through a membrane.*

8. The effect of electrolytes on colligative properties — *Calculate the freezing point or boiling points of solutions of electrolytes.*

SUMMARY OF SECTION 11-7

Questions: *When does water become a conductor of electricity? How do the properties of solutions differ from those of pure solvents?*

How can a clear solution be distinguished from the pure solvent? In many cases they appear identical to the naked eye. The answer is that the solution has many physical properties that distinguish it from the solvent. One difference involves the conduction of electricity. Although water itself is a **nonconductor** of electricity, a solution may or may not be a **conductor** depending on the nature of the solute in aqueous solution. It is the presence of ions in solution that allows water to become a conductor. Polar covalent compounds that dissolve in water without ion formation are known as **nonelectrolytes**. Compounds that produce ions are known as **electrolytes**. Ionic compounds and some polar covalent compounds (i.e., strong acids) are almost completely dissociated into ions in solution and are known as **strong electrolytes**. Some polar covalent compounds produce only a limited concentration of ions, and these are known as **weak electrolytes**.

The properties of a solution differ from a solvent in other ways. For example, the presence of a nonvolatile solute causes **vapor pressure lowering**, **boiling point elevation,** and **freezing point lowering** of the solution from the levels of the pure solvent. The magnitude of these effects depends on the amount of solute present in a given amount of solvent and not on the identity of the solute particles. Such a property is known as a **colligative property**.

The magnitude of the boiling point elevation (ΔT_b) and freezing point lowering (ΔT_f) are given by the equations

$$\Delta T_b = K_b m \qquad \Delta T_f = K_f m$$

where K_b and K_f are constants characteristic of the solvent and "m" is the **molality**. Molality is a unit of concentration defined as

$$m = \frac{\text{moles of solute}}{\text{kg of solvent}}$$

Freezing point lowering and boiling point elevation are often used in chemistry laboratories to determine the molar mass of an unknown pure compound. This procedure is illustrated by the following example.

Example B-1 Molar Mass by Freezing Point Lowering

A pure compound dissolves in water and is found to be a nonelectrolyte. When 50.0 g of this compound is dissolved in 473 g of water, the solution freezes at -2.13°C. What is the molar mass of the compound? For H_2O, $K_f = 1.86°C \cdot kg/mol$.

PROCEDURE

Calculate the value of ΔT and solve the equation for molality to obtain the molar mass (M.M.)

$$m = \frac{\text{mol Solute}}{\text{kg solvent}} = \frac{\frac{\text{g solute}}{\text{M.M.}}}{\frac{\text{g solvent}}{1000\ \text{g/kg}}}$$

Solving for M.M. we get $\quad \text{M.M.} = \dfrac{1000\ \text{g/kg x g solute}}{\text{g solvent x m}}$

SOLUTION

Since the freezing point of pure water is 0.00°C,

$\Delta T_f = 0.00 - (-2.13) = 2.13°C$ degrees

$$\Delta T_f = K_f m \qquad m = \frac{\Delta T}{K_f} = \frac{2.13\ \cancel{°C}}{1.86\ \frac{\cancel{°C} \cdot kg}{mol}} = 1.15\ \text{mol/kg}$$

$$\text{M.M.} = \frac{1000\ \text{g/}\cancel{\text{kg}}\ \text{x}\ 50.0\ \cancel{\text{g}}}{473\ \cancel{\text{g}}\ \text{x}\ 1.15\ \text{mol/}\cancel{\text{kg}}} = \underline{91.9\ \text{g/mol}}$$

Osmotic pressure, another colligative property, is important in many life processes. The process of **osmosis** concerns the unequal passage of solvent molecules through a semipermeable membrane separating solutions of different concentrations. We also discussed the effect of electrolytes on colligative properties. Since one mole of a solute such as NaCl produces two moles of particles (ions), the effect on colligative properties is about twice as much as the effect of one mole of a nonelectrolyte.

NEW TERMS

- Boiling point elevation
- Colligative property
- Conductor
- Electrolyte
- Freezing point lowering
- Molality
- Nonconductor
- Nonelectrolyte
- Osmosis
- Osmotic pressure
- Strong electrolyte
- Vapor pressure lowering
- Weak electrolyte

(3)

$$V = \frac{nRT}{P} = \frac{0.0794 \text{ mol} \times 0.0821 \frac{L \cdot atm}{K \cdot mol} \times 298 \text{ K}}{0.945 \text{ atm}} = \underline{2.06 \text{ L}}$$

B-1 Multiple Choice

1. **c** conduction of electricity (solute may be a nonelectrolyte)

2. **a** mass of solvent

3. **c** 100.512°C

4. **b** concentrating a dilute solution (Choices c and d would result when a solution is diluted.

5. **d** Li_2CO_3 (produces three moles of ions)

B-2 Problems

1. $m = \frac{\text{mol solute}}{\text{kg solvent}}$ $\qquad$ $\text{mol solute} = \frac{6.50 \text{ g}}{92.0 \text{ g/mol}} = 0.707 \text{ mol}$

$\text{kg solvent} = \frac{76.0 \text{ g}}{1000 \text{ g/kg}} = 0.0760 \text{ kg}$

$m = 0.0707 \text{ mol}/0.760 \text{ kg} = \underline{0.930 \text{ mol/kg}}$

2. $\Delta T_f = K_f m = 1.86^{\circ}C \cdot \text{kg/mol} \times 0.930 \text{ mol/kg} = 1.73^{\circ}C.$

Freezing point = $0.00^{\circ}C - 1.73^{\circ}C = \underline{-1.73^{\circ}C}$

$\Delta T_b = K_b m = 0.512^{\circ}C \cdot \text{kg/mol} \times 0.930 \text{ mol/kg} = 0.476^{\circ}C$

Boiling point = $100.000^{\circ}C + 0.476^{\circ}C = \underline{100.476^{\circ}C}$

C-1 Problems

1. $\Delta T_f = K_f m \quad m = \frac{\Delta T}{K_f} = \frac{2.98^{\circ}C}{1.86^{\circ}C \cdot \text{kg/mol}} = 1.60 \text{ mol/kg}$

Use the equation from the sample problem.

$$\text{M.M.} = \frac{1000 \text{ g/kg} \times \text{g solute}}{\text{g solvent} \times m} = \frac{1000 \text{ g/kg} \times 40.0 \text{ g}}{500 \text{ g} \times 1.60 \text{ mol/kg}} = 50.0 \text{ g/mol}$$

Since there are two ions per mole, the measured molar mass is the average of the ions. Twice this value is therefore the molar mass of the compound in solution. A molar mass of 100 g/mol corresponds to the compound BX. Therefore, AY precipitates.

SELF-TEST

B-1 Multiple Choice

____ 1. Which of the following properties of a solvent is not necessarily changed by the presence of a solute?

(a) boiling point
(b) freezing point
(c) conduction of electricity
(d) vapor pressure

____ 2. Given the mass of a certain solute, what other quantity is needed to calculate the molality?

(a) mass of solvent
(b) mass of solution
(c) volume of solution
(d) molar mass of solvent

____ 3. If the freezing point of an aqueous solution is -1.86°C, what is the normal boiling point? (For H_2O, $K_b = 0.512°C \cdot$ kg/mol.)

(a) 98.14°C (b) 101.8°C (c) 100.512°C (d) 99.488°C

____ 4. Which of the following can be accomplished by reverse osmosis?

(a) dilute a concentrated solution
(b) concentrate a dilute solution
(c) increase the freezing point of a solution
(d) decrease the boiling point of a solution

____ 5. Which of the following solutes is the most effective (per mole) in raising the osmotic pressure of water?

(a) KCl
(b) $C_3H_8O_3$ (a nonelectrolyte)
(c) HBr
(d) Li_2CO_3

B-2 Problems

1. What is the molality of a solution made by dissolving 6.50 g of glycerol in 76.0 g of water? Glycerol is a nonvolatile nonelectrolyte with the formula $C_3H_8O_3$.

2. What is the freezing point and boiling point of the solution in problem 1?

Review Section C

CHAPTER SUMMARY SELF-TEST

C-1 Problems

Two hypothetical ionic compounds with the formulas AX_2 and B_2Y are soluble in water. When solutions of these two compounds are mixed in stoichiometric amounts, however, a precipitate forms. The solution after filtration of the precipitate is found to contain 40.0 g of solute dissolved in 500 g of water. The freezing point of this solution was found to be -2.98°C. The atomic masses of the elements are as follows: A = 20, B = 30, X = 70, and Y = 40.

1. What is the precipitate, AY or BX? (Remember that the solution contains an electrolyte consisting of two ions. Therefore, the calculated molar mass must be multiplied by two to get the molar mass of the compound.)

2. The charge on the A ion is +2 and on the Y ion is -2. Write the molecular, total ionic, and net ionic equation illustrating the reaction that occurred.

3. What is the molarity of the solution containing the dissolved compound after the precipitate is removed? The density of the solution is 1.06 g/mL.

4. If 325 mL of an aqueous solution of AX_2 was originally mixed with a solution of B_2Y, what was the molarity of the original B_2Y solution? (Use the volume of the solution after mixing that was calculated in problem 3. Assume that the two solutions are mixed in exactly stoichiometric amounts.)

Answers to Self-Tests

A-1 Multiple Choice

1. **b** ion-dipole
2. **b** Only ionic compounds dissolve in water.
3. **a** 2 mol K^+ (and 1 mol SO_4^{2-})
4. **e** are miscible
5. **c** 20.0 g solute/100 g H_2O x 67 g H_2O = 13.4 g solute. If 12 g is present, the solution is unsaturated.

A-2 Problems

1. Mass of solution = mass of solute + mass of solvent = 0.0441 + 1.00 = 1.04 g

$$\frac{0.044 \text{ g}}{1.04 \text{ g}} \times 100\% = \underline{4.24\% \text{ solute}}$$

2. Let X = mass of solute: therefore the mass of solution is 1.00 + X

$$\frac{X}{1.00 + X} \times 100\% = 12.0\% \qquad X = \underline{0.136 \text{ g}}$$

3. (a) $M = \frac{n}{V} = \frac{0.117 \text{ mol}}{0.664 \text{ L}} = \underline{0.182 \text{ mol/L}}$

(b) $126 \text{ \sout{g H}}_3\text{\sout{PO}}_4 \times \frac{1 \text{ mol}}{97.99 \text{ \sout{g H}}_3\text{\sout{PO}}_4} = 1.29 \text{ mol} \quad M = \frac{1.29 \text{ mol}}{1.45 \text{ L}} = \underline{0.890 \text{ mol/L}}$

(c) $8.79 \times 10^{22} \text{ \sout{molecules}} \times \frac{1 \text{ mol}}{6.022 \times 10^{23} \text{ \sout{molecules}}} = 0.146 \text{ mol}$

$$M = \frac{0.146 \text{ mol}}{0.450 \text{ L}} = \underline{0.324 \text{ mol/L}}$$

4. First find moles of HNO_3: $18.0 \text{ \sout{g HNO}}_3 \times \frac{1 \text{ mol}}{63.02 \text{ \sout{g HNO}}_3} = 0.286 \text{ mol}$

Find requested volume: $M = \frac{n}{V} \quad V = \frac{n}{M} = \frac{0.286 \text{ mol}}{0.335 \text{ mol/L}} = \underline{0.854 \text{ L}}$

5. Assume exactly one liter (1000 mL) of solution.

mass of solution: 1000 mL x 1.23 g/mL = 1230 g

mass of NaOH in solution: 1230 g x 0.100 = 123 g

mol NaOH in solution: $123 \text{ g NaOH} \times \dfrac{1 \text{ mol}}{40.00 \text{ g NaOH}} = 3.08 \text{ mol}$

$$M = \frac{n}{V} = \frac{3.08 \text{ mol}}{1.00 \text{ L}} = \underline{3.08 \text{ mol/L}}$$

6. $$V_{con} = \frac{M_{dil} \times V_{dil}}{M_{con}} = \frac{500 \text{ mL} \times 0.330 \text{ M}}{6.00 \text{ M}} = \underline{27.5 \text{ mL}}$$

7. $$M_{dil} = \frac{M_{con} \times V_{con}}{V_{dil}} = \frac{2.00 \text{ M} \times 10.0 \text{ mL}}{115 \text{ mL}} = \underline{0.174 \text{ mol/L}}$$

8. (a) (1) Convert M and V of $Ca(NO_3)_2$ to moles $Ca(NO_3)_2$.
(2) Convert moles of $Ca(NO_3)_2$ to moles of Na_3PO_4.

$$\overset{(1)}{0.450 \text{ L} \times \frac{0.752 \text{ mol } Ca(NO_3)_2}{\text{L}}} \times \overset{(2)}{\frac{2 \text{ mol } Na_3PO_4}{3 \text{ mol } Ca(NO_3)_2}} = 0.226 \text{ mol } Na_3PO_4$$

(3)

$$V = \frac{n}{M} = \frac{0.226 \text{ mol}}{0.662 \text{ mol/L}} = \underline{0.341 \text{ L}}$$

(b) (1) Convert mass of $Ca_3(PO_4)_2$ to moles of $Ca_3(PO_4)_2$.
(2) Convert moles of $Ca_3(PO_4)_2$ to moles of $Ca(NO_3)_2$.
(3) Convert n and V of $Ca(NO_3)_2$ to M of $Ca(NO_3)_2$.

$$230 \text{ g } Ca_3(PO_4)_2 \times \overset{(1)}{\frac{1 \text{ mol } Ca_3(PO_4)_2}{310.2 \text{ g } Ca_3(PO_4)_2}} \times \overset{(2)}{\frac{3 \text{ mol } Ca(NO_3)_2}{1 \text{ mol } Ca_3(PO_4)_2}}$$
$$= 2.22 \text{ mol } Ca(NO_3)_2$$

(3)

$$M = \frac{n}{V} = \frac{2.22 \text{ mol}}{8.50 \text{ L}} = \underline{0.261 \text{ mol/L}}$$

9. (1) Convert V and M of HCl to mol of HCl.
(2) Convert mol of HCl to mol of CO_2.
(3) Convert n, T, and P of CO_2 to V of CO_2 using Ideal Gas Law.

$$1.27 \text{ L} \times \overset{(1)}{\frac{0.125 \text{ mol HCl}}{\text{L}}} \times \overset{(2)}{\frac{1 \text{ mol } CO_2}{2 \text{ mol HCl}}} = 0.0794 \text{ mol } CO_2.$$

(3)

$$V = \frac{nRT}{P} = \frac{0.0794 \text{ mol} \times 0.0821 \frac{\text{L} \cdot \text{atm}}{\text{K} \cdot \text{mol}} \times 298 \text{ K}}{0.945 \text{ atm}} = \underline{2.06 \text{ L}}$$

B-1 Multiple Choice

1. **c** conduction of electricity (solute may be a nonelectrolyte)
2. **a** mass of solvent
3. **c** 100.512°C
4. **b** concentrating a dilute solution (Choices c and d would result when a solution is diluted.
5. **d** Li_2CO_3 (produces three moles of ions)

B-2 Problems

1. $m = \frac{\text{mol solute}}{\text{kg solvent}}$

$$\text{mol solute} = \frac{6.50 \text{ g}}{92.0 \text{ g/mol}} = 0.707 \text{ mol}$$

$$\text{kg solvent} = \frac{76.0 \text{ g}}{1000 \text{ g/kg}} = 0.0760 \text{ kg}$$

m = 0.0707 mol/0.760 kg = <u>0.930 mol/kg</u>

2. $\Delta T_f = K_f m$ = 1.86°C · kg/mol x 0.930 mol/kg = 1.73°C.

Freezing point = 0.00°C - 1.73°C = <u>-1.73°C.</u>

$\Delta T_b = K_b m$ = 0.512°C · kg/mol x 0.930 mol/kg = 0.476°C

Boiling point = 100.000°C + 0.476°C = <u>100.476°C</u>

C-1 Problems

1. $\Delta T_f = K_f m \quad m = \frac{\Delta T}{K_f} = \frac{2.98 \ ^\circ\text{C}}{1.86 \ ^\circ\text{C} \cdot \text{kg/mol}} = 1.60 \text{ mol/kg}$

Use the equation from the sample problem.

$$\text{M.M.} = \frac{1000 \text{ g/kg} \times \text{g solute}}{\text{g solvent} \times m} = \frac{1000 \text{ g/kg} \times 40.0 \text{ g}}{500 \text{ g} \times 1.60 \text{ mol/kg}} = 50.0 \text{ g/mol}$$

Since there are two ions per mole, the measured molar mass is the average of the ions. Twice this value is therefore the molar mass of the compound in solution. A molar mass of 100 g/mol corresponds to the compound BX. Therefore, <u>AY precipitates</u>.

2. Molecular: $AX_2(aq) + B_2Y(aq) \rightarrow AY(s) + 2BX(aq)$

Total ionic: $A^{2+}(aq) + 2X^-(aq) + 2B^+(aq) + Y^{2-}(aq) \rightarrow$

$AY(s) + 2B^+(aq) + 2X^-(aq)$

Net ionic: $A^{2+}(aq) + Y^{2-}(aq) \rightarrow AY(s)$

3. Convert the total mass of the solution (500 g + 40.0 g = 540 g) into the equivalent volume.

$$540\ \text{g} \times \frac{1\ \text{mL}}{1.06\ \text{g}} = 509\ \text{mL} = 0.509\ \text{L} \quad \text{moles BX} = \frac{40.0\ \text{g}}{100\ \text{g/mol}} = 0.400\ \text{mol}$$

$M = n/V = 0.400\ \text{mol}/0.509\ \text{L} = \underline{0.786\ \text{mol/L}}$

4. Use the balanced molecular equation from (2) to calculate moles of B_2Y.

$$\text{g BX} \rightarrow \text{mol BX} \rightarrow \text{mol } B_2Y$$

$$40.0\ \text{g BX} \times \frac{1\ \text{mol BX}}{100\ \text{g BX}} \times \frac{1\ \text{mol } B_2Y}{2\ \text{mol BX}} = 0.200\ \text{mol } B_2Y$$

Volume of the B_2Y solution = 509 mL - 325 mL = 184 mL = 0.184 L

$M = n/V = 0.200\ \text{mol}/0.184\ \text{L} = \underline{1.09\ \text{mol/L}}$

Solutions to Black Text Problems

11-12 A specific amount of Li_2SO_4 will precipitate unless the solution becomes supersaturated.

$$(500\ \text{g } H_2O \times \frac{35\ \text{g}}{100\ \text{g } H_2O}) - (500\ \text{g } H_2O \times \frac{28\ \text{g}}{100\ \text{g } H_2O}) = 35\ \text{g } Li_2SO_4\ \text{(precipitate)}$$

11-13 Mass of solution = 650 g + 9.85 g = 660 g

$$\frac{9.85\ \text{g}}{660\ \text{g}} \times 100\% = \underline{1.49\%}$$

11-15 $150\ \text{g solution} \times \frac{10.0\ \text{g NaOH}}{100\ \text{g solution}} = 15.0\ \text{g NaOH}$

$$15.0\ \text{g NaOH} \times \frac{1\ \text{mol NaOH}}{40.00\ \text{g NaOH}} = \underline{0.375\ \text{mol NaOH}}$$

11-17 Solution is 23.2% KNO_3 and 76.8% H_2O

Let Y = Mass of the solution, then 0.768 x Y = 100 g Y = 130 g

Mass of solute = 130 g - 100 g = $\underline{30\ \text{g } KNO_3}$

11-18 M.M. (C_2H_5OH) = 46.07 g/mol   M.M. NaOH = 40.00 g/mol

Mass of solution = 40.00 g + (9 x 46.07 g) = 454.6 g

$\frac{40.00 \text{ g}}{454.6 \text{ g}} \times 100\% = \underline{8.799\% \text{ NaOH}}$

11-19 $\frac{10 \times 10^{-3} \cancel{\text{g}}}{100 \cancel{\text{g}}} \times 10^6 \text{ ppm} = \underline{100 \text{ ppm}}$

11-21 $\frac{1.00 \text{ g gold}}{x \text{ g sea water}} \times 10^9 \text{ ppb} = 1.2 \times 10^{-2} \text{ ppb}$

$x = 8.3 \times 10^{10} \cancel{\text{g sea water}} \times \frac{1 \cancel{\text{mL}}}{1.0 \cancel{\text{g}}} \times \frac{1 \text{ L}}{10^3 \cancel{\text{mL}}} = \underline{8.3 \times 10^7 \text{ L}}$

11-23 $\frac{2.44 \text{ mol}}{4.50 \text{ L}} = \underline{0.542 \text{ M}}$

11-24 (a) $\frac{n.}{V} = \frac{2.40 \text{ mol}}{2.75 \text{ L}} = \underline{0.873\text{M}}$

(b) $26.5 \cancel{\text{g}} \times \frac{1 \text{ mol}}{46.07 \cancel{\text{g}}} = 0.575 \text{ mol}$ $\frac{n}{V} = \frac{0.575 \text{ mol}}{0.410 \text{ L}} = \underline{1.40 \text{ M}}$

(c) $V = \frac{n}{M} = \frac{3.15 \cancel{\text{mol}}}{0.255 \cancel{\text{mol}}/\text{L}} = \underline{12.4 \text{ L}}$

(d) $n = M \times V = 0.625 \text{ mol}/\cancel{\text{L}} \times 1.25 \cancel{\text{L}} = 0.781 \text{ mol}$

$0.781 \cancel{\text{mol}} \times \frac{52.95 \text{ g}}{\cancel{\text{mol}}} = \underline{41.4 \text{ g}}$

(e) $M = \frac{n}{V} = \frac{0.250 \text{ mol}}{0.850 \text{ L}} = \underline{0.294 \text{ M}}$

(f) $n = M \times V = 0.054 \text{ mol}/\cancel{\text{L}} \times 0.45 \cancel{\text{L}} = \underline{0.024 \text{ mol}}$

(g) $14.7 \cancel{\text{g}} \times \frac{1 \text{ mol}}{138.2 \cancel{\text{g}}} = 0.106 \text{ mol};$ $V = \frac{0.106 \cancel{\text{mol}}}{0.345 \cancel{\text{mol}}/\text{L}} = 0.310 \text{ L} = \underline{307 \text{ mL}}$

(h) $n = 1.24 \text{ mol}/\cancel{\text{L}} \times 1.65 \cancel{\text{L}} = 2.05 \text{ mol}$ $2.05 \cancel{\text{mol}} \times \frac{23.95 \text{ g}}{\cancel{\text{mol}}} = \underline{49.1 \text{ g}}$

(i) $0.178 \cancel{\text{g}} \times \frac{1 \text{ mol}}{98.09 \cancel{\text{g}}} = 1.81 \times 10^{-3} \text{ mol}$

$V = \frac{n}{M} = \frac{1.81 \times 10^{-3} \cancel{\text{mol}}}{0.905 \cancel{\text{mol}}/\text{L}} = 0.00200 \text{ L} = \underline{2.00 \text{ mL}}$

11-28 $2.50 \times 10^{-4} \cancel{\text{g NaHCO}_3} \times \frac{1 \text{ mol NaHCO}_3}{84.01 \cancel{\text{g NaHCO}_3}} = 2.98 \times 10^{-6} \text{ mol}$

$\frac{2.98 \times 10^{-6} \text{ mol}}{2.54 \times 10^{-3} \text{ L}} = 1.17 \times 10^{-3} \text{ M}$

11-29 $13.5 \text{ g Ba(OH)}_2 \times \frac{1 \text{ mol Ba(OH)}_2}{171.3 \text{ g Ba(OH)}_2} = 0.0788 \text{ mol Ba(OH)}_2$

$\frac{0.0788 \text{ mol}}{0.475 \text{ L}} = 0.166 \text{ M [Ba(OH)}_2]$

$Ba(OH)_2 \rightarrow Ba^{2+}(aq) + 2OH^-(aq)$

$0.166 \text{ M Ba(OH)}_2 \times \frac{1 \text{ M Ba}^{2+}}{1 \text{ M Ba(OH)}_2} = \underline{0.166 \text{ M Ba}^{2+}}$

$0.166 \text{ M Ba(OH)}_2 \times \frac{2 \text{ M OH}^-}{1 \text{ M Ba(OH)}_2} = \underline{0.332 \text{ M OH}^-}$

11-31 Assume one liter (1000 mL) of solution $\quad 1000 \text{ mL} \times 1.21 \text{ g/mL} = 1210 \text{ g solution}$

$1210 \text{ g} \times \frac{25 \text{ g solute}}{100 \text{ g solution}} = 302.5 \text{ g solute} \quad 302.5 \text{ g} \times \frac{1 \text{ mol}}{164.1 \text{ g}} = 1.84 \text{ mol}$

$\frac{n}{V} = \frac{1.84 \text{ mol}}{1.00 \text{ L}} = \underline{1.84 \text{ M}}$

11-33 Assume one liter (1000 mL) of solution

$n = M \times V = 14.7 \text{ mol/L} \times 1.00 \text{ L} = 14.7 \text{ mol solute} \quad 14.7 \text{ mol} \times \frac{63.02 \text{ g}}{\text{mol}} = 926 \text{ g solute}$

Since 70% of the mass of the solution is solute,

$\text{mass of solution} \times \frac{70 \text{ g solute}}{100 \text{ g solution}} = 926 \text{ g solute} \quad \text{mass of solution} = 1320 \text{ g}$

$\text{Density} = \frac{1320 \text{ g}}{1000 \text{ mL}} = \underline{1.32 \text{ g/mL}}$

11-34 $M_d \times V_d = M_c \times V_c \quad V_c = \frac{1.50 \text{ M} \times 2.50 \text{ L}}{4.50 \text{ M}} = \underline{0.833 \text{ L}}$

11-36 First, find the volume of concentrated NaOH needed.

$Vc = \frac{1.00 \text{ L} \times 0.250 \text{ mol/L}}{0.800 \text{ mol/L}} = 0.313 \text{ L (313 mL)}$

Slowly add 313 mL of the 0.800 M NaOH to about 500 mL of water in a 1-L volumetric flask. Dilute to the 1-L mark with water.

11-38 $M_d = \frac{0.200 \text{ M} \times 3.50 \text{ L}}{5.00 \text{ L}} = \underline{0.140 \text{ M}}$

11-39 $V_d = \frac{M_c \times V_c}{M_d} = \frac{0.860 \text{ M} \times 1.25 \text{ L}}{0.545 \text{ M}} = 1.97 \text{ L} \quad 1.97 \text{ L} - 1.25 \text{ L} = 0.72 \text{ L} = \underline{720 \text{ mL}}$

11-41 **L → mol → g → mL**

$0.250 \text{ L} \times \frac{0.200 \text{ mol}}{\text{L}} \times \frac{60.05 \text{ g}}{\text{mol}} \times \frac{1.00 \text{ mL}}{1.05 \text{ g}} = \underline{2.86 \text{ mL}}$

11-42 Find the total moles and the total volume

$$n = V \times M = 0.150\ \cancel{L} \times 0.250\ \text{mol}/\cancel{L} = 0.0375\ \text{mol (solution 1)}$$

$$n = V \times M = 0.450\ \cancel{L} \times 0.375\ \text{mol}/\cancel{L} = 0.169\ \text{mol (solution 2)}$$

$V_T = 0.600$ L; $n_T = 0.206$ mol

$$\frac{n}{V} = \frac{0.206\ \text{mol}}{0.600\ \text{L}} = \underline{0.343\ \text{M}}$$

11-43 vol. KOH → mol KOH → mol $Cr(OH)_3$ → g $Cr(OH)_3$

$$0.500\ \cancel{L} \times \frac{0.250\ \cancel{\text{mol KOH}}}{\cancel{L}} \times \frac{1\ \cancel{\text{mol Cr(OH)}_3}}{3\ \cancel{\text{mol KOH}}} \times \frac{103.0\ \text{g Cr(OH)}_3}{\cancel{\text{mol Cr(OH)}_3}} = \underline{4.29\ \text{g Cr(OH)}_3}$$

11-45 vol. $Al_2(SO_4)_3$ → mol $Al_2(SO_4)_3$ → mol $BaSO_4$ → g $BaSO_4$

$$0.650\ \cancel{L} \times \frac{0.320\ \cancel{\text{mol Al}_2\text{(SO}_4)_3}}{\cancel{L}} \times \frac{3\ \cancel{\text{mol BaSO}_4}}{1\ \cancel{\text{mol Al}_2\text{(SO}_4)_3}} \times \frac{233.4\ \text{g BaSO}_4}{\cancel{\text{mol BaSO}_4}} = \underline{146\ \text{g BaSO}_4}$$

11-46 g $Al(OH)_3$ → mol $Al(OH)_3$ → mol $Ba(OH)_2$ → vol. $Ba(OH)_2$

$$265\ \cancel{\text{g Al(OH)}_3} \times \frac{1\ \cancel{\text{mol Al(OH)}_3}}{78.00\ \cancel{\text{g Al(OH)}_3}} \times \frac{3\ \text{mol Ba(OH)}_2}{2\ \cancel{\text{mol Al(OH)}_3}} = 5.10\ \text{mol Ba(OH)}_2$$

$$V = \frac{n}{M} = \frac{5.10\ \text{mol}}{1.25\ \text{mol/L}} = \underline{4.08\ \text{L}}$$

11-48 vol. $Ca(ClO_3)_2$ → mol $Ca(ClO_3)_2$ → mol Na_3PO_4 → vol. Na_3PO_4

$$0.580\ \cancel{L} \times \frac{3.75\ \cancel{\text{mol Ca(ClO}_3)_2}}{\cancel{L}} \times \frac{2\ \text{mol Na}_3\text{PO}_4}{3\ \cancel{\text{mol Ca(ClO}_3)_2}} = 1.45\ \text{mol Na}_3\text{PO}_4$$

$$V = \frac{n}{M} = \frac{1.45\ \cancel{\text{mol}}}{2.22\ \cancel{\text{mol}}/\text{L}} = \underline{0.653\ \text{L}}$$

11-50 $n(\text{NaOH}) = M \times V = 0.250\ \cancel{L} \times 0.240\ \text{mol}/\cancel{L} = 0.0600$ mol NaOH

$n(\text{MgCl}_2) = 0.400\ \cancel{L} \times 0.100\ \text{mol}/\cancel{L} = 0.0400$ mol $Mg(OH)_2$

$$\text{NaOH: } 0.0600\ \cancel{\text{mol NaOH}} \times \frac{1\ \text{mol Mg(OH)}_2}{2\ \cancel{\text{mol NaOH}}} = 0.0300\ \text{mol Mg(OH)}_2$$

$$\text{MgCl}_2\text{: } 0.0400\ \cancel{\text{mol MgCl}_2} \times \frac{1\ \text{mol Mg(OH)}_2}{1\ \cancel{\text{mol MgCl}_2}} = 0.0400\ \text{mol Mg(OH)}_2$$

Therefore, NaOH is the limiting reactant.

$$0.0300\ \cancel{\text{mol Mg(OH)}_2} \times \frac{58.33\ \text{g Mg(OH)}_2}{\cancel{\text{mol Mg(OH)}_2}} = \underline{1.75\ \text{g Mg(OH)}_2}$$

11-55 $n(\text{NaOH}) = 25.0 \text{ g NaOH} \times \frac{1 \text{ mol NaOH}}{40.00 \text{ g NaOH}} = 0.625 \text{ mol NaOH}$

$250 \text{ g} \times \frac{1 \text{ kg}}{10^3 \text{ g}} = 0.250 \text{ kg} \quad \frac{0.625 \text{ mol}}{0.250 \text{ kg}} = \underline{2.50 \text{ m}}$

The molality is the same in both solvents since the mass of solvent is the same.

11-57 $m = \frac{\text{n (solute)}}{\text{kg (solvent)}} \quad n = m \times \text{kg (solvent)}$

$0.550 \text{ kg} \times \frac{0.720 \text{ mol NaOH}}{\text{kg}} \times \frac{40.00 \text{ g NaOH}}{\text{mol NaOH}} = \underline{15.8 \text{ g NaOH}}$

11-59 $\Delta T_f = K_f m = 1.86 \times 0.20 = 0.37\ ^{\text{o}}\text{C} \quad \text{F.P.} = 0.00\ ^{\text{o}}\text{C} - 0.37\ ^{\text{o}}\text{C} = \underline{-0.37\ ^{\text{o}}\text{C}}$

11-65 Assume 1000 g of solution. There are then 100 g of $CaCl_2$ and 900 g of H_2O in the solution.

$100 \text{ g CaCl}_2 \times \frac{1 \text{ mol CaCl}_2}{111.0 \text{ g CaCl}_2} = 0.901 \text{ mol CaCl}_2 \qquad \frac{0.901 \text{ mol}}{0.900 \text{ kg}} = \underline{1.00 \text{ m}}$

11-67 $m = \Delta T_f/K_f = 5.0/1.86 = 2.7 \quad m = \frac{\text{n (solute)}}{\text{kg (solvent)}} = 2.7 \quad \frac{\text{n}}{5.00 \text{ kg}} = 2.7$

$n = 14 \text{ mol} \quad 14 \text{ mol} \times \frac{62.07 \text{ g}}{\text{mol}} = \underline{870 \text{ g glycol}}$

11-68 $\Delta T_b = K_b m = 0.512 \times 2.7 = 1.4^{\text{o}}\text{C} \quad \text{B.P.} = 100.0^{\text{o}}\text{C} + 1.4^{\text{o}}\text{C} = \underline{101.4^{\text{o}}\text{C}}$

11-70 $101.5^{\text{o}}\text{C} - 100.0^{\text{o}}\text{C} = 1.5^{\text{o}}\text{C} = \Delta T_b \qquad m = \Delta T_b/K_b = 1.5/0.512 = \underline{2.9\ m}$

11-73 $100 \text{ g CH}_3\text{OH} \times \frac{1 \text{ mol CH}_3\text{OH}}{32.04 \text{ g CH}_3\text{OH}} = 3.12 \text{ mol CH}_3\text{OH}$

$\Delta T_f = K_f m = 5.12 \times \frac{3.12 \text{ mol}}{0.800 \text{ kg}} = 20.0^{\text{o}}\text{C} \qquad \text{F.P.} = 5.5^{\text{o}}\text{C} - 20.0^{\text{o}}\text{C} = \underline{-14.5^{\text{o}}\text{C}}$

11-78 (a) $\Delta T_f = K_f m = 1.86 \times \frac{\frac{10.0 \text{ g}}{32.04 \text{ g/mol}}}{0.100 \text{ kg solvent}} = 5.8\ ^{\text{o}}\text{C} \quad \text{F.P.} = 0.0^{\text{o}}\text{C} - 5.8^{\text{o}}\text{C} = \underline{-5.8^{\text{o}}\text{C}}$

(b) $\Delta T_f = K_f m = 1.86 \times \frac{\frac{10.0 \text{ g}}{58.44 \text{ g/mol}}}{0.100 \text{ kg solvent}} = \underline{3.2^{\text{o}}\text{C}}$

$\text{NaCl} \longrightarrow \text{Na}^+(aq) + \text{Cl}^-(aq)$

The two ions per mole of NaCl cause the melting point to be lowered twice as much as one mole of a nonelectrolyte.

$2 \times 3.2^{\text{o}}\text{C} = 6.4^{\text{o}}\text{C} \qquad \text{F.P.} = 0.0^{\text{o}}\text{C} - 6.4^{\text{o}}\text{C} = -6.4^{\text{o}}\text{C}$

(c) $\Delta T_f = K_f m = 1.86 \times \frac{\frac{10.0 \text{ g}}{111.0 \text{ g/mol}}}{0.100 \text{ g solvent}} = \underline{1.68^{\text{o}}\text{C}}$

$CaCl_2 \rightarrow Ca^{2+}(aq) + 2Cl^-(aq)$

The three ions per mole of $CaCl_2$ cause the melting point to be lowered three times as much as one mole of a nonelectrolyte.

$3 \times 1.68^oC = 5.0^oC$ F.P. $= 0.0^oC - 5.0^oC = \underline{-5.0^oC}$

11-81 $150\ \cancel{mL} \times \frac{1.00\ g\ H_2O}{\cancel{mL}} = 150\ g\ H_2O$ mass of solution = 150 g + 10.0 g + 5.0 g = 165 g

$\frac{150\ g}{165\ g} \times 100\% = \underline{90.9\%\ H_2O}$; $\frac{10.0\ g}{165\ g} \times 100\% = \underline{6.06\ \%\ sugar}$

$100\% - (90.9 + 6.06)\% = \underline{3.0\%\ salt}$

11-83 $0.500\ \cancel{L} \times 0.20\ mol/\cancel{L} = 0.10\ mol\ Ag^+$ $0.500\ \cancel{L} \times 0.30\ mol/\cancel{L} = 0.15\ mol\ Cl^-$

0.10 mol of Ag^+ reacts with 0.10 mol of Cl^- to form AgCl leaving 0.05 mol of Cl^- in 1.00 L of solution. $[Cl^-] = 0.050\ mol/1.00\ L = \underline{0.05\ M}$

11-85 $0.225\ \cancel{L} \times \frac{0.196\ \cancel{mol\ HCl}}{\cancel{L}} \times \frac{1\ mol\ M}{2\ \cancel{mol\ HCl}} = 0.0221\ mol\ M$

$\frac{1.44\ g}{0.0221\ mol} = \underline{65.2\ g/mol\ (Zn)}$

11-87 $\Delta T = 5.67 - 2.22 = 3.45^oC$ $m = \frac{\Delta T}{K_f} = \frac{3.45}{8.10} = 0.426\ m$

$$\text{Molar mass} = \frac{1000 \times \text{mass solute}}{m \times \text{mass solution}} = \frac{1000 \times 3.07}{0.426 \times 120} = 60.1\ g/mol$$

C: $40.0\ \cancel{g\ C} \times \frac{1\ mol\ C}{12.01\ \cancel{g\ C}} = 3.33\ mol\ C$ H: $13.3\ \cancel{g\ H} \times \frac{1\ mol\ H}{1.008\ \cancel{g\ H}} = 13.2\ mol\ H$

N: $46.7\ \cancel{g\ N} \times \frac{1\ mol\ N}{14.01\ \cancel{g\ N}} = 3.33\ mol\ N$

H: $\frac{13.2}{3.33} = 4.0$ C & N: $\frac{3.33}{3.33} = 1.0$ empirical formula = CH_4N;

Emp mass = 30 g/emp unit $\frac{30.05\ \cancel{g}/emp\ unit}{60.1\ \cancel{g}/mol} = 2\ emp\ unit/mol$

Molecular formula = $\underline{C_2H_8N_2}$

11-89 0.30 m sugar < 0.12 m KCl (0.24 m ions) < 0.05 m $CrCl_3$ (0.20 m ions) < 0.05 m K_2CO_3 (0.15 m ions) < pure water

11-91 n(hydroxide) = M x V = 0.120 mol/L x 0.487 L = 0.0584 mol

$$n(H_2) = \frac{PV}{RT} = \frac{0.650\ \cancel{atm} \times 1.92\ \cancel{L}}{0.0821 \frac{\cancel{L} \cdot \cancel{atm}}{\cancel{K} \cdot mol} \times 298\ \cancel{K}} = 0.0292 \text{ mol } H_2$$

$$\frac{0.0584}{0.0292} = 2 \text{ (2 mol hydroxide : 1 mol } H_2)$$

Balanced Equations:

$2Na(s) + 2H_2O(l) \rightarrow 2NaOH(aq) + H_2(g)$ (2 mol hydroxide : 1 mol H_2)

$Ca(s) + 2H_2O(l) \rightarrow Ca(OH)_2(aq) + H_2(g)$ (1 mol hydroxide: 1 mol H_2)

The answer is sodium since it produces the correct ratio of hydrogen.

11-92 $NH_3(g) \rightarrow NH_3(aq)$

n(aq) = 0.450 mol/~~L~~ x 0.250 ~~L~~ = 0.113 mol

$$V(gas) = \frac{nRT}{P} = \frac{0.113\ \cancel{mol} \times 0.0821 \frac{L \cdot \cancel{atm}}{\cancel{K} \cdot \cancel{mol}} \times 298\ \cancel{K}}{0.951\ \cancel{atm}} = \underline{2.91\ L}$$

11-94 $NaHCO_3(aq) + HCl(aq) \rightarrow NaCl(aq) + H_2O + CO_2(g)$

n = M x V = 0.340 mol/~~L~~ x 1.00 ~~L~~ = 0.340 mol $NaHCO_3$

$$0.340\ \cancel{mol\ NaHCO_3} \times \frac{1 \text{ mol } CO_2}{1\ \cancel{mol\ NaHCO_3}} = 0.340 \text{ mol } CO_2$$

$$V(CO_2) = \frac{nRT}{P} = \frac{0.340\ \cancel{mol} \times 0.0821 \frac{L \cdot \cancel{atm}}{\cancel{K} \cdot \cancel{mol}} \times 308\ \cancel{K}}{1.00\ \cancel{atm}} = \underline{8.60\ L}$$

11-96 The formula must be PCl_3 with the Lewis structure

$$:\ddot{Cl}-\ddot{P}-\ddot{Cl}: \quad \text{(with } :\ddot{Cl}: \text{ bonded below P)}$$

which would be trigonal pyramidal.

The reaction is $PCl_3(g) + 3H_2O \rightarrow H_3PO_3(aq) + 3HCl(aq)$

$$0.750\,\cancel{L} \times \frac{1 \text{ mol } PCl_3}{22.4\,\cancel{L}} = 0.0335 \text{ mol } PCl_3$$

$$0.0335\,\cancel{\text{mol } PCl_3} \times \frac{3 \text{ mol HCl}}{1\,\cancel{\text{mol } PCl_3}} = 0.101 \text{ mol HCl}$$

$M = n/V = 0.101$ mol/0.250 L = <u>0.404 M HCl</u>

11-98 KCl (K^+Cl^-) $10.0\,\cancel{\text{g KCl}} \times \frac{1 \text{ mol KCl}}{74.55\,\cancel{\text{g KCl}}} = 0.134 \text{ mol KCl}$

$$\Delta T = 1.86 \times \frac{0.134 \text{ mol}}{\frac{100 \text{ g}}{1000 \text{ g/kg}}} \times 2 \text{ mol ions/mol KCl} = 4.98^{o}\text{C change}$$

Na_2S ($2Na^+S^{2-}$) $10.0\,\cancel{\text{g } Na_2S} \times \frac{1 \text{ mol } Na_2S}{78.05\,\cancel{\text{g } Na_2S}} = 0.128 \text{ mol } Na_2S$

$$\Delta T = 1.86 \times \frac{0.128 \text{ mol}}{\frac{100 \text{ g}}{1000 \text{ g/kg}}} \times 3 \text{ mol ions/mol } Na_2S = 7.14^{o}\text{C change}$$

$CaCl_2$ ($Ca^{2+}2Cl^-$) $10.0\,\cancel{\text{g } CaCl_2} \times \frac{1 \text{ mol } CaCl_2}{111.0\,\cancel{\text{g } CaCl_2}} = 0.0901 \text{ mol } CaCl_2$

$$\Delta T = 1.86 \times \frac{0.0901 \text{ mol}}{\frac{100 \text{ g}}{1000 \text{ g/kg}}} \times 3 \text{ mol ions/mol } CaCl_2 = 5.03^{o}\text{C change}$$

The answer is Na_2S since it would have a freezing point of -7.14^{o}C.

11-100 $\Delta T = 1.50°C$ $m = \Delta T/K_f = 1.50/1.86 = 0.806$

$n/1.00$ kg = 0.806 $n = 0.806$ mol   1 L = 1000 ~~mL~~ x $\frac{1.00 \text{ g}}{\text{mL}}$ = 1000 g (1 kg) H_2O

For dilute solution, Let X = kg of added water, then

$\Delta T = K_f m$ $1.15°C = 1.86 \times \frac{0.806}{1.00 + X}$

Solving for X, X = 0.30 kg   0.30 kg = 300 g = <u>300 mL of water added.</u>

12

Acids, Bases, and Salts

Review Section A *Acids, Bases, and the Formation of Salts*

OUTLINE / OBJECTIVES

12-1 Properties of Acids and Bases

1. The common characteristics of acids
2. The common characteristics of bases

Give the names and formulas of some common acids and bases derived from specified anions or cations.

3. The nature of the proton in water: the hydronium ion
4. The dissociation of acids in water

Apply the Arrhenius definition to identify compounds as acids or bases and write equations illustrating this behavior.

12-2 Bronsted-Lowry Acids and Bases

1. Bronsted-Lowry: acids and bases as proton donors and acceptors

Discuss the need for a broader definition.

2. Conjugate acid-base pairs

Determine the conjugate base of a specified acid or the conjugate acid of a specified base.

3. Water as an amphiprotic substance

Write specified proton exchange reactions by application of the Bronsted-Lowry definition.

12-3 Strengths of Acids and Bases

1. Complete ionization: strong acids
2. Partial ionization: weak acids

Distinguish between the behaviors of a strong acid and a weak acid in water.

3. Partial ionization and dynamic equilibrium
4. Strong bases and weak bases

Describe the dynamic equilibrium involved in the partial ionization of a weak acid or a weak base in water.

12-4 Neutralization and Salts

1. Molecular and net ionic equations of neutralization reactions

Write equations for neutralization reactions and identify the salts formed.

2. Ionic equations for strong base weak acid neutralizations

Write total and net ionic equations for the neutralization of a weak acid with a strong base.

3. Neutralization of a polyprotic acid
4. Acid salts from partial neutralization of polyprotic acids

Write equations for the step-wise neutralization of a polyprotic acid by a strong base.

SUMMARY OF SECTIONS 12-1 THROUGH 12-4

Questions: *What are the active ingredients in acids and in bases that cause their characteristic chemical behavior? How can acid-base nature be explained as a proton exchange reaction? Do all acids and all bases act the same? What happens when you mix acids and bases?*

Three ancient yet still important classes of compounds are known as **acids, bases**, and **salts**. This chapter deals exclusively with these compounds and the nature of their reactions in water. Some quantitative considerations of acids and bases will be presented in Chapter 14.

Although these types of compounds have been classified since antiquity, an early but still commonly used definition was advanced by Arrhenius in 1884. Acids are defined as substances that produce $H^+(aq)$ ions in water. Thus the acid nature of hydrobromic acid is illustrated as follows:

$$HBr(g) + H_2O \rightarrow \mathbf{H_3O^+(aq)} + Br^-(aq)$$

Acids are molecular compounds when pure. But, in the presence of polar water molecules, the acid molecules are **ionized**. It is the formation of the **hydronium ion** [H_3O^+ or $H^+(aq)$] that gives acids their unique character.

Bases obtain their unique character from the production of hydroxide (OH^-) ions in aqueous solution. The common strong bases are ionic solids that dissolve in water to form ions.

$$NaOH(s) \xrightarrow{H_2O} Na^+(aq) + \mathbf{OH^-(aq)}$$

To expand upon our concept of acids and bases in water and to extend this behavior to substances such as ionic compounds, it is helpful to employ a somewhat broader definition. In the **Bronsted-Lowry** definition, **acids** are identified as proton (H^+) donors and **bases** as proton acceptors. In this approach, the list of acids and bases is much more extensive than in the Arrhenius definition. An acid or a base is identified by the role (or potential role), it plays in a proton exchange reaction.

The reaction of a Bronsted-Lowry acid produces what is known as its **conjugate base**. The reaction of a base produces its **conjugate acid**. Conjugate acid-base pairs can be identified by the gain or loss of an H^+. An **amphiprotic** substance is one that has both a conjugate acid and a conjugate base. The relationships between acids and their conjugate bases are illustrated as follows:

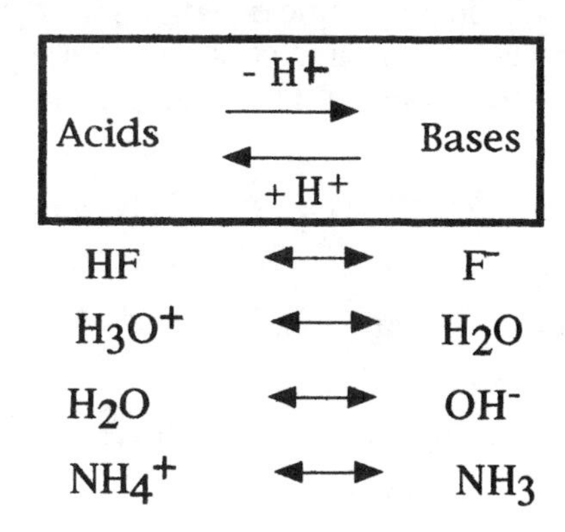

The strengths of acids vary according to what percentage of the acid molecules ionize to produce hydronium ions. **Strong acids** are 100% ionized in water and thus all have the same acid strength. **Weak acids**, on the other hand, are only partially ionized, which means that the following reaction goes to the right to a limited extent.

$$HNO_2(aq) + H_2O \rightleftharpoons H_3O^+(aq) + NO_2^-(aq)$$

The double arrows ($\rightleftharpoons$) indicate an incomplete reaction that reaches a state of dynamic equilibrium. In a reaction that reaches a state of equilibrium, the forward reaction ($\rightarrow$) and the reverse reaction ($\leftarrow$) are both occurring, but at the same rate, so that reactants and products coexist in the same solution in definite proportions. Thus weak acids (which are also weak electrolytes) produce small concentrations of H_3O^+ compared to the strong acids.

In a weak acid, most of the acid molecules are present as neutral molecules as shown on the left side of the equation. Likewise, **strong bases** are completely ionized, but **weak bases** such as ammonia produce a limited concentration of OH^- ions.

$$NH_3(aq) + H_2O \rightleftharpoons NH_4^+(aq) + OH^-(aq)$$

When solutions of acids and bases are mixed, a reaction known as a neutralization occurs. A **neutralization reaction** is a type of double replacement reaction similar to the precipitation reaction discussed in the previous chapter in which the cation and anion combine to form a solid precipitate. In a neutralization, the cation (of the acid) and the anion (of the base) combine to form the molecular compound water. The neutralization reaction between a strong acid and base was first illustrated in Chapter 7; an example of the balanced molecular, total ionic, and net ionic equations of such a reaction involving a **monoprotic acid** follows.

Molecular: $HNO_3(aq) + NaOH(aq) \rightarrow NaNO_3(aq) + H_2O(l)$

Total ionic: $H^+(aq) + NO_3^-(aq) + Na^+(aq) + OH^-(aq) \rightarrow Na^+(aq) + NO_3^-(aq) + H_2O(l)$

Net ionic: $H^+(aq) + OH^-(aq) \rightarrow H_2O$

The molecular, total ionic, and net ionic equations for the neutralization of a weak monoprotic acid and a strong base is shown below. In this case, the acid is not represented as ions in the ionic equations because most of the acid is present in solution as molecules.

Molecular: $HNO_2(aq) + KOH(aq) \rightarrow KNO_2(aq) + H_2O(l)$

Total ionic: $HNO_2(aq) + K^+(aq) + OH^-(aq) \rightarrow K^+(aq) + NO_2^-(aq) + H_2O(l)$

Net ionic: $HNO_2(aq) + OH^-(aq) \rightarrow NO_2^-(aq) + H_2O(l)$

Polyprotic acids have more than one replaceable hydrogen per molecule. Sulfurous acid, H_2SO_3, has two and is referred to as a **diprotic acid**. Phosphoric acid, H_3PO_4, has three and is an example of a **triprotic acid**. Polyprotic acids can react with bases in steps. For example, consider the reaction of one mole of sulfurous acid with one mole of KOH. The three appropriate equations representing the reaction are as follows:

Molecular: $H_2SO_3(aq) + KOH(aq) \rightarrow KHSO_3(aq) + H_2O(l)$

Total ionic: $H_2SO_3(aq) + K^+(aq) + OH^-(aq) \rightarrow K^+(aq) + HSO_3^-(aq) + H_2O(l)$

Net ionic: $H_2SO_3(aq) + OH^-(aq) \rightarrow HSO_3^-(aq) + H_2O(l)$

The potassium hydrogen sulfite ($KHSO_3$) is known as an **acid salt.** It is a salt whose anion contains at least one acid proton that can react with another mole of base. Acid salts are formed from the partial neutralization of polyprotic acids.

NEW TERMS

Acid
Amphiprotic
Conjugate acid (base)
Diprotic acid
Dynamic equilibrium
Hydronium ion
Monoprotic acid
Polyprotic acid
Weak acid
Weak base

SELF-TEST

A-1 Multiple Choice

___ 1. Which of the following is the formula of sulfurous acid?

(a) H_2S (b) HSO_3 (c) H_2SO_3 (d) HSO_2 (e) H_2SO_4

___ 2. Which of the following is the formula of strontium hydroxide?

(a) $Sr(OH)_2$ (b) $S(OH)_2$ (c) SrOH (d) $St(OH)_2$ (e) SrO

___ 3. Which of the following is the definition of a Bronsted-Lowry base?

(a) donates OH^- ion
(b) accepts H atoms
(c) accepts H^+ ions
(d) donates H^+ ions
(e) accepts OH^- ions

___ 4. Which of the following can act only as a Bronsted-Lowry base?

(a) S^{2-} (b) H_2S (c) HS^- (d) H^+ (e) HCl

___ 5. Which of the following can react as either a Bronsted acid or a Bronsted base?

(a) HS^- (b) Cl^- (c) NH_4^+ (d) H_2S (e) S^{2-}

___ 6. Which of the following is the conjugate base of $H_2PO_4^-$?

(a) HPO_4^{2-} (b) H_3PO_4 (c) HPO_4^- (d) PO_4^{3-} (e) OH^-

___ 7. Which of the following is a weak acid?

(a) HBr (b) H_2SO_3 (c) HNO_3 (d) HCl (e) HI

___ 8. Which of the following compounds is a strong base?

(a) NH_3 (b) $B(OH)_3$ (c) LiOH (d) HNO_2 (e) $CaCl_2$

___ 9. A 0.10 M solution of a substance has a H_3O^+ concentration of 0.01 M. The substance is:

(a) a weak base
(b) a strong base
(c) a strong acid
(d) a weak acid

___ 10. Which of the following is a salt formed by the complete neutralization of barium hydroxide with nitrous acid?

(a) $BaHNO_2$
(b) $Ba(NO_3)_2$
(c) $Ba(OH)_2$
(d) $BaNO_2$
(e) $Ba(NO_2)_2$

___ 11. Which of the following is an acid salt?

(a) NH_4Cl
(b) NH_4NO_3
(c) $BaHPO_4$
(d) $CaSO_3$
(e) H_2SO_3

___ 12. Which of the following is a product when one mole of H_3AsO_4 reacts with one mole of $Ca(OH)_2$?

(a) $Ca(H_2AsO_4)_2$
(b) $Ca(HAsO_4)$
(c) $Ca(AsO_4)_2$
(d) $Ca(HAsO_4)_2$
(e) CaH_3AsO_4

___ 13. Which of the following is a salt formed from the neutralization of a strong acid and a strong base?

(a) CaI_2 (b) $Ca(NO_2)_2$
(c) NH_4Br (d) NH_4NO_2
(e) KF

A-2 Problems

1. Identify each of the following as an acid or base when dissolved in water:

(a) HClO ______________ (b) $Mn(OH)_2$ ______________

(c) H_2Se ______________ (d) CH_3NH_2 ______________

(e) H_2SO_3 ______________ (f) CuOH ______________

2. Complete the following equations illustrating Bronsted-Lowry acid-base reactions:

Acid$_1$+ *Base$_2$* $\rightarrow$ *Acid$_2$* + *Base$_1$*

(a) $HClO_3$ + H_2O $\rightarrow$

(b) HSO_4^- + NH_3 $\rightarrow$

(c) H_2O + HCO_3^- $\rightarrow$

(d) HCO_3^- + H_2O $\rightarrow$

(e) H_2O + NH_2^- $\rightarrow$

(f) H_2O + CH_3NH_2 $\rightarrow$

(g) H_2S + H_2O $\rightarrow$

(h) $H_2PO_4^-$ + CH_3O^- $\rightarrow$

3. Complete and balance the following equations illustrating complete neutralizations:

(a) HNO_3 + $Ca(OH)_2$ $\rightarrow$ $Ca(NO_3)_2$ + ________

(b) $HC_2H_3O_2$ + NaOH $\rightarrow$ ________ + ______

(c) ________ + KOH → K_2CO_3 + H_2O

(d) H_2S + $Mg(OH)_2$ → ________ + H_2O

(e) $HClO$ + NH_3 → ________

(f) ________ + ________ → $Fe_2(SO_4)_3$ + ________

4. Write the total ionic and the net ionic equation for 3(a), (b), and (c).

5. Write equations representing the preparation of the following salts or acid salts from the reaction of the appropriate acids and bases:

(a) $RbHS$

(b) $Ca(H_2PO_4)_2$

(c) NiS

6. Write the net ionic equations for 5 (a) and (b).

Review Section B *Measurement of Acid Strength*

OUTLINE	OBJECTIVES
12-5 Equilibrium of Water	
1. The autoionization of water	
2. K_w: the ion product of water	*Calculate $[OH^-]$ from a specified $[H_3O^+]$ and vice versa by use of K_w.*
3. Acidic, basic, and neutral solutions in terms of $[H_3O^+]$ and $[OH^-]$	*Distinguish among acidic, basic, and neutral solutions in terms of $[OH^-]$ and $[H_3O^+]$.*

12-6 pH Scale

1. The definition of pH and pOH — *Convert pH to [H_3O^+] and vice versa.*

2. Acidic, basic, and neutral solutions in terms of pH and pOH — *Distinguish among strongly acidic, weakly acidic, neutral, weakly basic, and strongly basic solutions in terms of pH.*

SUMMARY OF SECTIONS 12-5 AND 12-6

Questions: *How does the H_3O^+ concentration relate to the OH^- concentration in water? How is the degree of acidity or basicity measured?*

We are now ready to expand our view of acids and bases in water based on the fact that there is a small but important equilibrium concentration of H_3O^+ and OH^- present in pure water. This occurs because of the **autoionization** of water, illustrated as follows.

$$2H_2O \rightleftharpoons H_3O^+ + OH^-$$

The product of the molar concentrations of H_3O^+ and OH^- is known as the **ion product** of water and has a constant value of 1.0×10^{-14} at 25°C. Thus in pure water

$$K_w = [H_3O^+][OH^-] = 1.0 \times 10^{-14} \qquad [H_3O^+] = [OH^-] = 1.0 \times 10^{-7}$$

In this light, we can redefine an acid as a substance that increases [H_3O^+]. Since there is an inverse relation between [H_3O^+] and [OH^-], an increase in [H_3O^+] is the same as a decrease in [OH^-]. (As an analogy, pushing one side of a see-saw down is the same as pushing the other up.) A base does the opposite; it increases [OH^-] while decreasing [H_3O^+].

The [H_3O^+] in a solution is more conveniently expressed on a logarithmic scale rather than in scientific notation. This scale, known as **pH**, is defined as follows:

$$pH = -\log [H_3O^+] \quad (\text{or } \mathbf{pOH} = -\log [OH^-])$$

Some example calculations follow.

Example B-1 Conversion of $[H_3O^+]$ to pH

What is the pH of a solution if $[H_3O^+] = 8.3 \times 10^{-4}$?

PROCEDURE

$pH = -\log [H_3O^+]$ Take the log of 8.3×10^{-4}.

Using a calculator, enter 8.3, EXP, 4, +/-, and take the log.
The answer on the calculator is changed to a positive number.
Using a log table, the answer is 4 (minus the log of 10^{-4}) - log 8.3

SOLUTION

$pH = -\log (8.3 \times 10^{-4}) = \underline{3.08}$

Example B-2 Conversion of pH to $[H_3O^+]$

What is $[H_3O^+]$ if pH = 10.43?

PROCEDURE

$-\log [H_3O^+] = 10.43 \qquad \log [H_3O^+] = -10.43$
Using a calculator, enter 10.43, +/-, and INV (or shift), and log.

Using a log table, change -10.43 to 0.57 – 11.
Find the antilog of 0.57 which is then multiplied by 10^{-11}.

(Review the discussion of logarithms in Appendix C in the text if necessary.)

SOLUTION

Inverse log (antilog) of $-10.43 = [H_3O^+] = 3.7 \times 10^{-11}$

Acidic, basic, and neutral solutions can be defined in terms of $[H_3O^+]$, $[OH^-]$, pH, and pOH as follows:

	$[H_3O^+]$	$[OH^-]$	pH	pOH
Neutral	1.0×10^{-7}	1.0×10^{-7}	7.00	7.00
Acidic	$> 1.0 \times 10^{-7}$	$< 1.0 \times 10^{-7}$	<7.00	>7.00
Basic	$< 1.0 \times 10^{-7}$	$> 1.0 \times 10^{-7}$	>7.00	<7.00

NEW TERMS

Acidic solution
Autoionization
Basic solution
Ion product of water
pH
pOH

SELF-TEST

B-1 Multiple Choice

___ 1. Which of the following is an acidic solution?

(a) $[H_3O^+] = 10^{-1}$
(b) $[OH^-] = 10^{-7}$
(c) $[OH^-] = 10^{-4}$
(d) $[OH^-] = 10^{-14}$

___ 2. When $[H_3O^+] = 10^{-3}$

(a) $[OH^-] = 10^{-11}$
(b) $[OH^-] = 10^{-3}$
(c) $[OH^-] = 10^{-7}$
(d) $[OH^-] = 10^{-9}$

___ 3. If there is 10^{-4} mol of H_3O^+ in 10 L of water, what is $[OH^-]$?

(a) 10^{-5} (b) 10^{11} (c) 10^{-4} (d) 10^{-9} (e) 10^{-3}

___ 4. What is the pH of the solution in the preceding problem?

(a) 4 (b) 5 (c) 9 (d) 11 (e) 3

___ 5. When $[H_3O^+] = 1.0 \times 10^{-9}$, pOH equals which of the following?

(a) 5.0 (b) 7.0 (c) 4.0 (d) 9.0 (e) 10.0

___ 6. Which of the following is the most acidic solution?

(a) pH = 7.0
(b) pOH = 1.2
(c) pOH = 13.8
(d) pH = 4.3
(e) pH = 11.2

B-2 Problems

1. What is the $[H_3O^+]$ in the following solutions?

(a) $[OH^-] = 1.0 \times 10^{-12}$

(b) $[OH^-] = 5.9 \times 10^{-4}$

(c) pH = 9.0

(d) pH = 7.85

(e) pOH = 4.18

2. What is the pH of the following solutions?

   (a) pOH = 4.51

   (b) $[OH^-] = 1.0 \times 10^{-7}$

   (c) $[H_3O^+] = 3.6 \times 10^{-4}$

   (d) $[H_3O^+] = 48 \times 10^{-10}$

Review Section C *Salts and Oxides as Acids and Bases*

OUTLINE	OBJECTIVES
12-7 The Effects of Salts on pH - Hydrolysis	
1. Reactions of ions with water	*Distinguish between the anions that undergo hydrolysis and those that don't.*
2. Solutions of salts in water	*Predict the relative acidity of a given salt when dissolved in water*

12-8 Control of pH - Buffer Solutions

1. Buffer solutions: a solution of a weak acid and a salt containing its conjugate base

Describe how a solution acts as a buffer and which solutions exhibit this behavior.

12-9 Oxides as Acids and Bases

1. The reactions of oxides with water: acid and base anhydrides

Write equations illustrating the reactions of oxides as acids or bases.

SUMMARY OF SECTIONS 12-7 THROUGH 12-9

Questions: *Can salts act as acids or bases? How is acidity controlled? How can rain become acidic?*

Besides molecular compounds, solutions of certain salts may be acidic or basic as well as neutral. Cations of strong bases (i.e., Na^+ and Ca^{2+}) do not affect the pH of water. Anions of strong acids (Cl^-, Br^-, I^-, NO_3^-, ClO_4^- but not HSO_4^-) also do not affect the pH of water. Salts containing any combination of such cations and anions form neutral aqueous solutions (pH = 7.0). Anions of weak acids behave as weak bases in water and cations of weak acids behave as weak acids in water. Such reactions are known as **hydrolysis** reactions. The reaction of an anion as a base, illustrated as follows:

$$X^- + H_2O \rightleftharpoons HX + OH^-$$

Thus solutions of salts such as $KC_2H_3O_2$, $Sr(NO_2)_2$, and RbClO are weakly basic. Cations of weak bases also hydrolyze in water to form acidic solutions illustrated by the following equation:

$$MH^+ + H_2O \rightleftharpoons M + H_3O^+$$

Thus solutions of salts such as NH_4Br and $N_2H_5^+Cl^-$ are weakly acidic.

The partial ionization of a weak acid can be utilized in the formation of a **buffer** solution. A buffer solution is one that can absorb limited quantities of added H_3O^+ or OH^- so that the pH of a solution remains relatively constant. Obviously, such a solution contains two substances - one that is a potential base to remove added H_3O^+ and one that is a potential acid to remove added OH^-. A solution of a weak acid and a salt of its conjugate base (or a weak base and a salt of its conjugate acid) serves this purpose. The equilibrium of a buffer solution is shown as follows:

$$\underset{\textbf{HCN}}{0.25\text{ M}} + H_2O \rightleftharpoons H_3O^+ + \underset{\textbf{CN}^-}{0.25\text{ M}}$$

Thus a solution that is 0.25 M in both HCN and CN^- has a reservoir of species that are available to react with either H_3O^+ or OH^- added to the solution, as illustrated by the following two equations.

$$HCN + OH^- \rightarrow CN^- + H_2O$$

$$CN^- + H_3O^+ \rightarrow HCN + H_2O$$

Buffers are important in many chemical and life processes to control pH within strict limits.

The final topic concerns an important environmental issue. This is the role of oxides as acids or bases. Generally, nonmetal oxides react with water to form acids and are thus called **acid anhydrides.**

$$SO_2(g) + H_2O(l) \rightarrow H_2SO_3(aq)$$

$$Cl_2O_5(aq) + H_2O(l) \rightarrow 2HClO_3(aq)$$

Many metal oxides (generally, ionic metal oxides) are **base anhydrides** and react with water to form a hydroxide.

$$K_2O(s) + H_2O \rightarrow 2KOH(aq)$$

$$BaO(s) + H_2O \rightarrow Ba(OH)_2(aq)$$

NEW TERMS

Acid anhydride
Base anhydride
Buffer
Hydrolysis

SELF-TEST

C-1 Multiple Choice

___ 1. Which of the following anions does not hydrolyze?

(a) I^- (b) F^- (c) ClO_2^- (d) CHO_2^- (e) NO_2^-

___ 2. A solution of which of the following salts produces a basic solution?

(a) NH_4Cl
(b) NH_4NO_3
(c) KBr
(d) $Mg(ClO_4)_2$
(e) Rb_2S

___ 3. Which of the following solutions is a buffer solution?

(a) 0.20 M HNO_2 and 0.20 M KNO_3
(b) 0.20 M HNO_3 and 0.20 M KNO_2
(c) 0.20 M HNO_3 and 0.20 M KNO_3
(d) 0.20 M HNO_2 and 0.20 M KNO_2
(e) 0.20 M HNO_2 and 0.20 M HNO_3

___ 4. Given one mole of NaOH in 10 L of water, which of the following makes a buffer solution when added to this solution?

(a) one mole of HCl
(b) two moles of $HC_2H_3O_2$
(c) one mole of $HC_2H_3O_2$
(d) one mole of $NaC_2H_3O_2$
(e) two moles of HCl

___ 5 Which of the following is the acid formed when P_4O_{10} reacts with water?

(a) H_3PO_3 (b) HPO_2 (c) H_3PO_4 (d) $H_2P_4O_{11}$ (e) H_2SO_4

___ 6. Which of the following is the anhydride of $Mn(OH)_3$?

(a) Mn_2O_7 (b) MnO (c) MnO_2 (d) Mn_2O_3 (e) MnO_2H

C-2 Problems

1. Indicate whether solutions of the following salts in water are acidic, basic, or neutral. If a solution is acidic or basic, write the equation illustrating this behavior.

   (a) $MgCl_2$

   (b) NH_4I

   (c) K_2CO_3

2. Complete the following equations concerning anhydrides:

   (a) $SO_2 + H_2O \rightarrow$ __________

   (b) $2HClO_4 \rightarrow$ __________ $+ H_2O$

   (c) $2Fe(OH)_3 \rightarrow$ __________ $+ 3\ H_2O$

Review Section D

CHAPTER SUMMARY SELF-TEST

D-1 Matching I

___ a strong acid

___ a weak base

___ an acid salt

___ a salt formed from a strong acid and a strong base

___ the conjugate acid of CH_3NH_2

___ the anhydride of $HClO_3$

___ the anhydride of KOH

(a) $NaNO_3$

(b) $HClO_2$

(c) NH_4Br

(d) Cl_2O_3

(e) $CH_3NH_3^+$

(f) KO

(g) $KHSO_3$

(h) $HClO_4$

(i) Cl_2O_5

(j) K_2O

(k) CH_3NH^-

(l) CH_3NH_2

D-2 Matching II

For each of the following solutions pick a pH, and any acid, base, salt, or oxide listed whose solution could produce the appropriate acidity. (Assume that a significant amount of each compound is in solution.)

weakly acidic ______________ strongly acidic ______________

weakly basic ______________ strongly basic ______________

neutral ______________

pH	Acid or base	Salt	Soluble oxide
1.0	NH_3	K_2SO_3	SO_3
5.0	HNO_3	NH_4NO_3	SO_2
7.0	H_2SO_3	KI	BaO
9.0	KOH		
13.0			

D-3 Problems

1. Write equations illustrating reactions of the following with water. Where no reaction occurs write "NR".

(a) NH_3 + H_2O

(b) HClO + H_2O

(c) ClO^- + H_2O

(d) I^- + H_2O

(e) $N_2H_5^+$ + H_2O

(f) CN^- + H_2O

(g) H_2S + H_2O

(h) Ca^{2+} + H_2O

(i) HCN + H_2O

Answers to Self-Tests

A-1 Multiple Choice

1. **c** H_2SO_3
2. **a** $Sr(OH)_2$
3. **c** accepts H^+ ions
4. **a** S^{2-}
5. **a** HS^- [It has a conjugate base (S^{2-}) and a conjugate acid (H_2S)]
6. **a** HPO_4^{2-}
7. **b** H_2SO_3
8. **c** LiOH
9. **d** a weak acid (It is 10% ionized.)

10. e $Ba(NO_2)_2$ [$2HNO_2 + Ba(OH)_2 \rightarrow Ba(NO_2)_2 + 2H_2O$]

11. c $BaHPO_4$

12. b $CaHAsO_4$ [$H_3AsO_4 + Ca(OH)_2 \rightarrow CaHAsO_4 + 2H_2O$]

13. a CaI_2 [formed from $Ca(OH)_2$ and HI]

A-2 Problems

1. (a) acid (b) base (c) acid (d) base (e) acid (f) base

2.

	Acid$_1$	+	Base$_2$	→	Acid$_2$	+	Base$_1$
(a)	$HClO_3$	+	H_2O	→	H_3O^+	+	ClO_3^-
(b)	HSO_4^-	+	NH_3	→	NH_4^+	+	SO_4^{2-}
(c)	H_2O	+	HCO_3^-	→	H_2CO_3	+	OH^-
(d)	HCO_3^-	+	H_2O	→	H_3O^+	+	CO_3^{2-}
(e)	H_2O	+	NH_2^-	→	NH_3	+	OH^-
(f)	H_2O	+	CH_3NH_2	→	$CH_3NH_3^+$	+	OH^-
(g)	H_2S	+	H_2O	→	H_3O^+	+	HS^-
(h)	$H_2PO_4^-$	+	CH_3O^-	→	CH_3OH	+	HPO_4^{2-}

3. (a) $2HNO_3 + Ca(OH)_2 \rightarrow Ca(NO_3)_2 + \mathbf{2H_2O}$

(b) $HC_2H_3O_2 + NaOH \rightarrow \mathbf{NaC_2H_3O_2} + \mathbf{H_2O}$

(c) $\mathbf{H_2CO_3} + 2KOH \rightarrow K_2CO_3 + 2H_2O$

(d) $H_2S + Mg(OH)_2 \rightarrow \mathbf{MgS} + 2H_2O$

(e) $HClO + NH_3 \rightarrow \mathbf{NH_4ClO}$

(f) $\mathbf{3H_2SO_4 + 2Fe(OH)_3} \rightarrow Fe_2(SO_4)_3 + \mathbf{6}H_2O$

4. (a) $2H^+(aq) + 2NO_3^-(aq) + Ca^{2+}(aq) + 2OH^-(aq) \rightarrow Ca^{2+}(aq) + 2NO_3^-(aq) + 2H_2O(l)$

$H^+(aq) + OH^-(aq) \rightarrow H_2O(l)$

(b) $HC_2H_3O_2 + Na^+(aq) + OH^-(aq) \rightarrow Na^+(aq) + C_2H_3O_2^-(aq) + H_2O(l)$

$HC_2H_3O_2(aq) + OH^-(aq) \rightarrow C_2H_3O_2^-(aq) + H_2O(l)$

(c) $H_2CO_3(aq) + 2K^+(aq) + 2OH^-(aq) \rightarrow 2K^+(aq) + CO_3^{2-}(aq) + 2H_2O(l)$

$H_2CO_3(aq) + 2OH^-(aq) \rightarrow CO_3^{2-}(aq) + 2H_2O(l)$

5. (a) $H_2S + RbOH \rightarrow RbHS + H_2O$

(b) $2H_3PO_4 + Ca(OH)_2 \rightarrow Ca(H_2PO_4)_2 + 2H_2O$

(c) $H_2S + Ni(OH)_2 \rightarrow NiS + 2H_2O$

6. (a) $H_2S(aq) + OH^-(aq) \rightarrow HS^-(aq) + H_2O(l)$

(b) $H_3PO_4(aq) + OH^-(aq) \rightarrow H_2PO_4^-(aq) + H_2O(l)$

B-1 Multiple Choice

1. **d** $[OH^-] = 10^{-14}$

2. **a** $[OH^-] = 10^{-11}$

3. **d** $[H_3O^+] = \frac{10^{-4}\text{ mol}}{10\text{ L}} = 10^{-5}$ M, $[OH^-] = \frac{K_w}{[H_3O^+]} = \frac{10^{-14}}{10^{-5}} = 10^{-9}$

4. **b** $pH = -\log [H_3O^+] = -\log (10^{-5}) = 5$

5. **a** pOH = 5 (pOH = 14 - pH = 14.0 - 9.0 = 5.0)

6. **c** pOH = 13.8 (pH = 14.0 - pOH = 14.0 - 13.8 = 0.2)

B-2 Problems

1. (a) $[H_3O^+] = \underline{1.0 \times 10^{-2}}$ (b) $[H_3O^+] = \underline{1.7 \times 10^{-11}}$

(c) $[H_3O^+] = \underline{1.0 \times 10^{-9}}$ (d) $[H_3O^+] = \underline{1.4 \times 10^{-8}}$

(e) pOH = 4.18; pH = 14 - 4.18 = $\underline{9.82}$

$-\log [H_3O^+] = \underline{9.82}$ $[H_3O^+] = \underline{1.5 \times 10^{-10}}$

2. (a) pH = $\underline{9.49}$

(b) $[H_3O^+] = 1.0 \times 10^{-8}$; pH = $\underline{8.00}$

(c) pH = 4.00 - log 3.6 = 4.00 - 0.56 = $\underline{3.44}$

(d) $[H_3O^+] = 48 \times 10^{-10} = 4.8 \times 10^{-9}$; pH = $\underline{8.32}$

C-1 Multiple Choice

1. **a** I^- (I^- is the anion or conjugate base of the strong acid HI.)

2. **e** Rb_2S (The S^{2-} ion hydrolyzes according to the equation
   $S^{2-} + H_2O \rightleftharpoons HS^- + OH^-$.)

3. **d** 0.20 M HNO_2 and 0.20 M KNO_2. (This is the only solution shown of a weak acid and a salt containing its conjugate base.)

4. **b** two moles of $HC_2H_3O_2$ (One mole of NaOH neutralizes one mole of $HC_2H_3O_2$ to produce one mole of $NaC_2H_3O_2$,
   i.e., $NaOH + HC_2H_3O_2 \rightarrow NaC_2H_3O_2 + H_2O$
   The solution now contains one remaining mole of $HC_2H_3O_2$ and one mole of $NaC_2H_3O_2$ and is thus a buffer solution.)

5. **c** H_3PO_4 ($P_4O_{10} + 6H_2O \rightarrow 4H_3PO_4$)

6. **d** Mn_2O_3 ($2Mn(OH)_3 \rightarrow Mn_2O_3 + 3H_2O$)

C-2 Problem

1. (a) neutral (Neither ion hydrolyzes).
   (b) acidic (cation hydrolysis) $NH_4^+ + H_2O \rightleftharpoons NH_3 + H_3O^+$
   (c) basic (anion hydrolysis) $CO_3^{2-} + H_2O \rightleftharpoons HCO_3^- + OH^-$

2. (a) $SO_2 + H_2O \rightarrow H_2SO_3$
   (b) $2HClO_4 \rightarrow Cl_2O_7 + H_2O$
   (c) $2Fe(OH)_3 \rightarrow Fe_2O_3 + 3H_2O$

D-1 Matching I

h $HClO_4$ is a strong acid.

g $KHSO_3$ is an acid salt.

e $CH_3NH_3^+$ is the conjugate acid of CH_3NH_2.

i Cl_2O_5 is the anhydride of $HClO_3$.

l CH_3NH_2 is a weak base.

a $NaNO_3$ is formed from a strong acid (HNO_3) and a strong base (NaOH).

j K_2O is the anhydride of KOH.

D-2 Matching II

weakly acidic	pH = 5.0, H_2SO_3, NH_4NO_3, SO_2 (forms H_2SO_3 in water)
strongly acidic	pH = 1.0, HNO_3, SO_3 (forms H_2SO_4 in water)
neutral	pH = 7.0, KI
weakly basic	pH = 9.0, NH_3, K_2SO_3
strongly basic	pH = 13.0, KOH, BaO

D-3 Problems

(a) $NH_3 + H_2O \rightleftharpoons NH_4^+ + OH^-$

(b) $HClO + H_2O \rightleftharpoons H_3O^+ + ClO^-$

(c) $ClO^- + H_2O \rightleftharpoons HClO + OH^-$

(d) $I^- + H_2O \rightarrow$ NR

(e) $N_2H_5^+ + H_2O \rightleftharpoons N_2H_4 + H_3O^+$

(f) $CN^- + H_2O \rightleftharpoons HCN + OH^-$

(g) $H_2S + H_2O \rightleftharpoons H_3O^+ + HS^-$

$HS^- + H_2O \rightleftharpoons H_3O^+ + S^{2-}$

(h) $Ca^{2+} + H_2O \rightarrow$ NR

(i) $HCN + H_2O \rightleftharpoons H_3O^+ + CN^-$

Solutions to Black Text Problems

12-22 HX is a weak acid $\quad HX + H_2O \rightleftharpoons H_3O^+ + X^-$

The concentration of H_3O^+ must equal the concentration of the HX that ionized.

$\frac{0.010}{0.100}$ x 100% = 10% ionized

12-24 Since $HClO_4$ is one of the strong acids, it is completely dissociated.

Thus $[H_3O^+]$ = 0.55 M

12-25 3.0% of the original HX concentration is ionized to form H_3O^+.

Thus, $[H_3O^+]$ = 0.030 x 0.55 = 0.017 M

12-27 $H_2SO_4 + H_2O \xrightarrow{100\%} H_3O^+ + HSO_4^-$

From the first ionization: $[H_3O^+] = 0.354$ M

$HSO_4^- + H_2O \rightleftharpoons H_3O^+ + SO_4^{2-}$ (25% to the right)

The concentration of HSO_4^- from the first ionization is equal to 0.354 M. Of that, 25% dissociates.

$0.25 \times 0.354 = 0.089$ M $[H_3O^+]$ from the second ionization.

The total $[H_3O^+] = 0.354 + 0.089 = \underline{0.443\ M}$

12-28 The concentration of H_3O^+ is equal to the concentration of the ionized acid.

Therefore, $\frac{0.050\ M}{1.0\ M} \times 100\% = \underline{5.0\%\ \text{ionized}}$

12-50 The system would not be at equilibrium if $[H_3O^+] = [OH^-] = 10^{-2}$ M. Therefore, H_3O^+ reacts with OH^- until the concentration of each is reduced to 10^{-7} M. This is a neutralization reaction.

i.e., $H_3O^+ + OH^- \rightarrow 2H_2O$

12-51 (a) $[H_3O^+] = \frac{K_w}{[OH^-]} = \frac{10^{-14}}{10^{-12}} = 10^{-2}$ M

(b) $[H_3O^+] = \frac{10^{-14}}{10} = 10^{-15}$ M

(c) $[OH^-] = \frac{K_w}{[H_3O^+]} = \frac{1.0 \times 10^{-14}}{2.0 \times 10^{-5}} = 5.0 \times 10^{-10}$ M

12-53 $HClO_4 + H_2O \rightarrow H_3O^+ + ClO_4^-$ (100% to the right)

Since all $HClO_4$ ionizes, $[H_3O^+] = 0.250$ mol/10.0 L = $\underline{0.0250\ M}$

$$[OH^-] = \frac{K_w}{[H_3O^+]} = \frac{1.0 \times 10^{-14}}{2.50 \times 10^{-2}} = 4.0 \times 10^{-13}\ M$$

12-59 (a) $-\log(1.0 \times 10^{-6}) = \underline{6.00}$

(b) $-\log(1.0 \times 10^{-9}) = \underline{9.00}$

(c) pOH = $-\log(1.0 \times 10^{-2}) = 2.00$ pH = 14.00 - pOH = $\underline{12.00}$

(d) pOH = $-\log(2.5 \times 10^{-5}) = 4.60$ pH = 14.00 - 4.60 = $\underline{9.40}$

(e) pH = $-\log(6.5 \times 10^{-11}) = \underline{10.19}$

12-61 (a) $0.0001 = 10^{-4}$ $\underline{pH = 4.0,\ pOH = 10.0}$
(b) $0.00001 = 10^{-5}$ $\underline{pOH = 5.0,\ pH = 9.0}$
(c) $0.020 = 2.0 \times 10^{-2}$ $\underline{pH = 1.70,\ pOH = 12.30}$
(d) $0.000320 = 3.20 \times 10^{-4}$ $\underline{pOH = 3.495,\ pH = 10.505}$

12-63 (a) $-\log[H_3O^+] = 3.00$ $[H_3O^+] = 1.0 \times 10^{-3}$ M
(b) $-\log[H_3O^+] = 3.54$ $[H_3O^+] = \text{antilog}(-3.54)$ $[H_3O^+] = 2.9 \times 10^{-4}$ M
(c) $-\log[OH^-] = 8.00$ $[OH^-] = 1.0 \times 10^{-8}$ M $[H_3O^+] = 1.0 \times 10^{-6}$ M
(d) $pH = 14.00 - 6.38 = 7.62$ $-\log[H_3O^+] = 7.62$ $[H_3O^+] = 2.4 \times 10^{-8}$ M
(e) $-\log[H_3O^+] = 12.70$ $[H_3O^+] = 2.0 \times 10^{-13}$ M

12-67 If pH = 3.0, $[H_3O^+] = 10^{-3}$ $\frac{10^{-3}}{100} = 10^{-5}$ $\underline{pH = 5.0}$ (less acidic)
$10^{-3} \times 10 = 10^{-2}$ $\underline{pH = 2.0}$ (more acidic)

12-69 $HNO_3 + H_2O \rightarrow H_3O^+ + NO_3^-$ (100% to the right)
Since this is a strong acid $[H_3O^+] = 0.075$ M $-\log(7.5 \times 10^{-2}) = \underline{1.12}$

12-71 $Ca(OH)_2$ is a strong base producing 2 mol of OH^- per mol of $Ca(OH)_2$.
$[OH^-] = 2 \times 0.018 = 0.036 = 3.6 \times 10^{-2}$ M
$pOH = -\log(3.6 \times 10^{-2}) = 1.44$ $pH = 14.00 - 1.44 = \underline{12.56}$

12-72 $HX \rightleftharpoons H_3O^+ + X^-$ $[H_3O^+] = 10\%$ of $[HX]$
$[H_3O^+] = 0.100 \times 0.10 = 0.010 = 1.0 \times 10^{-2}$ M $-\log(1.0 \times 10^{-2}) = \underline{2.00}$

12-114 kg coal → kg FeS_2 → g FeS_2 → mol FeS_2 → mol SO_2 → mol SO_3 → mol H_2SO_4 → g H_2SO_4

$$100\ \cancel{\text{kg coal}} \times \frac{5.0\ \cancel{\text{kg FeS}_2}}{100\ \cancel{\text{kg coal}}} \times \frac{10^3\ \cancel{\text{g FeS}_2}}{\cancel{\text{kg FeS}_2}} \times \frac{1\ \cancel{\text{mol FeS}_2}}{120.0\ \cancel{\text{g FeS}_2}} \times \frac{8\ \cancel{\text{mol SO}_2}}{4\ \cancel{\text{mol FeS}_2}}$$

$$\times \frac{2\ \cancel{\text{mol SO}_3}}{2\ \cancel{\text{mol SO}_2}} \times \frac{1\ \cancel{\text{mol H}_2\text{SO}_4}}{1\ \cancel{\text{mol SO}_3}} \times \frac{98.09\ \text{g H}_2\text{SO}_4}{\cancel{\text{mol H}_2\text{SO}_4}} = 8200\ \text{g} = \underline{8.2\ \text{kg of H}_2\text{SO}_4}$$

12-115 $2.50\ \text{g HCl} \times \frac{1\ \text{mol HCl}}{36.46\ \text{g HCl}} = 0.0686\ \text{mol HCl}$

$[H_3O^+] = \frac{0.0686\ \text{mol}}{0.245\ \text{L}} = 0.280\ \text{mol/L}$ pH = -log(0.280) = <u>0.553 (con)</u>

$[H_3O^+] = \frac{0.0686\ \text{mol}}{0.890\ \text{L}} = 0.0771\ \text{mol/L}$ pH = -log(0.0771) = <u>1.113 (dilute)</u>

12-117 $10.0\ \text{g HCl} \times \frac{1\ \text{mol HCl}}{36.46\ \text{g HCl}} = 0.274\ \text{mol HCl}$

$10.0\ \text{g NaOH} \times \frac{1\ \text{mol NaOH}}{40.00\ \text{g NaOH}} = 0.250\ \text{mol NaOH}$

$HCl + NaOH \longrightarrow NaCl + H_2O$ (1 mol HCl reacts with 1 mol NaOH)

0.274 mol HCl - 0.250 mol NaOH = 0.024 mol HCl remaining

$[HCl] = [H_3O^+] = 0.024\ \text{mol}/1.00\ \text{L}$ pH = 1.62

12-119 $2HNO_3 + Ca(OH)_2 \longrightarrow Ca(NO_3)_2 + 2H_2O$

$0.500\ \text{L} \times 0.10\ \text{mol/L} = 0.050\ \text{mol}\ HNO_3 = 0.0500\ \text{mol}\ [H_3O^+]$
$0.500\ \text{L} \times 0.10\ \text{mol/L} = 0.050\ \text{mol}\ Ca(OH)_2$

$0.050\ \text{mol}\ HNO_3 \times \frac{1\ \text{mol}\ Ca(OH)_2}{2\ \text{mol}\ HNO_3} = 0.025\ \text{mol}\ Ca(OH)_2$ reacts

$0.050 - 0.025 = 0.025\ \text{mol}\ Ca(OH)_2$ remaining in 1.00 L

$[OH^-] = 0.025\ \text{mol}\ Ca(OH)_2 \times \frac{2\ \text{mol}\ OH^-}{\text{mol}\ Ca(OH)_2} = 0.050\ \text{mol/L}$

pOH = 1.30   <u>pH = 12.70</u>

13

Oxidation-Reduction Reactions

Review Section A *Redox Reactions - The Exchange of Electrons*

OUTLINE / OBJECTIVES

13-1 The Nature of Oxidation and Reduction

1. Oxidation, reduction, and redox reactions — *Understand the definition of oxidation and reduction in terms of electron exchange.*
2. Oxidizing and reducing agents — *Define oxidizing and reducing agents.*

13-2 Oxidation States

1. Electron bookkeeping: oxidation states — *Determine the oxidation state of each element in a compound from a set of rules.*

13-3 Balancing Redox Equations: Oxidation State Method

1. Balancing equations using oxidation states

13-4 Balancing Redox Equations: The Ion-Electron Method

1. Balancing equations with half-reactions in acidic solutions
2. Balancing equations with half-reactions in basic solutions — *Balance redox equations by either the oxidation state method or the ion-electron method in acidic or basic solutions.*

SUMMARY OF SECTIONS 13-1 THROUGH 13-4

Questions: *What is meant by electron exchange? How can we keep track of electrons in electron exchange reactions? How do we balance complex electron exchange reactions?*

Another important class of chemical reactions is known as **oxidation-reduction** reactions or simply **redox** reactions. In this type of reaction (unlike the double-replacement reactions studied in the previous two chapters) electrons are exchanged between two reactants. Electrons are assigned to atoms in a molecule by assuming all atoms in the molecule are present as ions. This is known as the **oxidation state** (or oxidation number) of the element in a compound. Two examples of the calculation of oxidation states follow. Redox reactions involve changes in oxidation states of at least two elements in two different reactants. One atom undergoes an increase in oxidation state which means it has lost electrons in going from reactant to product. An atom in another reactant does the opposite: it undergoes a decrease in oxidation state, which means that it has gained electrons in going from reactant to product.

The molecule or ion containing the atom that loses electrons is said to undergo **oxidation** and is also known as the **reducing agent**. The molecule or ion containing the atom that gains electrons is said to undergo **reduction** and is also known as the **oxidizing agent.**

Example A-1 Calculation of Oxidation States

What is the oxidation state of (a) P in H_3PO_3 and (b) the S in $K_2S_2O_6$?

(a) **PROCEDURE**

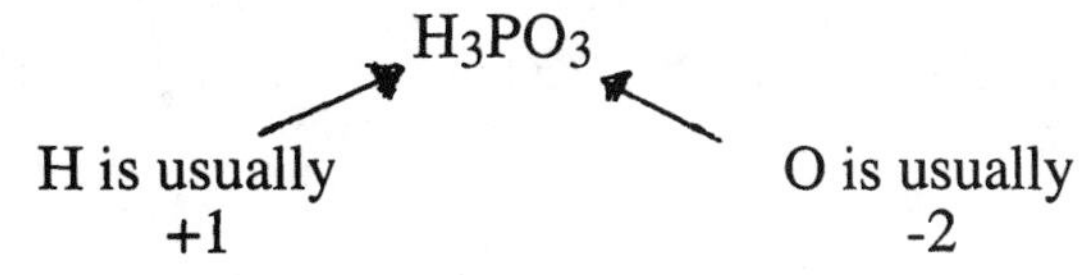

SOLUTION

Since all oxidation states in a neutral compound add to zero

$$3(H) + P + 3(O) = 0$$
$$3(+1) + P + 3(-2) = 0$$
$$P = \underline{+3}$$

(b) **PROCEDURE**

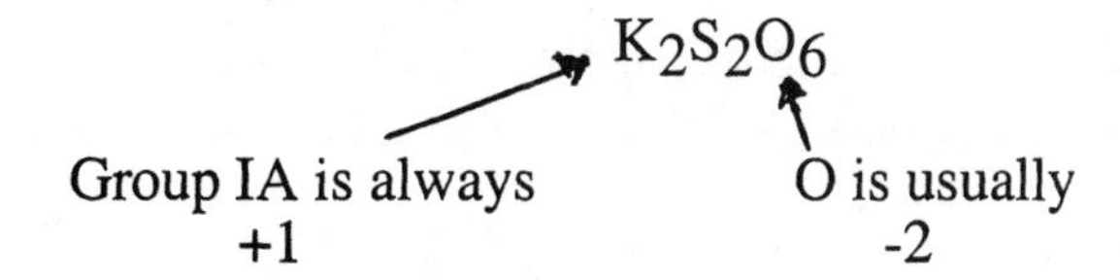

SOLUTION

$$2(K) + 2(S) + 6(O) = 0$$
$$2(+1) + 2(S) + 6(-2) = 0$$
$$S = \underline{+5}$$

In order to identify oxidizing and reducing agents in a reaction containing several reactants and products, it is necessary to calculate oxidation states. For example, calculation of oxidation states of the atoms of all of the elements in the following equation allows us to identify the atoms that undergo a change.

$KMnO_4$ is reduced (oxidizing agent)

$$\overset{+1\ \underline{+7}\ -2}{KMnO_4} + \overset{+1\ +6\ -2}{H_2SO_4} + \overset{\underline{-3}\ +1}{PH_3} \rightarrow \overset{+1\ +6\ -2}{K_2SO_4} + \overset{\underline{+2}\ +6\ -2}{MnSO_4} + \overset{+1\ -2}{H_2O} + \overset{+1\ \underline{+5}\ -2}{H_3PO_4}$$

PH_3 is oxidized (reducing agent)

Balancing redox reactions such as illustrated by the previous unbalanced equation can be extremely tedious if done by inspection. Two methods are introduced in this text to balance equations. Both methods are based on the principle that "electrons gained (by the oxidizing agent) equal electrons lost (by the reducing agent)." In the **oxidation state method** one focuses only on the elements that change. Coefficients are introduced so as to equalize the electron exchange.

Example A-2 Balancing an Equation by the Oxidation State Method

Balance the following equation.

$$MnO_2 + PbO_2 + HNO_3 \rightarrow Pb(NO_3)_2 + HMnO_4 + H_2O$$

PROCEDURE

(a) The two atoms that change are Mn and Pb. The Mn is oxidized from the +4 state in MnO_2 to +7 in $HMnO_4$. The Pb is reduced from the +4 state in PbO_2 to the +2 state in $Pb(NO_3)_2$.

(b) Since MnO_2 gains three electrons and PbO_2 loses two, the MnO_2 and its product are multiplied by two and the PbO_2 and its product are multiplied by three so that exactly six electrons are exchanged.

(c) Since there are six NO_3^- ions on the right, there must be six HNO_3 molecules on the left. Since there are now six H's on the left, there must be two H_2O's on the right to balance the hydrogens. Notice that a check of the O's indicates that the equation is balanced.

SOLUTION

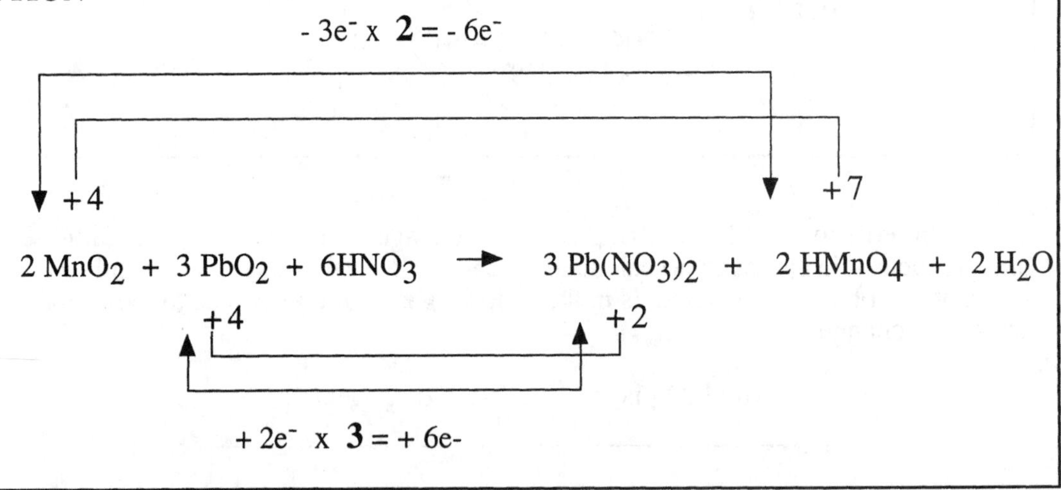

In the **ion-electron method** the focus is on the entire oxidizing or reducing agent, not just the one element that changes. It is necessary to recognize only the species that change not the actual oxidation states. In this method, the total reaction is divided into two **half-reactions** that are balanced separately. An example of balancing an equation in acid solution follows.

Example A-3 Balancing an Equation by the Ion-Electron Method

Balance the following by the ion-electron method.

$$MnO_4^- + NO_2 + H_2O \rightarrow Mn^{2+} + NO_3^- + H^+$$

PROCEDURE

(a) Notice that the MnO_4^- ion and the NO_2 molecule undergo changes in oxidation states. Balance each half-reaction separately. In each half-reaction balance the atom that changes first, oxygens second by adding H_2O in acidic solution. Next, balance the hydrogens by adding H^+. The last step, and most important, is to balance the charge by adding electrons to the more positive side of the equation. To do this, calculate the charge on each side by multiplying the coefficient of the ion by its charge. Neutral molecules do not contribute to the charge.

(b) The oxidation reaction is multiplied by five and then added to the reduction reaction so that each process involves five electrons. When the equations are added, the five electrons subtract out as do 4 H_2O's and 8 H^+'s.

(c) In basic solution the rules are modified as follows: First balance the half-reaction as in acid solution. Convert each H^+ to H_2O by adding an OH^- ion to both sides of the equation. (The H^+ converts to H_2O by means of the neutralization reaction $H^+ + OH^- \rightarrow H_2O$.) Simplify if necessary by canceling any H_2O's on both sides of the equation.

SOLUTION

(a)

$$5e^- + 8H^+ + MnO_4^- \rightarrow Mn^{2+} + 4H_2O$$

$$H_2O + NO_2 \rightarrow NO_3^- + 2H^+ + e^-$$

(b)

$$5H_2O + 5NO_2 \rightarrow 5NO_3^- + 10\ H^+ + 5e^-$$

Adding the oxidation and the reduction reaction, we have

$$5e^- + 8H^+ + MnO_4^- + 5H_2O + 5NO_2 \rightarrow Mn^{2+} + 4H_2O + 5NO_3^- + 10\ H^+ + 5e^-$$

The final balanced net ionic equation is

$$MnO_4^- + 5NO_2 + H_2O \rightarrow Mn^{2+} + 5NO_3^- + 2H^+$$

NEW TERMS

Half-reaction
Ion-electron method
Oxidation
Oxidation-reduction (redox)
Oxidation state (number)
Oxidation state method
Oxidizing agent
Reducing agent
Reduction

SELF-TEST

A-1 Multiple Choice

____ 1. What is the oxidation state of the Br in KBrO?

(a) +5 (b) +3 (c) -7 (d) -6 (e) none of these

____ 2. What is the oxidation state of the Co in $Co_2(SO_4)_3$?

(a) +2 (b) +6 (c) -4 (d) +3 (e) none of these

____ 3. Which of the following is an oxidation-reduction reaction?

(a) $K + O_2 \rightarrow KO_2$

(b) $CaO + CO_2 \rightarrow CaCO_3$

(c) $KCl + AgNO_3 \rightarrow AgCl(s) + KNO_3$

(d) $HBr + NaOH \rightarrow NaBr + H_2O$

(e) $H_2CO_3 \rightarrow H_2O + CO_2$

____ 4. Which of the following unbalanced equations represents an impossible reaction?

(a) $Zn + 2H^+ \rightarrow Zn^{2+} + H_2$

(b) $Bi + HNO_3 \rightarrow Bi(NO_3)_3 + H_2O$

(c) $Zn^{2+} + Ni^{2+} \rightarrow Zn + Ni$

(d) $NaI + Br_2 \rightarrow NaBr + I_2$

(e) $Mg + N_2 \rightarrow Mg_3N_2$

____ 5. In the following unbalanced equation, what is the oxidizing agent?

$$Fe^{2+} + H^+ + MnO_4^- \rightarrow Fe^{3+} + Mn^{2+} + H_2O$$

(a) Fe^{2+} (b) Fe^{3+} (c) Mn^{2+} (d) H^+ (e) MnO_4^-

____ 6. In the following equation, how many electrons have been gained ($+e^-$) or lost ($-e^-$) by one molecule of O_3?

$$6H^+ + O_3 \rightarrow 3H_2O$$

(a) $+ 2e^-$ (b) $- 2e^-$ (c) $+ 6e^-$ (d) $+ 4e^-$ (e) $- 6e^-$

A-2 Problems

1. Give the oxidation states of the following:

   (a) the Mn in $Ca(MnO_4)_2$ ________

   (b) the Mn in K_2MnO_4 ________

   (c) the Sn in $Sn(SO_4)_2$ ________

   (d) the O in H_2O_2 ________

   (e) the C in CH_4O ________

   (f) the Cr in CrC_2O_4 ________

2. Fill in the blanks concerning the following unbalanced equations.

   (a) $Cu + HNO_3 \rightarrow Cu^{2+} + NO + H_2O$

   (b) $H_2S + KMnO_4 + HCl \rightarrow S + KCl + MnCl_2 + H_2O$

   (c) $Bi(OH)_3 + K_2Sn(OH)_4 \rightarrow Bi + K_2Sn(OH)_6$

reaction	(a)	(b)	(c)
species oxidized	________	________	________
electrons lost	________	________	________
species reduced	________	________	________
electrons gained*	________	________	________
oxidizing agent	________	________	________
reducing agent	________	________	________

*per atom or formula unit

3. Balance the following equations by the oxidation state method.

(a) $Cu + HNO_3 \rightarrow Cu(NO_3)_2 + NO + H_2O$

(b) $H_2S + KMnO_4 + HCl \rightarrow S + KCl + MnCl_2 + H_2O$

(c) $Bi(OH)_3 + K_2Sn(OH)_4 \rightarrow Bi + K_2Sn(OH)_6$

4. Balance the following half-reactions by the ion-electron method.

(a) $S_2O_3^{2-} \rightarrow SO_4^{2-}$ (acid)

(b) $CrO_2^- \rightarrow CrO_4^{2-}$ (base)

(c) $ClO_3^- \rightarrow Cl^-$ (acid)

(d) $H_2O_2 \rightarrow O_2$ (acid)

(e) $O_2^{2-} \rightarrow OH^-$ (base)

(f) $As \rightarrow AsO_4^{3-}$ (acid)

(g) $CN^- \rightarrow CNO^-$ (base)

(h) $PH_3 \rightarrow H_3PO_4$ (acid)

5. Balance the following redox equations by the ion-electron method.

(a) $Fe^{2+} + H_2O_2 + H^+ \rightarrow Fe^{3+} + H_2O$

(b) $OH^- + Cl_2 + IO_3^- \rightarrow IO_4^- + Cl^- + H_2O$

(c) $I_2 + H^+ + NO_3^- \rightarrow IO_3^- + NO_2 + H_2O$

(d) $H_2O_2 + ClO_2 + OH^- \rightarrow O_2 + ClO_2^- + H_2O$

(e) $Cr_2O_7^{2-} + H_2C_2O_4 + H^+ \rightarrow Cr^{3+} + CO_2 + H_2O$

Review Section B *Spontaneous and Nonspontaneous Redox Reactions*

OUTLINE / OBJECTIVES

13-5 Predicting Spontaneous Redox Reactions

1. Determining the comparative strength of oxidizing agents

   Rank the strength of two competing oxidizing agents and reducing agents from observance of the direction of a spontaneous reaction.

2. The inverse relationship between an oxidizing agent and its product

   Describe the inverse relationship between the oxidizing strength of a metal ion and the reducing strength of the corresponding metal.

3. Favorable redox reactions

   Predict the occurrence and write the equations for spontaneous redox reactions by use of an appropriate table.

13-6 Voltaic Cells

1. Voltaic cells: spontaneous redox reactions

   Describe how the potential energy of reactants and products relates to the reactions occurring in a voltaic cell.

2. Electrodes: anodes and cathodes
3. The Daniell cell, lead storage battery, dry cell, and fuel cell

   Write the anode, cathode, and overall reactions occurring in specified voltaic cells, including the common batteries or cells.

13-7 Electrolytic Cells

1. Electrolyte cells: nonspontaneous redox reactions
2. Electrolysis and electroplating

   Give some practical applications of electrolytic cells.

SUMMARY OF SECTION 13-5 THROUGH 13-7

Questions: *Can we predict spontaneous redox reactions? How do spontaneous reactions produce electricity? What are the applications of nonspontaneous reactions?*

What determines the direction of a spontaneous reaction? In the previous chapter recall that we found that a stronger acid reacts with a stronger base to produce a weaker acid and base (in the Bronsted-Lowry concept). A similar situation exists for redox reactions. That is, the stronger oxidizing agent reacts with the stronger reducing agent to produce weaker agents. For example, in the following reaction, (b) is the reverse of (a). How do we determine which is the spontaneous reaction and which is the nonspontaneous reaction?

$$\text{(a)}\ Zn + Cu^{2+} \rightarrow Zn^{2+} + Cu$$

$$\text{(b)}\ Cu + Zn^{2+} \rightarrow Cu^{2+} + Zn$$

The answer lies in the observations of a laboratory experiment. Since Zn forms a coating of Cu when immersed in Cu^{2+} solution, we can conclude that reaction (a) is spontaneous. Thus we conclude that Cu^{2+} is a stronger oxidizing agent than Zn^{2+}. Since there is an inverse relationship between the strength of an oxidizing agent (e.g., Cu^{2+}) and the reducing strength of its product (e.g., Cu), we can also say that Zn is a stronger reducing agent than Cu. With even more experiments, a table, such as Table 13-1 in the text, can be constructed with the strongest oxidizing agent at the top. Because of the inverse relationship, the species on the right are listed with the strongest reducing agent at the bottom. A spontaneous chemical reaction is predicted between an oxidizing agent and any reducing agent lying below it in the table. For example, the following reaction is predicted to be spontaneous.

$$Br_2 + Ni \rightarrow Ni^{2+} + 2\,Br^-$$

Table 13-1 also implies that water can be either oxidized or reduced. Certain nonmetals (e.g., F_2 and Cl_2) oxidize water to form an acid solution and oxygen. On the other hand, water oxidizes (is reduced by) several metals (e.g., Na and Mg) to form a hydroxide solution and hydrogen. Other metals (e.g., Fe and Ni) are not affected by water but are oxidized by strong acid (H^+) solutions. Some metals (e.g., Cu and Ag) are not affected by either water or most strong acid solutions. (Nitric acid oxidizes copper and silver due to the oxidizing action of the nitrate ion.)

In a spontaneous reaction, the products have lower potential energy than the reactants. (In most cases, anyway.) This difference in potential energy is what accounts for the heat, light, and electricity that is liberated by the reaction. In order for the energy to be evolved as electrical energy, we need a redox reaction that can be physically separated into its two half-reactions. The electron lost in the oxidation process can then be made to travel in an external circuit (e.g., a wire) where it eventually ends up in the reduction process. As the electrons are pushed through the wire they can be made to do work such as make a light glow, a car start, or a stereo blast. A **voltaic cell** converts chemical energy directly into electrical energy. Surfaces where half-reactions take place are called **electrodes**. The electrode where oxidation occurs is called the **anode**, and the electrode where reduction occurs is called the **cathode**. One or more voltaic cells arranged in a series is known as a **battery**. Examples of important batteries are the lead-acid battery used in the automobile, the dry cell, the Daniell cell, and the fuel cell. The half-reactions and the total reaction in the lead-acid battery are as follows.

anode: $Pb(s) + H_2SO_4(aq) \rightarrow PbSO_4(s) + 2H^+(aq) + 2e^-$

cathode: $PbO_2(s) + 2H^+(aq) + H_2SO_4(aq) + 2e^- \rightarrow PbSO_4(s) + 2H_2O$

total: $Pb(s) + PbO_2(s) + 2H_2SO_4(aq) \rightarrow 2PbSO_4(aq) + 2H_2O$

Some chemical reactions can be forced in the nonspontaneous direction if energy is supplied from an outside source. For example, the lead-acid battery in the automobile is especially useful because it can be forced in the reverse direction if electrical energy is supplied. After the engine starts, it supplies the energy that regenerates the battery. When electrical energy is thus converted to chemical energy, the cell is known as an **electrolytic cell.** Electrolytic cells are useful in processes such as **electrolysis** and electroplating. Also, large amounts of several elements such as aluminum and chlorine are freed from their compounds by electrolytic means.

NEW TERMS

Anode
Battery
Cathode
Daniell cell
Dry cell
Electrode
Electrolysis
Electrolytic cell
Fuel cell
Lead-acid cell
Voltaic cell

SELF-TEST

B-1 Multiple Choice

____ 1. A strong oxidizing agent

(a) is also a strong reducing agent.
(b) produces a strong reducing agent.
(c) produces a weak reducing agent.
(d) cannot be easily reduced.

____ 2. Which of the following elements does not react with pure water?

(a) Cl_2 (b) Na (c) Mg (d) Br_2

____ 3. Which of the following metals does not react with pure water but is oxidized by strong acid solution?

(a) Zn (b) Cu (c) Ag (d) Al

____ 4. A voltaic cell

(a) converts chemical energy to electrical energy
(b) proceeds in a nonspontaneous direction
(c) does not utilize a redox reaction
(d) is used to silverplate base metals

____ 5. In the following voltaic cell what reaction occurs at the anode?

$$Cu^{2+} + Fe \rightarrow Fe^{2+} + Cu$$

(a) Fe^{2+} is oxidized.
(b) Cu^{2+} is reduced.
(c) Fe is reduced.
(d) Fe is oxidized.
(e) Cu is oxidized.

____ 6. When molten $MgCl_2$ is electrolyzed which of the following occurs?

(a) H_2 is produced at the cathode.
(b) Mg is produced at the cathode.
(c) Cl_2 is produced at the cathode.
(d) Mg is produced at the anode.
(e) O_2 is produced at the cathode.

B-2 Problems

1. A partial list from Table 13-1 is shown below. The strongest oxidizing agent (on the left) is at the top and the strongest reducing agent (on the right) is at the bottom.

$$Br_2 + 2e^- \rightleftarrows 2Br^-$$

$$Cu^{2+} + 2e^- \rightleftarrows Cu$$

$$Fe^{2+} + 2e^- \rightleftarrows Fe$$

$$Mg^{2+} + 2e^- \rightleftarrows Mg$$

(a) Write a spontaneous reaction involving Cu and Br.

(b) A piece of copper (Cu) is placed in a solution of Fe^{2+} ion. Write the spontaneous reaction, if any.

(c) A piece of iron (Fe) is placed in a solution of Cu^{2+} ion. Write the spontaneous reaction, if any.

(d) A piece of iron is placed in a solution of Br^- ion. Write the spontaneous reaction, if any.

(e) Lead metal (Pb) forms a coating of Cu when placed in a Cu^{2+} solution but does not form a coating of Fe in a Fe^{2+} solution. Where does Pb^{2+} rank in the table as an oxidizing agent?

(f) Write the spontaneous reaction between the strongest oxidizing agent and the strongest reducing agent from the partial table.

2. Write balanced equations representing:

(a) the oxidation of water by F_2.

(b) the reduction of water by Al

(c) the reduction of an H^+ solution by Fe

Review Section C

CHAPTER SUMMARY SELF-TEST

C-1 Problems

1. The species below are arranged in a sequence such as in Table 13-1 in the text.

strongest oxidizing agent → $ClO_3^- \rightleftarrows Cl_2$

$MnO_2 \rightleftarrows Mn^{2+}$

$N_2O_4 \rightleftarrows NO$ ← strongest reducing agent

(a) Balance all three of the preceding half-reactions (acid solution).

(b) Write the balanced equation representing the spontaneous reaction involving the N_2O_4 - NO and the MnO_2 - Mn^{2+} half-reactions.

(c) In the reaction from part (b) indicate the following:

species oxidized ______ oxidizing agent ______

species reduced ______ reducing agent ______

(d) Write balanced equations representing two other spontaneous reactions.

Answers to Self-Tests

A-1 Multiple Choice

1. **e** none of these. (Since K is +1 and O is -2, the Br is +1.)
2. **d** +3 (Since the charge on the sulfate is SO_4^{2-}, Co is +3.)
3. **a** $K + O_2 \rightarrow KO_2$
4. **c** $Zn^{2+} + Ni^{2+} \rightarrow Zn + Ni$ (Both changes are reductions.)
5. **e** MnO_4^- (It is also reduced.)
6. **c** $+ 6e^-$ $(6e^- + 6H^+ + O_3 \rightarrow 3H_2O)$

A-2 Problems

1. (a) Mn +7
(b) Mn +6
(c) Sn +4
(d) O -1
(e) C -2
(f) Cr +2

2.

reaction	a	b	c
species oxidized	Cu	H_2S	$K_2Sn(OH)_4$
electrons lost	2	2	2
species reduced	HNO	$KMnO_4$	$Bi(OH)_3$
electrons gained*	3	5	3
oxidizing agent	HNO_3	$KMnO_4$	$Bi(OH)_3$
reducing agent	Cu	H_2S	$K_2Sn(OH)_4$

3. (a)

$$\overset{-2e^- \times \mathbf{3} = -6e^-}{\overbrace{\overset{0}{3Cu} + 8HNO_3 \rightarrow \overset{+2}{3Cu}(NO_3)_2}}$$

$$3Cu + \underset{+5}{8HNO_3} \rightarrow 3Cu(NO_3)_2 + \underset{+2}{2NO} + 4H_2O$$

$$+3e^- \times \mathbf{2} = +6\ e^-$$

In this equation six HNO_3 molecules are needed in addition to the two that are reduced. These provide NO_3^- ions for the Cu^{2+}.

(b) $-2e^- \times \mathbf{5} = -10\ e^-$ (S: $-2 \rightarrow 0$)

$$5H_2S + 2KMnO_4 + 6HCl \rightarrow 5S + 2KCl + 2MnCl_2 + 8H_2O$$

(Mn: $+7 \rightarrow +2$) $+\ 5e^- \times \mathbf{2} = +10\ e^-$

After balancing the oxidized and reduced species, you can balance the K's on the right. Then from the total Cl's, balance the HCl on the left. Finally, from the total H's on the left balance the H_2O's on the right.

(c) $-\ 3e^- \times \mathbf{2} = -\ 6\ e^-$ (Bi: $+3 \rightarrow 0$)

$$2Bi(OH)_3 + 3K_2Sn(OH)_4 \rightarrow 2Bi + 3K_2Sn(OH)_6$$

(Sn: $+2 \rightarrow +4$) $+\ 2e^- \times \mathbf{3} = +\ 6\ e^-$

4. (a) $5H_2O + S_2O_3^{2-} \rightarrow 2SO_4^{2-} + 10H^+ + 8e^-$

(b) $2H_2O + CrO_2^- \rightarrow CrO_4^{2-} + 4H^+ + 3e^-$ (acid)

$\mathbf{4OH^-} + 2H_2O + CrO_2^- \rightarrow CrO_4^{2-} + (4H^+ + \mathbf{4OH^-}) + 3e^-$

$4OH^- + 2H_2O + CrO_2^- \rightarrow CrO_4^{2-} + 4H_2O + 3e^-$

$4OH^- + CrO_2^- \rightarrow CrO_4^{2-} + 2H_2O + 3e^-$ (base)

(c) $6e^- + 6H^+ + ClO_3^- \rightarrow Cl^- + 3H_2O$

(d) $H_2O_2 \rightarrow O_2 + 2H^+ + 2e^-$

(e) $2e^- + 2H_2O + O_2^{2-} \rightarrow 4OH^-$

(f) $4H_2O + As \rightarrow AsO_4^{3-} + 8H^+ + 5e^-$

(g) $H_2O + CN^- \rightarrow CNO^- + 2H^+ + 2e^-$ (acid)

$\mathbf{2OH^-} + H_2O + CN^- \rightarrow CNO^- + (2H^+ + \mathbf{2OH^-}) + 2e^-$

$2OH^- + H_2O + CN^- \rightarrow CNO^- + 2H_2O + 2e^-$

$2OH^- + CN^- \rightarrow CNO^- + H_2O + 2e^-$ (base)

(h) $4H_2O + PH_3 \rightarrow H_3PO_4 + 8H^+ + 8e^-$

5. (a)

$Fe^{2+} \rightarrow Fe^{3+} + e^-$ | x 2

$2e^- + 2H^+ + H_2O_2 \rightarrow 2H_2O$ | x 1

$2Fe^{2+} + 2H^+ + H_2O_2 \rightarrow 2Fe^{3+} + 2H_2O$

(b)

$2OH^- + IO_3^- \rightarrow IO_4^- + H_2O + 2e^-$ | x 1

$2e^- + Cl_2 \rightarrow 2Cl^-$ | x 1

$2OH^- + Cl_2 + IO_3^- \rightarrow IO_4^- + 2Cl^- + H_2O$

(c)

$6H_2O + I_2 \rightarrow 2IO_3^- + 12H^+ + 10e^-$ | x 1

$e^- + 2H^+ + NO_3^- \rightarrow NO_2 + H_2O$ | x 10

$6H_2O + I_2 + 20\ H^+ + 10NO_3^- \rightarrow 2IO_3^- + 12H^+ + 10NO_2 + 10H_2O$

$I_2 + 8H^+ + 10NO_3^- \rightarrow 2IO_3^- + 10NO_2 + 4H_2O$

(d)

$2OH^- + H_2O_2 \rightarrow O_2 + 2H_2O + 2e^-$ | x 1

$e^- + ClO_2 \rightarrow ClO_2^-$ | x 2

$H_2O_2 + 2ClO_2 + 2OH^- \rightarrow 2ClO_2^- + O_2 + 2H_2O$

(e)

$6e^- + 14H^+ + Cr_2O_7^{2-} \rightarrow 2Cr^{3+} + 7H_2O$ | x 1

$H_2C_2O_4 \rightarrow 2CO_2 + 2H^+ + 2e^-$ | x 3

$8H^+ + Cr_2O_7^{2-} + 3H_2C_2O_4 \rightarrow 2Cr^{3+} + 7H_2O + 6CO_2$

B-1 Multiple Choice

1. **c** produces a weak reducing agent
2. **d** Br_2
3. **a** Zn
4. **a** converts chemical energy into electrical energy
5. **d** Fe is oxidized

6. **b** Mg is produced at the cathode (from reduction of Mg^{2+})

B-2 Problems

1. (a) $Cu + Br_2 \rightarrow Cu^{2+} + 2Br^-$
   (b) No spontaneous reaction
   (c) $Fe + Cu^{2+} \rightarrow Fe^{2+} + Cu$
   (d) No spontaneous reaction
   (e) $Pb^{2+} + 2e^- \rightarrow Pb$ half-reaction belongs between Cu^{2+} and Fe^{2+}.
   (f) $Br_2 + Mg \rightarrow Mg^{2+} + 2Br^-$

2. (a) $2F_2(g) + 2H_2O \rightarrow 4HF(aq) + O_2(g)$
   (Since HF is a weak acid, it is written in molecular form rather than as H^+ and F^-.)
   (b) $2Al(s) + 6H_2O(l) \rightarrow 2Al(OH)_3(s) + 3H_2(g)$
   [$Al(OH)_3$ is insoluble in water.]
   (c) $Fe(s) + 2H^+(aq) \rightarrow Fe^{2+}(aq) + H_2(g)$

C-1 Problems

1. (a) $10e^- + 12H^+ + 2ClO_3^- \rightleftarrows Cl_2 + 6H_2O$

   $2e^- + 4H^+ + MnO_2 \rightleftarrows Mn^{2+} + 2H_2O$

   $4e^- + 4H^+ + N_2O_4 \rightleftarrows 2NO + 2H_2O$

   (b) $4H^+ + 2MnO_2 + 2NO \rightarrow N_2O_4 + 2Mn^{2+} + 2H_2O$

   (c) NO is oxidized and is also the reducing agent.
   MnO_2 is reduced and is also the oxidizing agent.

   (d) $2ClO_3^- + 5Mn^{2+} + 4H_2O \rightarrow Cl_2 + 5MnO_2 + 8H^+$

   $4H^+ + 4ClO_3^- + 10NO \rightarrow 2Cl_2 + 2H_2O + 5N_2O_4$

Solutions to Black Text Problems

13-6 (a) $3(+1) + P + 4(-2) = 0$ — $P = +5$
(b) $2(+1) + 2C + 4(-2) = 0$ — $C = +3$
(c) $Cl + 4(-2) = -1$ — $Cl = +7$
(d) $(+2) + 2Cr + 7(-2) = 0$ — $Cr = +6$
(e) $S + 6(-1) = 0$ — $S = +6$
(f) $(+1) + N + 3(-2) = 0$ — $N = +5$
(g) $(+1) + Mn + 4(-2) = 0$ — $Mn = +7$

13-16

$-5e^- \times 4 = -20e^-$ (N: $-3 \rightarrow +2$)

(a) $4NH_3 + 5O_2 \rightarrow 4NO + 6H_2O$

$+4e^- \times 5 = +20e^-$ (O: $0 \rightarrow -2$)

$-4e^- \times 1 = -4e^-$ (Sn: $0 \rightarrow +4$)

(b) $Sn + 4HNO_3 \rightarrow SnO_2 + 4NO_2 + 2H_2O$

$+1e^- \times 4 = +4e^-$ (N: $+5 \rightarrow +4$)

(c) Before the number of electrons lost is calculated, notice that a temporary coefficient of "2" is needed for the Na_2CrO_4 in the products since there are 2 Cr's in Cr_2O_3 in the reactants.

$+2e^- \times 3 = +6e^-$ (N: $+5 \rightarrow +3$)

$Cr_2O_3 + 2Na_2CO_3 + 3KNO_3 \rightarrow 2CO_2 + 2Na_2CrO_4 + 3KNO_2$

$-6e^- \times 1 = -6e^-$ (Cr: $+6 \rightarrow +12$)

$-4e^- \times 3 = -12e^-$ (Se: $0 \rightarrow +4$)

(d) $3Se + 2BrO_3^- + 3H_2O \rightarrow 3H_2SeO_3 + 2Br^-$

$+6e^- \times 2 = +12e^-$ (Br: $+5 \rightarrow -1$)

13-20 (a)

$$S^{2-} \rightarrow S + 2e^- \quad \times 3$$
$$3e^- + 4H^+ + NO_3^- \rightarrow NO + 2H_2O \quad \times 2$$
$$3S^{2-} + 8H^+ + 2NO_3^- \rightarrow 3S + 2NO + 4H_2O$$

(b)

$$2S_2O_3^{2-} \rightarrow S_4O_6^{2-} + 2e^- \quad \times 1$$
$$2e^- + I_2 \rightarrow 2I^- \quad \times 1$$
$$2S_2O_3^{2-} + I_2 \rightarrow S_4O_6^{2-} + 2I^-$$

(c)

$$H_2O + SO_3^{2-} \rightarrow SO_4^{2-} + 2H^+ + 2e^- \quad \times 3$$
$$6e^- + 6H^+ + ClO_3^- \rightarrow Cl^- + 3H_2O \quad \times 1$$
$$3SO_3^{2-} + ClO_3^- \rightarrow Cl^- + 3SO_4^{2-}$$

(d)

$$Fe^{2+} \rightarrow Fe^{3+} + e^- \quad \times 2$$
$$2e^- + 2H^+ + H_2O_2 \rightarrow 2H_2O \quad \times 1$$
$$2H^+ + 2Fe^{2+} + H_2O_2 \rightarrow 2Fe^{3+} + 2H_2O$$

(e)

$$2I^- \rightarrow I_2 + 2e^- \quad \times 1$$
$$2e^- + 2H^+ + AsO_4^{3-} \rightarrow AsO_3^{3-} + H_2O \quad \times 1$$
$$AsO_4^{3-} + 2I^- + 2H^+ \rightarrow I_2 + AsO_3^{3-} + H_2O$$

(f)

$$Zn \rightarrow Zn^{2+} + 2e^- \quad \times 4$$
$$8e^- + 10H^+ + NO_3^- \rightarrow NH_4^+ + 3H_2O \quad \times 1$$
$$4Zn + NO_3^- + 10H^+ \rightarrow 4Zn^{2+} + NH_4^+ + 3H_2O$$

13-22 (a) $H_2O + SnO_2^{2-} \rightarrow SnO_3^{2-} + 2H^+ + 2e^-$ (acid)

$\underline{2OH^-} + H_2O + SnO_2^{2-} \rightarrow SnO_3^{2-} + (2H^+ + \underline{2OH^-} = 2H_2O) + 2e^-$

$2OH^- + SnO_2^{2-} \rightarrow SnO_3^{2-} + H_2O + 2e^-$

(b) $6e^- + 8H^+ + 2ClO_2^- \rightarrow Cl_2 + 4H_2O$ (acid)

$6e^- + (8H^+ + \underline{8OH^-} = 8H_2O) + 2ClO_2^- \rightarrow Cl_2 + 4H_2O + \underline{8OH^-}$

$6e^- + 4H_2O + 2ClO_2^- \rightarrow Cl_2 + 8OH^-$

(c) $3H_2O + Si \rightarrow SiO_3^{2-} + 6H^+ + 4e^-$ (acid)

$\underline{6OH^-} + 3H_2O + Si \rightarrow SiO_3^{2-} + (6H^+ + \underline{6OH^-} = 6H_2O) + 4e^-$

$6OH^- + Si \rightarrow SiO_3^{2-} + 3H_2O + 4e^-$

(d) $8e^- + 9H^+ + NO_3^- \rightarrow NH_3 + 3H_2O$ (acid)

$8e^- + (9H^+ + \underline{9OH^-} = 9H_2O) + NO_3^- \rightarrow NH_3 + 3H_2O + \underline{9OH^-}$

$8e^- + 6H_2O + NO_3^- \rightarrow NH_3 + 9OH^-$

13-24 (a)

$8OH^- + S^{2-} \rightarrow SO_4^{2-} + 4H_2O + 8e^-$ | x 1

$I_2 + 2e^- \rightarrow 2I^-$ | x 4

$S^{2-} + 8OH^- + 4I_2 \rightarrow SO_4^{2-} + 8I^- + 4H_2O$

(b)

$8OH^- + I^- \rightarrow IO_4^- + 4H_2O + 8e^-$ | x 1

$e^- + MnO_4^- \rightarrow MnO_4^{2-}$ | x 8

$8OH^- + I^- + 8MnO_4^- \rightarrow 8MnO_4^{2-} + IO_4^- + 4H_2O$

(c) $2OH^- + SnO_2^{2-} \rightarrow SnO_3^{2-} + H_2O + 2e^-$ | x 1

$2e^- + 3H_2O + BiO_3^- \rightarrow Bi(OH)_3 + 3OH^-$ | x 1

$2H_2O + SnO_2^{2-} + BiO_3^- \rightarrow SnO_3^{2-} + Bi(OH)_3 + OH^-$

(d) $32OH^- + CrI_3 \rightarrow CrO_4^{2-} + 3IO_4^- + 16H_2O + 27e^-$ | x 2

$2e- + Cl_2 \rightarrow 2Cl^-$ | x 27

$2CrI_3 + 64OH^- + 27Cl_2 \rightarrow 2CrO_4^{2-} + 6IO_4^- + 32\ H_2O + 54Cl^-$

13-26 (a)

$$H_2 \rightarrow 2H^+ + 2e^- \qquad \times 2$$
$$4e^- + 4H^+ + O_2 \rightarrow 2H_2O \qquad \times 1$$
$$2H_2 + O_2 \rightarrow 2H_2O$$

$$2OH^- + H_2 \rightarrow 2H_2O + 2e^- \qquad \times 2$$
$$4e^- + 2H_2O + O_2 \rightarrow 4OH^- \qquad \times 1$$
$$2H_2 + O_2 \rightarrow 2H_2O$$

(b)

$$H_2O_2 \rightarrow O_2 + 2H^+ + 2e^- \qquad \times 1$$
$$2e^- + 2H^+ + H_2O_2 \rightarrow 2H_2O \qquad \times 1$$
$$2H_2O_2 \rightarrow O_2 + 2H_2O$$

$$2OH^- + H_2O_2 \rightarrow O_2 + 2H_2O + 2e^- \qquad \times 1$$
$$2e^- + H_2O_2 \rightarrow 2OH^- \qquad \times 1$$
$$2H_2O_2 \rightarrow O_2 + 2H_2O$$

13-40 The spontaneous reaction is:

$$Pb(NO_3)_2(aq) + Fe(s) \rightarrow Fe(NO_3)_2(aq) + Pb(s)$$

Anode: $Fe \rightarrow Fe^{2+} + 2e^-$

cathode: $Pb^{2+} + 2e^- \rightarrow Pb$

In the anode compartment a piece of Fe is immersed in a $Fe(NO_3)_2$ solution. In the cathode compartment a piece of Pb is immersed in a $Pb(NO_3)_2$ solution. The electrodes are connected to the motor (or light bulb, etc.) by wires and the two compartments are connected by a salt bridge.

13-52

$$Zn \rightarrow Zn^{2+} + 2e^- \qquad \times 5$$
$$10e^- + 12H^+ + 2NO_3^- \rightarrow N_2 + 6H_2O \qquad \times 1$$
$$12H^+(aq) + 5Zn(s) + 2NO_3^-(aq) \rightarrow 5Zn^{2+}(aq) + N_2(g) + 6H_2O(l)$$

g N_2 → mol N_2 → mol Zn → g Zn

$$0.658 \text{ g } N_2 \times \frac{1 \text{ mol } N_2}{28.02 \text{ g } N_2} \times \frac{5 \text{ mol Zn}}{1 \text{ mol } N_2} \times \frac{65.39 \text{ g}}{\text{mol Zn}} = \underline{7.68 \text{ g Zn}}.$$

13-54

$$3e^- + 4H^+ + NO_3^- \rightarrow NO + 2H_2O \quad \times 2$$
$$2H^+ + Cu_2O \rightarrow 2Cu^{2+} + H_2O + 2e^- \quad \times 3$$
$$14H^+(aq) + 2NO_3^-(aq) + 3Cu_2O(s) \rightarrow 6Cu^{2+}(aq) + 2NO(g) + 7H_2O(l)$$

g Cu_2O → mol Cu_2O → mol NO → vol. NO

$$10.0\ \text{g } Cu_2O \times \frac{1\ \text{mol } Cu_2O}{143.1\ \text{g } Cu_2O} \times \frac{2\ \text{mol NO}}{3\ \text{mol } Cu_2O} \times \frac{22.4\ \text{L(STP)}}{\text{mol NO}} = \underline{1.04\ \text{L of NO}}\ \text{(STP)}$$

13-56

$$4OH^- + Zn \rightarrow Zn(OH)_4^{2-} + 2e^- \quad \times 4$$
$$8e^- + 6H_2O + NO_3^- \rightarrow NH_3 + 9OH^- \quad \times 1$$
$$7OH^-(aq) + 4Zn(s) + 6H_2O(l) + NO_3^-(aq) \rightarrow NH_3(g) + 4Zn(OH)_4^{2-}(aq)$$

g Zn → mol Zn → mol NH_3 → vol. NH_3

$$6.54\ \text{g Zn} \times \frac{1\ \text{mol Zn}}{65.39\ \text{g Zn}} \times \frac{1\ \text{mol } NH_3}{4\ \text{mol Zn}} = 0.0250\ \text{mol } NH_3$$

$$V = \frac{nRT}{P} = \frac{0.0250\ \text{mol} \times 0.0821 \frac{\text{L} \cdot \text{atm}}{\text{K} \cdot \text{mol}} \times 300\ \text{K}}{1.25\ \text{atm}} = \underline{0.493\ \text{L}}$$

13-57 (a)

$$MnO_4^- + 8H^+ + 5e^- \rightarrow Mn^{2+} + 4H_2O \quad \times 1$$
$$Fe^{2+} \rightarrow Fe^{3+} + e^- \quad \times 5$$
$$MnO_4^- + 8H^+ + 5Fe^{2+} \rightarrow 5Fe^{3+} + Mn^{2+} + 4H_2O$$

(b)

$$MnO_4^- + 8H^+ + 5e^- \rightarrow Mn^{2+} + 4H_2O \quad \times 2$$
$$2Br^- \rightarrow Br_2 + 2e^- \quad \times 5$$
$$2MnO_4^- + 16H^+ + 10Br^- \rightarrow 5Br_2 + 2Mn^{2+} + 8H_2O$$

(c)

$$MnO_4^- + 8H^+ + 5e^- \rightarrow Mn^{2+} + 4H_2O \quad \times 2$$
$$C_2O_4^{2-} \rightarrow 2CO_2 + 2e^- \quad \times 5$$
$$2MnO_4^- + 16H^+ + 5C_2O_4^{2-} \rightarrow 10CO_2 + 2Mn^{2+} + 8H_2O$$

13-58 **g $FeCl_2$ → mol $FeCl_2$ (mol Fe^{2+}) → mol MnO_4^- (mol $KMnO_4$) → vol. $KMnO_4$**

$$25.0 \text{ g } FeCl_2 \times \frac{1 \text{ mol } FeCl_2}{126.8 \text{ g } FeCl_2} \times \frac{1 \text{ mol } KMnO_4}{5 \text{ mol } FeCl_2} = 0.0394 \text{ mol } KMnO_4$$

$$V = \frac{n}{M} = \frac{0.0394 \text{ mol}}{0.220 \text{ mol/L}} = 0.179 \text{ L} = \underline{179 \text{ mL}}$$

13-59 **vol. $KMnO_4$ → mol $KMnO_4$(MnO_4^-) → mol Br^-(KBr) → vol. KBr**

$$0.125 \text{ L} \times \frac{0.220 \text{ mol } KMnO_4}{\text{L}} \times \frac{10 \text{ mol KBr}}{2 \text{ mol } KMnO_4} \times \frac{1 \text{ L KBr}}{0.450 \text{ mol KBr}} = 0.306 \text{ L} = \underline{306 \text{ mL}}$$

14

Reaction Rates and Equilibrium

Review Section A *Collisions of Molecules and Reactions at Equilibrium*

OUTLINE

OBJECTIVES

14-1 How Reactions Take Place

1. The collision theory of chemical reactions
2. Conditions necessary for a reaction to take place

Describe the two conditions necessary for reactants to be transformed into products.

3. Activation energy

Show how an activated complex forms and how activation energy relates to exothermic and endothermic reactions.

14-2 Rates of Chemical Reactions

1. Factors that affect the rate of a reaction
   (a) activation energy
   (b) temperature

Discuss the two factors that increase the rate of a reaction.

   (c) concentration of reactants
   (d) particle size
   (e) catalysts

Describe how temperature, concentration, particle size, and catalysts affect the rate of a reaction.

14-3 Reactions at Equilibrium

1. Reversible reactions and dynamic equilibrium

Illustrate how a reversible reaction demonstrates dynamic equilibrium in terms of reaction rates.

2. Examples of systems at equilibrium
   (a) gas phase reactions
   (b) liquid-vapor
   (c) saturated solutions of a salt
   (d) weak acids and bases

Explain how equilibrium is demonstrated in previous examples cited in the text.

14-4 Point of Equilibrium and Le Chatelier's Principle

1. Le Chatelier's principle and changing the position of equilibrium
   (a) Changing concentration by addition or removal of a component
   (b) Changing the volume and pressure
   (c) Changing the temperature of the reaction mixture

Apply the principle of Le Chatelier to show how changes in concentrations, temperature, and pressure affect a system at equilibrium.

   (d) Adding a catalyst

Describe the role of a catalyst in a reaction that reaches a point of equilibrium.

SUMMARY OF SECTIONS 14-1 THROUGH 14-4

Questions: *How do reactants become transformed into products? What is meant by the energy barrier for a reaction? Do all reactions go to completion? What affects how fast a reaction takes place? Can we affect the point at which a reaction seems to stop?*

Collision theory tells us that reactants are transformed through collisions of reactant molecules or ions with each other. All collisions between reactants do not lead to products, however, since collisions must have the proper orientation and a minimum amount of energy known as the

activation energy. This is the minimum kinetic energy that colliding molecules must have in order for old bonds to break so that new ones can form. The lower the activation energy, the faster the reaction. At maximum impact, colliding molecules form an **activated complex**, which then rearranges to form products. The activation energy is equal to the difference in energy between reactants and the activated complex.

Chemical reactions take place at a wide range of rates. The chemical reaction in an explosion is almost instantaneous, whereas the reaction in the aging of a fine wine takes years. The rate at which a particular reaction takes place is affected by five factors. They are as follows:

(1) *The activation energy.* This depends on the particular reaction.
(2) *The temperature.* The temperature relates to the average kinetic energy of the molecules. Thus, the higher the temperature, the more molecules that will have the minimum kinetic energy (the activation energy) for a reaction to occur. The rate of a reaction is also proportional to the frequency of collisions between reacting molecules. Since a higher temperature increases the average velocity of the molecules, collisions become more frequent at higher temperatures which also increases the rate.
(3) *The concentration of reactants..* Concentration also relates to the frequency of collisions. The more concentrated the reactants, the more frequent the collisions and the faster is the rate of the reaction.
(4) *Particle size.* In a heterogeneous reaction, the more surface area for a particular reactant, the more frequent the collisions and the faster the reaction.
(5) *Catalysts.* A **catalyst** provides an alternate reaction pathway with a lower activation energy and thus increases the rate of the reaction.

The hypothetical reaction discussed in this text

$$A_2(g) + B_2(g) \rightleftharpoons 2AB(g)$$

is a **reversible reaction** that reaches a **point of equilibrium** where both the forward and reverse reactions occur under the same conditions. This phenomenon is sometimes referred to as a **dynamic equilibrium** since the identity of reactants and products is constantly changing although the relative concentrations are constant.

Equilibrium phenomena have been discussed several times in previous chapters. For example, we discussed gas phase equilibria in Chapter 8, liquid-vapor equilibria in Chapter 10, the equilibrium between a salt and its saturated solution in Chapter 11, and finally, the equilibria of weak acids and bases in Chapter 12.

A system at equilibrium remains at equilibrium as long as there is no change in conditions. Should any condition change, such as temperature, volume of the container, pressure, or the concentration of a reactant or product, the point of equilibrium changes. According to **Le Chatelier's principle**, a system at equilibrium shifts to counteract any stress. For example, consider the commercially important process used to make ammonia for fertilizers.

$$N_2(g) + 3H_2(g) \rightleftharpoons 2NH_3(g) + \text{heat}$$

The following conditions increase the concentration of NH_3 present at equilibrium.

1. A large concentration of N_2 and/or H_2

   Increased concentration of substances on one side of an equation leads eventually to an increase on the other.

2. Compression of the reaction mixture

   Compression of the reaction mixture means that the reaction occurs at a higher pressure and a lower volume. A low-volume container favors the side with the fewer moles of gas. In this case, four moles of reactants combine to form two moles of products. Therefore, the equilibrium shifts to the right when the reaction mixture is compressed.

3. A low temperature

   Heat can be considered as a reactant in endothermic reactions or as a product in exothermic reactions. Addition or removal of heat affects the point of equilibrium just like addition or removal of any other component. In this case, removal of heat by lowering the temperature causes the equilibrium to shift to the right.

4. Addition of a catalyst

   It takes time for equilibrium to be established (the lower the temperature the more time it takes). A catalyst is a substance that increases the rate of the reaction so that equilibrium is achieved faster. (It does not affect the eventual distribution of reactants and products.) In the industrial process just discussed, the presence of a catalyst counteracts the decreased rate of the reaction brought on by the lower temperature. Still, the reaction is run at about 400°C.

NEW TERMS

Activated complex
Activation energy
Chemical kinetics
Collision theory
Dynamic equilibrium
Le Chatelier's principle
Rate of reaction
Reversible reaction

SELF-TEST

A-1 Multiple Choice

____ 1. Which of the following is not an example of a reaction that reaches a point of equilibrium?

(a) the reaction of N_2 and H_2 to produce NH_3
(b) the ionization of a weak acid
(c) the ionization of a strong acid
(d) the formation of crystals in a saturated solution of a salt

____ 2. Which of the following is not an assumption of collision theory?

(a) Molecules react through collisions with each other.
(b) Colliding molecules must have a minimum energy to react.
(c) Molecules react because they collide with the sides of the container.
(d) Colliding molecules must have the proper orientation.

____ 3. A reaction at equilibrium

(a) goes to completion.
(b) goes so far to the right then stops.
(c) does not go to the right.
(d) goes to the right and to the left simultaneously.
(e) evolves a gaseous product.

____ 4. When two reactants are mixed and come to a point of equilibrium, when does the rate of the reverse reaction reach a maximum?

(a) at the beginning of the reaction
(b) when the rate of the forward reaction is at a maximum
(c) at equilibrium
(d) when the rate of the forward reaction is decreasing

____ 5. Phosphorus burns spontaneously in air, but coal must be ignited (heated) before it begins to burn. What conclusion can we draw?

(a) The combustion of phosphorus has a higher activation energy than the combustion of coal.
(b) The concentration of phosphorus is greater than that of coal.
(c) More phosphorus molecules have the proper orientation during collisions at room temperature.
(d) The combustion of coal has a higher activation energy than the combustion of phosphorus.

____ 6. Increasing the temperature of a reaction mixture

(a) increases the average velocity of the molecules.
(b) increases the average energy of colliding molecules.
(c) both (a) and (b).
(d) increases the average mass of the colliding molecules.
(e) has little effect on the rate of a reaction.

____ 7. Increasing the concentrations of reactant molecules

(a) increases the average velocity of the molecules.
(b) increases the frequency of collisions between molecules.
(c) increases the average energy of the colliding molecules.
(d) increases the temperature.
(e) decreases the volume of the container.

____ 8. Hot iron metal burns rapidly in pure oxygen but slowly in air. This difference implies

(a) the activation energy is lower in pure oxygen.
(b) the temperature is higher in pure oxygen.
(c) the concentration of oxygen is important in the rate of the reaction.
(d) collisions in pure oxygen are more energetic.

____ 9. Which of the following does not contribute to a fast rate of reaction?

(a) a low activation energy
(b) a low temperature
(c) a high concentration of reactants
(d) a high temperature

____ 10. Given the gaseous equilibrium: $H_2(g) + I_2(g) \rightleftharpoons 2HI(g)$, which of the following happens at equilibrium if the pressure on the system is increased at constant temperature?

(a) shifts to the right (b) shifts to the left (c) no effect

A-2 Problem

1. Assume the following system at equilibrium:

$$\text{heat} + 4HCl(g) + O_2(g) \rightleftharpoons 2Cl_2(g) + 2H_2O(g)$$

Determine what effect the following changes will have on the amount of O_2 present at equilibrium.

(a) heat the reaction mixture ____________________

(b) increase [HCl] ____________________

(c) decrease $[Cl_2]$ ____________________

(d) add a catalyst ____________________

(e) compress the reaction mixture ____________________

(f) add some $H_2O(g)$ ____________________

(g) increase the volume of the container ____________________

Review Section B *The Quantitative Aspects of Reactions at Equilibrium*

OUTLINE

14-5 Point of Equilibrium and the Law of Mass Action

1. The position of equilibrium: the law of mass action

OBJECTIVES

Write the law of mass action for any homogeneous reaction.

2. The magnitude of the equilibrium constant	*Describe how the magnitude of the equilibrium constant expression predicts the point of equilibrium.*
3. Calculation of the value of K_{eq}	*Calculate the value of the equilibrium constant or the concentration of a component at equilibrium given appropriate experimental data.*

14-6 Equilibria of Weak Acids and Weak Bases in Water

1. The law of mass action for weak acids and bases	*Describe how the values of K_a and K_b relate to the strength of a weak acid or weak base.*
2. Calculation of K_a and K_b	*Use equilibrium concentrations of all species or percent ionization to calculate K_a or K_b.*
3. pH of a weak acid or a weak base solution	*Use K_a or K_b and initial concentrations to calculate the pH of a solution.*
4. pH of a buffer solution	*Use the Henderson-Hasselbalch equation to calculate the pH of a buffer solution.*

14-7 Solubility Equilibria

1. The Solubility Product, K_{sp}	*Calculate the value of K_{sp}. form molar solubility data.*
2. Solubility data from K_{sp}.	*Calculate molar solubility from the value of K_{sp}.*
3. The concentration of specific ions.	*Calculate the concentration of one ion from the concentration of the other ion and K_{sp}.*

SUMMARY OF SECTION 14-5 AND 14-6

Questions: *Can the concentration of reactants and products in a given reaction at equilibrium be predicted? Can the acidity of a given weak acid or a weak base be predicted? How does the solubility of ionic compounds relate to equilibrium?*

It has long been known that reactants and products are distributed in proportions at equilibrium as predicted by the **law of mass action**. For the gaseous reaction

$$H_2(g) + I_2(g) \rightleftharpoons 2HI(g)$$

the law of mass action can be written as follows:

$$K_{eq} = \frac{[HI]^2}{[H_2][I_2]} \qquad [X] = \text{mol/L}$$

K_{eq} is known as the **equilibrium constant.** Its numerical value gives some indication of the distribution of reactants and products at equilibrium *at a specified temperature*. A large value tells us that products are favored, whereas a small value tells us that reactants are favored. The fraction to the right of the equal sign is known as the **equilibrium constant expression**. It is written by placing the concentrations of products in the numerator, with each product raised to the power of the coefficient of that product, and by placing reactants in the denominator, also raised to the power of their respective coefficients. For example, the mass action expression for the hypothetical equilibrium

$$3A(g) + B(g) \rightleftharpoons C(g) + 2D(g)$$

is

$$K_{eq} = \frac{[C][D]^2}{[A]^3[B]}$$

The value of K_{eq} at a specified temperature is found from a measurement of the concentrations of reactants and products at equilibrium.

Example B-1 Calculation of the value of K_{eq}

What is the value of K_{eq} for the following reaction?

$$3A(g) + B(g) \rightleftharpoons C(g) + 2D(g)$$

If at equilibrium [A] = 0.0112, [B] = 0.0314, [C] = 0.0667, and [D] = 0.0432.

PROCEDURE

Write the equilibrium constant expression and substitute appropriate concentrations.

SOLUTION

$$K_{eq} = \frac{[C][D]^2}{[A]^3[B]} = \frac{(0.0667)(0.0432)^2}{(0.0112)^3(0.0314)} = \underline{2.82 \times 10^3}$$

Sometimes the concentrations of all species at equilibrium are implied by information about the initial and final concentrations of one reactant or product. An example follows.

Example B-2 Calculation of the value of K_{eq}

Assume the hypothetical equilibrium $2E(g) + F(g) \rightleftharpoons G(g) + 2H(g)$.
Initially, only E and F are present, both at a concentration of 1.00 mol/L. At equilibrium [F] = 0.60 mol/L. What is the value of K_{eq}?

PROCEDURE

Since [F] is known at equilibrium, the concentrations of all other species are implied by the stoichiometry of the reaction. The concentration of the F that *reacted* is

$$[F]_{reacted} = 1.00 - 0.60 = 0.40 \text{ mol/L.}$$

From the stoichiometry of the reaction, notice that for every one mole of F that reacts, one mole of G and two moles of H are formed.
Therefore, at equilibrium 0.40 mol/L of F yields the following:

$$[G] = 0.40 \text{ mol/L} \qquad 0.40 \text{ mol/L F} \times \frac{2 \text{ mol/L H}}{1 \text{ mol/L F}} = 0.80 \text{ mol/L H}$$

Now we must find [E] at equilibrium. Again from the stoichiometry of the reaction, if 0.40 mol/L of F reacted, 0.80 mol/L of E reacted (two moles of E per mole of F). Therefore, at equilibrium

$$[E] = 1.00 - 0.80 = 0.20 \text{ mol/L}$$

SOLUTION

$$K_{eq} = \frac{[G][H]^2}{[E]^2[F]} = \frac{(0.40)(0.80)^2}{(0.20)^2(0.60)} = \underline{11}$$

When the equilibrium constant is known, the concentration of a substance can be calculated if the concentrations of all other species in the reaction are known or implied.

Example B-3 Calculation of a component using the value of K_{eq}

In the hypothetical reaction discussed in problem B-2, what is the concentration of F at equilibrium if [E] = 0.40, [G] = 0.55, and [H] = 0.18?

PROCEDURE

$$K_{eq} = \frac{[G][H]^2}{[E]^2[F]} \qquad \text{Solving for [F]} \qquad [F] = \frac{[G][H]^2}{[E]^2 K_{eq}}$$

SOLUTION

$$[F] = \frac{(0.55)(0.18)^2}{(0.40)^2 \cdot 11} = \underline{0.010 \text{ mol/L}}$$

We can now discuss the equilibrium involved in weak acids and weak bases. $\mathbf{K_a}$ represents an **acid ionization constant** and $\mathbf{K_b}$ represents a **base ionization constant.** Examples of the equilibria, and the laws of mass action follow.

weak acid $\quad HNO_2 + H_2O \rightleftharpoons H_3O^+ + NO_2^- \qquad K_a = \frac{[H_3O^+][NO_2^-]}{[HNO_2]}$

weak base $\quad NH_3 + H_2O \rightleftharpoons NH_4^+ + OH^- \qquad K_b = \frac{[NH_4^+][OH^-]}{[NH_3]}$

The values of the respective constants tell us how far to the right the equilibrium lies. The smaller the constant the more the equilibrium lies to the left, and thus the weaker the acid or base. The values of the constants are determined from experiments as illustrated in the following example.

Example B-4 Calculation of the value of K_a

The pH of a 0.10 M solution of HNO_2 is 2.17. What is K_a for HNO_2?

PROCEDURE

$$HNO_2 + H_2O \rightleftharpoons H_3O^+ + NO_2^- \qquad K_a = \frac{[H_3O^+][NO_2^-]}{[HNO_2]}$$

pH = 2.17 or $-\log [H_3O^+] = 2.17$ $\quad \log [H_3O^+] = -2.17$ $\quad [H_3O^+] = 6.8 \times 10^{-3}$

In this solution at equilibrium

$[H_3O^+] = [NO_2^-] = 6.8 \times 10^{-3}$ $\quad [HNO_2] = 0.10 - 0.0068 \approx 0.$

(0.0068 is negligible compared to 0.10)

SOLUTION

$$K_a = \frac{(6.8 \times 10^{-3})(6.8 \times 10^{-3})}{(0.10)} = \underline{4.6 \times 10^{-4}}$$

Once the constants are known, they can be used to calculate the pH of a solution of any original concentration of acid as follows.

Example B-5 Calculation of the pH of a weak acid solution

What is the pH of a 0.027 M solution of HNO_2?

PROCEDURE

In this case, the unknown is $[H_3O^+]$ which we say is X. Therefore,

$[H_3O^+] = [NO_2^-] = X$ and $[HNO_2] = 0.027 - X$

(If X mol/L of H_3O^+ is formed, then X mol/L of HNO_2 reacted. Therefore, the HNO_2 left at equilibrium is the original concentration minus X.)

Since X is probably small compared to 0.027, we can attempt the following approximation.

$$[HNO_2] = 0.027 - X \approx 0.027$$

SOLUTION

$$\frac{[H_3O^+]\,[NO_2^-]}{[HNO_2]} = \frac{X \cdot X}{0.027} = Ka = 4.6 \times 10^{-4}$$

$X^2 = 0.12 \times 10^{-4}$ $\quad X = 3.5 \times 10^{-3}$ $\quad pH = -\log (3.5 \times 10^{-3}) = \underline{2.46}$

[Notice X (0.0012) is less than 10% of the original concentration (0.027).
The approximation is therefore valid.

Buffer solutions contain a weak acid (or weak base) and a salt providing its conjugate base (or acid). Calculation of the pH of buffers is simplified since one can consider the initial concentrations of acid and conjugate base to be essentially unchanged by any ionization. By rearranging the mass acition expression for a weak acid and using logarithms, we can derive the **Henderson-Hasselbalch equation** which is used specifically for buffer solutions.

$$pH = pK_a + \log \frac{[\text{base}]}{[\text{acid}]} \quad \text{or} \quad pOH = pK_b + \log \frac{[\text{acid}]}{[\text{base}]}$$

Example B-6 Calculation of the pH of a buffer solution

What is the pH of a solution made by dissoving 0.024 mol of hydrofluoric acid (HF) and 0.082 mol of sodium fluoride (NaF) in 2.0 L of water? For HF, $K_a = 6.7 \times 10^{-4}$

PROCEDURE

$HF + H_2O \rightleftharpoons H_3O^+ + F^-$ $\quad pK_a = -\log K_a = -\log(6.7 \times 10^{-4}) = 3.17$

$NaF(s) \rightarrow Na^+(aq) + F^-(aq)$ (The Na^+ is a spectator ion.)

HF is the acid species so [acid] = 0.024 mol/2.0L
F^- is the base species so [base] = 0.082 mol/2.0L
For a buffer solution use $pH = pK_a + \log \frac{[\text{base}]}{[\text{acid}]}$

SOLUTION

$$pH = 3.17 + \log \frac{0.082 \cancel{\text{mol/2.0 L}}}{0.024 \cancel{\text{mol/2.0 L}}} = 3.17 + 0.53 = \underline{3.70}$$

The solubility of an ionic compound in water is also a dynamic equilibrium. The equilibrium constant which is known as the **solubility product (K_{sp})** is simply equal to the concentrations of the two ions each raised to a power equal to their coefficients. For example, the solubility of the compound $Ba_3(PO_4)_2$ is set up as follows.

$$Ba_3(PO_4)_2(s) \rightleftharpoons 3Ba^{2+}(aq) + 2PO_4^{3-}(aq) \qquad K_{sp} = [Ba^{2+}]^3[PO_4^{3-}]^2$$

By knowing the molar solubility, the value of K_{sp} can be calculated. On the other hand, if the value of K_{sp} is known the molar solubility can be calculated. Also, if K_{sp} and the concentration of one ion is known, the concentration of the other can be calculated. The solubility product can also be used to determine whether a precipitate will or will not form from given concentrations of the two ions. If the concentrations of ions are substituted into the equilibrium constant expression and the result is greater than the value of K_{sp} then a precipitate forms.

Example B-7 Calculation of K_{sp} from the Molar Solubility.

The molar solubility of Ag_3PO_4 is 4.4 x 10^{-5} mol/L. What is the value of K_{sp}?

PROCEDURE

The equilibrium involved is $Ag_3PO_4(s) \rightleftharpoons 3Ag^{+}(aq) + PO_4^{3-}(aq)$

Notice for each mole of Ag_3PO_4 that dissolves three moles of Ag^{+} and one mole of PO_4^{3-} are present in solution. Therefore, in solution

$$[Ag^{+}] = 3 \text{ x solubility} = 3 \text{ x } 4.4 \text{ x } 10^{-5} \text{ mol/L} = 1.32 \text{ x } 10^{-4} \text{ mol/L}$$
$$[PO_4^{3-}] = \text{solubility} = 4.4 \text{ x } 10^{-5} \text{ mol/L}$$

SOLUTION

$$K_{sp} = [Ag^{+}]^3[PO_4^{3-}] = [1.32 \text{ X } 10^{-4}]^3[4.4 \text{ X } 10^{-5}] = \underline{1.0 \text{ x } 10^{-17}}$$

Example B-8 Calculation of Molar Solubility from the value of K_{sp}.

What is the molar solubility of $NiCO_3$? For $NiCO_3$, K_{sp} = 1.45 x 10^{-7}

PROCEDURE

The equilibrium involved is $NiCO_3(s) \rightleftharpoons Ni^{2+}(aq) + CO_3^{2-}(aq)$

If we let X = the molar solubility, then at equilibrium

$[Ni^{2+}] = [CO_3^{2-}] = X$

SOLUTION

$$K_{sp} = [Ni^{2+}][CO_3^{2-}] = [X][X] = X^2 = 1.45 \text{ X } 10^{-7}$$

$X = \underline{3.81 \text{ x } 10^{-4} \text{ mol/L}}$

NEW TERMS

Acid ionization constant (Ka)
Base ionization constant (K_b)
Equilibrium constant (K_{eq})
Equilibrium constant expression
Henderson-Hasselbalch equation
Ionization constant
Law of mass action
Solubility Product (K_{sp})

SELF-TEST

B-1 Multiple Choice

____ 1. For the hypothetical reaction: A + B $\rightleftharpoons$ C + D, $K_{eq} = 10^{-5}$. At equilibrium which of the following is true?

(a) [A][B] > [C][D]
(b) [C][D] > [A][B]
(c) [C][B] > [A][D]
(d) [B][D] > [A][C]
(e) cannot tell

____ 2. For the hypothetical reaction: 2A + B $\rightleftharpoons$ C + D, what are the units of K_{eq}?

(a) mol/L
(b) L/mol
(c) mol^2/L^2
(d) L^2/mol^2
(e) other

____ 3. Assume the hypothetical reaction 2A + B $\rightleftharpoons$ 3C.
If one starts with 0.50 mol/L of A and the concentration of A at equilibrium is 0.20 mol/L, what is the concentration of C at equilibrium?

(a) 0.50 mol/L
(b) 0.30 mol/L
(c) 0.75 mol/L
(d) 0.45 mol/L
(e) 0.20 mol/L

____ 4. For HCH_2O (formic acid) $K_a = 1.8 \times 10^{-4}$. Which of the following statements is true?

(a) $HCHO_2$ is a strong base.
(b) $HCHO_2$ is a strong acid.
(c) $HCHO_2$ is mostly ionized in water.
(d) Most $HCHO_2$ is not ionized in water.
(e) $HCHO_2$ is a weak base.

____ 5. Which of the following approximations is not valid?

(a) $0.10 - 0.003 \approx 0.10$
(b) $1.23 - 0.002 \approx 1.23$
(c) $0.010 - 0.005 \approx 0.010$
(d) $0.020 - 0.001 \approx 0.20$

____ 6. Which of the changes on a system at equilibrium changes the value of K_{eq}?

(a) the pressure
(b) the volume
(c) the concentration of a reactant or product
(d) addition of a catalyst
(e) the temperature

____ 7. Which of the following do not form a buffer solution when the two compounds are dissolved in water?

(a) HNO_3 and $Ca(NO_3)_2$
(b) CH_3NH_2 and $CH_3NH_3^+Br^-$
(c) H_2S and KHS
(d) $Mg(NO_2)_2$ and HNO_2

____ 8. For $CaCrO_4$ the value of $K_{sp} = 1.0 \times 10^{-4}$. If the concentration of Ca^{2+} in a solution is equal to 1.0×10^{-3} mol/L, the concentration of CrO_4^{2-} in that solution can be no more than

(a) 1.0×10^{-4} mol/L
(b) 1.0×10^{-3} mol/L
(c) 1.0×10^{-1} mol/L
(d) 1.0×10^{-5} mol/L

B-2 Problems

1. Write the law of mass action for each of the following:

(a) $P_4(g) + 6H_2(g) \rightleftharpoons 4PH_3(g)$

(b) $2NOCl(g) \rightleftharpoons 2NO(g) + Cl_2(g)$

(c) $SO_2(g) + NO_2(g) \rightleftharpoons SO_3(g) + NO(g)$

(d) $HMnO_4(aq) + H_2O \rightleftharpoons H_3O^+(aq) + MnO_4^-(aq)$

(e) $Cr_2(C_2O_4)_3(s) \rightleftharpoons 2Cr^{3+}(aq) + 3C_2O_4^{2-}(aq)$

2. Assume the following equilibrium: $2NOCl(g) \rightleftharpoons 2NO(g) + Cl_2(g)$. At a specified temperature, at equilibrium $[NO] = 1.32 \times 10^{-3}$ mol/L, $[Cl_2] = 3.76 \times 10^{-3}$ mol/L, and $[NOCl] = 0.542$ mol/L. What is the value of K_{eq}?

3. Assume the following equilibrium: $COBr_2(g) \rightleftharpoons CO(g) + Br_2(g)$. If one starts with 1.00 mole of $COBr_2$ in a 10.0-L container, it is later found that 8.50% of the $COBr_2$ dissociates (reacts) at equilibrium. What is the value of K_{eq}?

4. Assume the following equilibrium: $2NO(g) + O_2(g) \rightleftharpoons 2NO_2(g)$. If one starts with the 0.050 moles of both NO and O_2 in a 1.0-L container, the concentration of NO_2 at equilibrium is found to be 0.022 mol/L. What is K_{eq}?

5. Assume the following equilibrium:
$SO_2(g) + NO_2(g) \rightleftharpoons SO_3(g) + NO(g)$. The value of K_{eq} at a specified temperature is 7.5×10^{-2}. What is $[SO_2]$ at equilibrium if $[SO_3] = [NO] =$ 0.045 mol/L and $[NO_2] = 0.13$ mol/L at equilibrium?

6. Consider the same equilibrium and constant as in problem 5. If one starts with 2.00 moles of NO_2 and some SO_2 in a 10.0-L container and no products, it is later found that 0.038 mol/L of NO_2 reacted. How many moles of SO_2 are present at equilibrium?

7. For the following equilibrium: $2CO(g) + O_2(g) \rightleftharpoons 2CO_2(g)$, $K_{eq} = 2.0 \times 10^3$ at a specified temperature. What is the concentration of CO at equilibrium if $[CO_2] = 2.0 \times 10^{-3}$ and $[O_2] = 6.7 \times 10^{-5}$ mol/L?

8. What is K_a for the hypothetical acid HX of a 0.25 M solution of HX if the pH of the solution is 3.58?

9. What is K_b for a hypothetical nitrogen base like ammonia (NX) if 0.075 M solution is 0.40% ionized in aqueous solution?

10. What is the pH of a 0.54 M solution of hypobromous acid (HBrO)? For HBrO, $K_a = 2.1 \times 10^{-9}$.

11. Novocaine (Nv) is an ammonia-like nitrogen base with $K_b = 6.6 \times 10^{-6}$. What is the pH of a 0.62 M solution of novocaine that also is 0.35 M in the salt, NvH^+Cl^-?

12. What is the value of K_{sp} for silver(I) oxalate ($Ag_2C_2O_4$) if the molar solubility of the compound equals 1.2×10^{-6} mol/L?

Review Section C

CHAPTER SUMMARY SELF-TEST

C-1 Matching

____ collision theory

____ activated complex

____ an equilibrium constant expression

____equilibrium

____ catalyst

____ activation energy

____ an equilibrium constant for a partial ionization

____ a law of mass action

(a) $\dfrac{[NH_3]^2}{[N_2][H_2]^3}$

(b) the minimum kinetic energy needed for reactant molecules to be transformed into products

(c) reactants are transformed into products by means of collisions of molecules

(d) $K_a = 4.5 \times 10^4$

(e) a substance that shifts the point of equilibrium to the right

(f) the reactant molecules at the instant of maximum compression.

(g) a substance that does not affect the point of equilibrium but does affect the rate of the reaction

(h) a system at equilibrium shifts to counteract stress

(i) a situation where the rate of the forward and reverse reactions are equal

(j) a situation where a reaction stops before it is complete

(k) $K_{eq} = \dfrac{[PCl_5]}{[PCl_3][Cl_2]}$

(l) $K_b = 1.5 \times 10^{-4}$

Answers to Self-Tests

A-1 Multiple Choice

1. **c** the ionization of a strong acid (The reaction is essentially complete.)

2. **c** Molecules react because they collide with the sides of the container. (Actually, they react because they collide with each other.)

3. **d** A reaction at equilibrium goes to the right and to the left simultaneously.

4. **c** at equilibrium

5. **d** The combustion of coal has a higher activation energy than the combustion of phosphorus.

6. **c** both (a) and (b)

7. **b** increases the frequency of collisions between molecules

8. **c** The concentration of oxygen is important in the rate of the reaction.

9. **b** a low temperature

10. **c** no effect. (There are the same number of moles of gas on each side of the equation.)

A-2 Problem

1. (a) Heating decreases the amount of O_2 present at equilibrium.

   (b) Increasing [HCl] decreases O_2.

   (c) Decreasing [Cl_2] decreases O_2.

   (d) Adding a catalyst has no effect.

   (e) Compressing the reaction mixture decreases the amount of O_2.

   (f) Adding $H_2O(g)$ increases O_2.

   (g) Increasing the volume increases the amount of O_2.

B-1 Multiple Choice

1. **a** [A][B] > [C][D] The reactants are favored. The equilibrium lies far to the left.

2. **b** L/mol $K_{eq} = \frac{\cancel{[mol/L]}\cancel{[mol/L]}}{\cancel{[mol/L]^2}[mol/L]} = \frac{1}{mol/L} = L/mol$

3. **d** 0.45 mol/L If 0.50 - 0.20 = 0.30 mol/L of A reacted,

$$0.30 \cancel{\text{mol A}} \times \frac{3 \text{ mol C}}{2 \cancel{\text{mol A}}} = 0.45 \text{ mol A/L is formed.}$$

4. **d** The constant indicates that the equilibrium lies to the left, which means that most $HCHO_2$ is not ionized in water.

5. **c** 0.010 - 0.005 ≈ 0.010 (Notice that the second number is 50% of the first number.)

6. **e** the temperature (K_{eq} is larger at higher temperatures for an endothermic reaction and smaller for an exothermic reaction.)

7. **a** HNO_3 and $Ca(NO_3)_2$ (HNO_3 is a strong acid.)

8. **c** $K_{sp} = [Ca^{2+}][CrO_4^{2-}]$ $[CrO_4^{2-}] = K_{sp}/[Ca^{2+}] = 1.0 \times 10^{-1}$ mol/L

B-2 Problems

1. (a) $K_{eq} = \frac{[PH_3]^4}{[P_4]\,[H_2]^6}$ (b) $K_{eq} = \frac{[NO]^2[Cl_2]}{[NOCl]^2}$

(c) $K_{eq} = \frac{[SO_3]\,[NO]}{[SO_2]\,[NO_2]}$ (d) $K_a = \frac{[H_3O^+]\,[MnO_4^-]}{[HMnO_4]}$

(e) $K_{sp} = [Cr^{3+}]^2[C_2O_4^{2-}]^3$

2. $K_{eq} = \frac{[NO]^2[Cl_2]}{[NOCl]^2} = \frac{(1.32 \times 10^{-3})^2(3.76 \times 10^{-3})}{(0.542)^2} = \underline{2.23 \times 10^{-8}}$

3. The amount of $COBr_2$ that reacts is 0.0850 x 1.00 mol = 0.0850 mol. From the stoichiometry, notice that if 0.0850 moles of $COBr_2$ react, then 0.0850 moles of both CO and Br_2 are formed. The concentrations are

$[Br_2] = [CO] = 0.0850 \text{ mol}/10.0 \text{ L} = 8.50 \times 10^{-3}$ mol/L

The $COBr_2$ left at equilibrium is the initial amount minus the amount that reacts.

1.00 - 0.085 = 0.91 mol $[COBr_2] = 0.91 \text{ mol}/10.0 \text{ L} = 0.091$ mol/L

$$K_{eq} = \frac{[CO][Br_2]}{[COBr_2]} = \frac{(8.50 \times 10^{-3})(8.50 \times 10^{-3})}{(0.091)} = \underline{7.9 \times 10^{-4}}$$

4. (a) Find the concentrations of NO and O_2 that reacted to form 0.022 mol/L of NO_2.
   (b) From the initial concentrations of NO and O_2 find the concentration of each remaining at equilibrium (the initial minus the amount that reacts).

$$0.022 \text{ mol } NO_2 \times \frac{2 \text{ mol NO}}{2 \text{ mol } NO_2} = 0.022 \text{ mol/L NO reacted}$$

$$0.022 \text{ mol } NO_2 \times \frac{1 \text{ mol } O_2}{2 \text{ mol } NO_2} = 0.011 \text{ mol/L } O_2 \text{ reacted}$$

At equilibrium: [NO] = 0.050 - 0.022 = 0.028 mol/L
$[O_2]$ = 0.050 - 0.011 = 0.039 mol/L

$$K_{eq} = \frac{[NO_2]^2}{[NO]^2[O_2]} = \frac{(0.022)^2}{(0.028)^2(0.039)} = \underline{16}$$

5. $$K_{eq} = \frac{[SO_3][NO]}{[SO_2][NO_2]} \quad \text{Solving for } [SO_2],\ [SO_2] = \frac{[SO_3]\,[NO]}{[NO_2]\cdot K_{eq}}$$

$$[SO_2] = \frac{(0.045)(0.045)}{(0.13)(7.5 \times 10^{-2})} = \underline{0.21 \text{ mol/L}}$$

6. (a) Find $[NO_2]$ at equilibrium from the initial concentration minus the concentration that reacted.
   (b) From the concentration of NO_2 that reacted find the concentration of SO_3 and NO present at equilibrium.
   (c) Substitute and solve for $[SO_2]$. Then find the total amount in 10.0 L.

$[NO_2]_{eq}$= (2.00 mol/10.0 L) - 0.038 mol/L = 0.162 mol/L

From the stoichiometry, $[SO_3]$ = [NO] = 0.038 mol/L at equilibrium

$$\text{Solving for } [SO_2] \quad [SO_2] = \frac{[SO_3][NO]}{[NO_2]\cdot K_{eq}} = \frac{(0.038)(0.038)}{(0.162)(7.5 \times 10^{-2})} = 0.12$$

0.12 mol/L x 10.0 L = $\underline{1.2 \text{ mol } SO_2}$

7. Solve for [CO]

$$K_{eq} = \frac{[CO_2]^2}{[CO]^2[O_2]} \quad [CO]^2 = \frac{[CO_2]^2}{[O_2]\cdot K_{eq}} = \frac{(2.0 \times 10^{-3})^2}{(6.7 \times 10^{-5})(2.0 \times 10^{3})}$$

$[CO]^2 = 30 \times 10^{-6}$ $\quad$ $[CO] = 5.5 \times 10^{-3}$ mol/L

8. $HX + H_2O \rightleftharpoons H_3O^+ + X^-$ $K_a = \dfrac{[H_3O^+][X^-]}{[HX]}$

pH = 3.58; $\log [H_3O^+] = -3.58$ $[H_3O^+] = 2.6 \times 10^{-4}$

In this solution $[X^-] = [H_3O^+]$ and $[HX] = 0.25 - (2.6 \times 10^{-4}) \approx 0.25$

$$K_a = \frac{(2.6 \times 10^{-4})(2.6 \times 10^{-4})}{(0.25)} = \underline{2.7 \times 10^{-7}}$$

9. $NX + H_2O \rightleftharpoons NXH^+ + OH^-$ $K_b = \dfrac{[NXH^+][OH^-]}{[NX]}$

[NX] that ionizes is $0.0040 \times 0.075 = 3.0 \times 10^{-4}$.
Since each mole of NX that ionizes produces one mole of NXH^+ and one mole of OH^-, we have at equilibrium

$$[NXH^+] = [OH^-] = 3.0 \times 10^{-4}$$

The [NX] left at equilibrium is $0.075 - (3.0 \times 10^{-4}) \approx 0.075$

$$K_b = \frac{(3.0 \times 10^{-4})(3.0 \times 10^{-4})}{(0.075)} = \underline{1.2 \times 10^{-6}}$$

10. $HBrO + H_2O \rightleftharpoons H_3O^+ + BrO^-$ $K_a = \dfrac{[H_3O^+][BrO^-]}{[HBrO]} = 2.1 \times 10^{-9}$

Let $X = [H_3O^+] = [BrO^-]$; then at equilibrium $[HBrO] = 0.54 - X \approx 0.54$
(the value of the constant tells us that X is small compared to 0.54).

$$K_a = \frac{X \cdot X}{0.54} = 2.1 \times 10^{-9} \quad X^2 = 1.13 \times 10^{-9} = 11.3 \times 10^{-10}$$

$$X = 3.4 \times 10^{-5} \quad pH = -\log(3.4 \times 10^{-5}) = \underline{4.47}$$

11. $Nv + H_2O \rightleftharpoons NvH^+ + OH^-$ This is a buffer solution.

$NvH^+Cl^-(s) \rightarrow NvH^+(aq) + Cl^-(aq)$ Cl^- is a spectator ion.
$pK_b = -\log(6.6 \times 10^{-6}) = 5.18$
Nv = base so [base] = 0.62 M; NvH^+ = acid so [acid] = 0.35 M
$pOH = pK_b + \log\dfrac{[acid]}{[base]}$ $pOH = 5.18 + \log\dfrac{0.35\ M}{0.62\ M} = 5.18 - 0.25 = 4.93$

pH = 14.00 - 4.93 = $\underline{9.07}$

12. The equilibrium involved is $Ag_2C_2O_4(s) \rightleftharpoons 2Ag^+(aq) + C_2O_4^{2-}(aq)$

$K_{sp} = [Ag^+]^2[C_2O_4^{2-}]$

At equilibrium, $[Ag^+] = 2 \times$ solubility and $[C_2O_4^{2-}]$ = solubility
Thus $[Ag^+] = 2 \times 1.2 \times 10^{-6} = 2.4 \times 10^{-6}$ mol/L $[C_2O_4^{2-}] = 1.2 \times 10^{-6}$ mol/L

$K_{sp} = [2.4 \times 10^{-6}]^2[1.2 \times 10^{-6}] = \underline{6.9 \times 10^{-18}}$

C-1 Matching

c Collision theory tells us that reactants are transformed into products by means of collisions.

f An activated complex is formed at the point of maximum compression of reactant molecules.

a An equilibrium constant expression is $\frac{[NH_3]^2}{[N_2][H_2]^3}$.

i Equilibrium is a situation where the rates of the forward and reverse reactions are equal.

g A catalyst is a substance that does not affect the point of equilibrium but does affect the rate of the reaction.

b Activation energy is the minimum kinetic energy needed for reactant molecules to be transformed into products.

l $K_b = 1.5 \times 10^{-4}$ is an equilibrium constant for a partial ionization of a weak base.

k $K_{eq} = \frac{[PCl_5]}{[PCl_3][Cl_2]}$ represents a law of mass action.

Solutions to Black Text Problems

14-20 $K_{eq} = \frac{[O_3]^2}{[O_2]^3} = \frac{(0.12)^2}{(0.35)^3} = \underline{0.34}$

14-21 $K_{eq} = \frac{[NO_2]^2}{[N_2][O_2]^2} = \frac{(6.20 \times 10^{-4})^2}{(1.25 \times 10^{-3})(2.50 \times 10^{-3})^2} = \underline{49.2}$

14-23 $K_{eq} = \frac{[CO_2][H_2]^4}{[CH_4][H_2O]^2} = \frac{(2.20/30.0)(4.00/30.0)^4}{(6.20/30.0)(3.00/30.0)^2} = \underline{0.0112}$

14-25 (a) $0.60 \text{ mol HI} \times \frac{1 \text{ mol } H_2}{2 \text{ mol HI}} = 0.30 \text{ mol } H_2$ $[H_2] = [I_2] = \underline{0.30 \text{ mol/L}}$

(b) As in (a) $[H_2] = \underline{0.30 \text{ mol/L}}$ $[I_2] = 0.20 + 0.30 = \underline{0.50 \text{ mol/L}}$

(c) $[HI] = [HI]_{initial} - [HI]_{reacted} = 0.60 - 0.20 = \underline{0.40 \text{ mol/L}}$

$0.20 \text{ mol HI} \times \frac{1 \text{ mol } H_2}{2 \text{ mol HI}} = \underline{0.10 \text{ mol /L } H_2}$ $[I_2] = [H_2] = \underline{0.10 \text{ mol/L}}$

(d) $K_{eq} = \frac{[H_2][I_2]}{[HI]^2} = \frac{(0.10)(0.10)}{(0.40)^2} = \underline{0.063}$

(e) $K_r = \frac{1}{K_{eq}} = \frac{1}{0.063} = \underline{16}$

This is a smaller value than that used in Table 14-1. This indicates that the equilibrium in this problem was established at a different temperature than that of Table 14-1.

14-27 (a) $[N_2]_{reacted} = 0.50 - 0.40 = 0.10 \text{ mol/L}$

$0.10 \text{ mol } N_2 \times \frac{3 \text{ mol } H_2}{1 \text{ mol } N_2} = 0.30 \text{ mol/L } H_2$ (reacts)

$[H_2]_{eq} = 0.50 - 0.30 = \underline{0.20 \text{ mol/L}}$

$0.10 \text{ mol } N_2 \times \frac{2 \text{ mol } NH_3}{1 \text{ mol } N_2} = 0.20 \text{ mol/L } NH_3$ formed

$[NH_3]_{eq} = 0.50 + 0.20 = \underline{0.70 \text{ mol/L}}$

(b) $K_{eq} = \frac{(0.70)^2}{(0.20)^3(0.40)} = \underline{150}$

14-28 (a) The concentration of O_2 that reacts is

$0.25 \text{ mol } NH_3 \times \frac{5 \text{ mol } O_2}{4 \text{ mol } NH_3} = \underline{0.31 \text{ mol/L } O_2}$

(b) $[NH_3]_{eq} = [[NH_3]_{init.} - [NH_3]_{reacted} = 1.00 - 0.25 = \underline{0.75 \text{ mol/L}}$

$[O_2]_{eq} = 1.00 - 0.31 = \underline{0.69 \text{ mol/L } O_2}$

$0.25 \text{ mol } NH_3 \times \frac{4 \text{ mol NO}}{4 \text{ mol } NH_3} = \underline{0.25 \text{ mol/L NO}}$ (formed)

$0.25 \text{ mol } NH_3 \times \frac{6 \text{ mol } H_2O}{4 \text{ mol } NH_3} = 0.38 \text{ mol/L } H_2O$

(c) $K_{eq} = \frac{[NO]^4[H_20]^6}{[NH_3]^4[O_2]^5} = \frac{(0.25)^4(0.38)^6}{(0.75)^4(0.69)^5}$

14-30 $K_{eq} = \dfrac{[PCl_5]}{[PCl_3][Cl_2]}$ $\quad [PCl_5] = K_{eq}[PCl_3][Cl_2] = (0.95)(0.75)(0.40) = \underline{0.28\text{ mol/L}}$

14-32 $[H_2O]^2 = \dfrac{[CO_2][H_2]^4}{K_{eq}[CH_4]} = \dfrac{(0.24)(0.20)^4}{0.0112(0.50)} = 6.86 \times 10^{-2}$ $\quad [H_2O] = \underline{0.26\text{ mol/L}}$

14-33 Let X = [HCl] = $[CH_3Cl]$ $\quad K_{eq} = \dfrac{[HCl][CH_3Cl]}{[CH_4][Cl_2]} = \dfrac{X^2}{(0.20)(0.40)} = 56$

$X^2 = 4.5 \quad X = \underline{2.1\text{ mol/L}}$

14-39 $HOCN + H_2O \rightleftharpoons H_3O^+ + OCN^-$

(a) $[HOCN]_{eq} = [HOCN]_{initial} - [HOCN]_{ionized}$

From the equation $[HOCN]_{ionized} = [H_3O^+] = [OCN^-]$

$[HOCN]_{eq} = 0.20 - 0.0062 = \underline{0.19}$

(b) $K_a = \dfrac{[H_3O^+][OCN^-]}{[HOCN]} = \dfrac{(6.2 \times 10^{-3})(6.2 \times 10^{-3})}{(0.19)} = \underline{2.0 \times 10^{-4}}$

(c) $-\log(6.2 \times 10^{-3}) = \underline{2.21}$

14-40 (a) $HX + H_2O \rightleftharpoons H_3O^+ + X^-$ $\quad K_a = \dfrac{[H_3O^+][X^-]}{[HX]}$

(b) From the equation $[H_3O^+] = [X^-] = 10.0\%$ of $[HX]_{init}$

$0.100 \times 0.58 = [H_3O^+] = [X^-] \quad [HX]_{eq} = 0.58 - 0.058 = 0.52$

(c) $K_a = \dfrac{(0.058)(0.058)}{(0.52)} = 6.5 \times 10^{-3}$

(d) pH = $-\log(5.8 \times 10^{-2}) = \underline{1.24}$

14-43 $Nv + H_2O \rightleftharpoons NvH^+ + OH^-$

pH = 11.46 $\quad$ pOH = 14.00 - 11.46 = 2.54 $\quad [OH^-] = 2.88 \times 10^{-3}$

Since $[NvH^+] = [OH^-] = 2.88 \times 10^{-3}$ $\quad [Nv]_{eq} = [Nv]_{init} - [OH^-] = 1.25 - 0.00288 = 1.25$

$K_b = \dfrac{[NvH^+][OH^-]}{[Nv]} = \dfrac{(2.88 \times 10^{-3})^2}{(1.25)} = \underline{6.6 \times 10^{-6}}$

14-44 $HC_2Cl_3O_2 + H_2O \rightleftharpoons H_3O^+ + C_2Cl_3O_2^-$

$[H_3O^+] = [HC_2Cl_3O_2]_{ionized} = [HC_2Cl_3O_2]_{initial} - [HC_2Cl_3O_2]_{eq}$

$= 0.300 - 0.277 = 0.023$ $\quad$ $pH = -\log(2.3 \times 10^{-2}) = \underline{1.64}$

$$K_a = \frac{(0.023)^2}{0.277} = \underline{1.9 \times 10^{-3}}$$

14-45 $HBrO + H_2O \rightleftharpoons H_3O^+ + BrO^-$

Let $X = [H_3O^+] = [BrO^-]$ $\quad$ Then, $[HBrO]_{eq} = [HBrO]_{init} - [HBrO]_{ionized}$

Thus, $[HBrO]_{eq} = 0.50 - X$. Since K_a is small, X is small and

$[HBrO] \approx 0.50$ $\quad K_a = \frac{(X)(X)}{0.50} = 2.1 \times 10^{-9}$ $\quad [X] = [H_3O^+] = 3.2 \times 10^{-5}$ $\quad \underline{pH = 4.43}$

14-47 $NH_3 + H_2O \rightleftharpoons NH_4^+ + OH^-$

Let $X = [OH^-] = [NH_4^+]$ $\quad [NH_3] = 0.55 - X \approx 0.55$

$$K_b = \frac{[NH_4^+][OH^-]}{[NH_3]} = \frac{X^2}{0.55} = 1.8 \times 10^{-5} \qquad X = [OH^-] = \underline{3.2 \times 10^{-3}}$$

14-49 $(CH_3)_2NH + H_2O \rightleftharpoons (CH_3)_2NH_2^+ + OH^-$

Set $X = [OH^-] = [(CH_3)_2NH_2^+]$ $\quad [(CH_3)_2NH] = 1.00 - X \approx 1.00$

$$K_b = \frac{[(CH_3)_2NH_2^+][OH^-]}{[(CH_3)_2NH]} = \frac{X^2}{1.00} = 7.4 \times 10^{-4}$$

$X = 2.7 \times 10^{-2}$ $\quad pOH = 1.57$ $\quad pH = \underline{12.43}$

14-51 $HCN + H_2O \rightleftharpoons H_3O^+ + CN^-$ $\quad [HCN] = \text{acid}, [CN^-] = \text{base}$

$pK_a = -\log(4.0 \times 10^{-10}) = 9.40$ $\quad pH = 9.40 + \log \frac{0.45\ \cancel{\text{mol/2.50 L}}}{0.45\ \cancel{\text{mol/2.50 L}}}$

$pH = pKa = \underline{9..40}$

14-53 $HBrO + H_2O \rightleftharpoons H_3O^+ + BrO^-$

$pK_a = -\log(2.1 \times 10^{-9}) = 8.68 \quad pH = 8.68 + \log \dfrac{0.20\ \cancel{mol/0.850\ L}}{0.60\ \cancel{mol/0.850\ L}} = 8.68 - 0.48 = \underline{8.20}$

14-56 $N_2H_4 + H_2O \rightleftharpoons N_2H_5^+ + OH^-$

$1.50\ g\ N_2H_4 \times \dfrac{1\ mol\ N_2H_4}{32.05\ g\ N_2H_4} = 0.0468\ mol\ N_2H_4$

$1.97\ g\ N_2H_5Cl \times \dfrac{1\ mol\ N_2H_5Cl}{68.51\ g\ N_2H_5Cl} = 0.0288\ mol\ N_2H_5Cl\ (N_2H_5^+)$

$pK_b = -\log(9.8 \times 10^{-7}) = 6.01 \quad pOH = 6.01 + \log \dfrac{0.0288\ \cancel{mol/2.00\ L}}{0.0468\ \cancel{mol/2.00\ L}}$

$pOH = 6.01 - 0.21 = 5.80 \quad \underline{pH = 8.20}$

14-62 $CuI(s) \rightleftharpoons Cu^+(aq) + I^-(aq) \quad K_{sp} = [Cu^+][I^-]$

At equilibrium $[Cu^+] = [I^-] = 2.2 \times 10^{-6}$ mol/L

$K_{sp} = [2.2 \times 10^{-6}][2.2 \times 10^{-6}] = \underline{4.8 \times 10^{-12}}$

14-64 $Ca(OH)_2(s) \rightleftharpoons Ca^{2+}(aq) + 2OH^-(aq) \qquad K_{sp} = [Ca^{2+}][OH^-]^2$

At equilibrium $[Ca^{2+}] = 1.3 \times 10^{-2}$ mol/L $[OH^-] = 2 \times (1.3 \times 10^{-2})$mol/L $= 2.6 \times 10^{-2}$ mol/L

$K_{sp} = [1.3 \times 10^{-2}][2.6 \times 10^{-2}]^2 = \underline{8.8 \times 10^{-6}}$

14-66 $AgI(s) \rightleftharpoons Ag^+(aq) + I^-(aq) \quad K_{sp} = [Ag^+][I^-] = 8.3 \times 10^{-17}$

Let x = molar solubility of AgI, then $[Ag^+] = [I^-] = x$

$[x][x] = 8.3 \times 10^{-17} \qquad x = \underline{9.1 \times 10^{-9}\ mol/L}$

14-68 $AgBr(s) \rightleftharpoons Ag^+(aq) + Br^-(aq) \quad K_{sp} = [Ag^+][Br^-] = 7.7 \times 10^{-13}$

$[7 \times 10^{-6}][8 \times 10^{-7}] = 6 \times 10^{-12}$

This is a larger number than K_{sp} so a precipitate of AgBr does form.

14-70 (a) $3Cl_2(g) + NH_3(g) \rightleftharpoons NCl_3(g) + 3HCl(g)$

(b) $K_{eq} = \dfrac{[NCl_3][HCl]^3}{[NH_3][Cl_2]^3}$ (c) No effect (d) decreases $[NH_3]$

(e) reactants (f) $\dfrac{(2.0 \times 10^{-4})(1.0 \times 10^{-3})^3}{[NH_3](0.10)^3} = 2.4 \times 10^{-9}$ $[NH_3] = \underline{0.083 \text{ mol/L}}$

14-72 (a) $2N_2O(g) \rightleftharpoons 2N_2(g) + O_2(g)$

(b) $K_{eq} = \dfrac{[N_2]^2[O_2]}{[N_2O]^2}$ (c) decreases $[N_2]$ (d) decreases $[N_2O]$

(e) $[N_2O]_{eq} = 0.10 - (0.015 \times 0.10) = 0.10$ mol/L; $[N_2] = 0.015 \times 0.10 = 1.5 \times 10^{-3}$ mol/L;

$$[O_2] = \frac{1.5 \times 10^{-3}}{2} = 7.5 \times 10^{-4} \text{ mol/L}$$

$$K_{eq} = \frac{(1.5 \times 10^{-3})^2(7.5 \times 10^{-4})}{(0.10)^2} = \underline{1.7 \times 10^{-7}}$$

14-76 $HC_2H_3O_2 + H_2O \rightarrow H_3O^+ + C_2H_3O_2^-$

$0.265 \cancel{L} \times 0.22 \text{ mol}/\cancel{L} = 0.058 \text{ mol } HC_2H_3O_2$

$$0.375 \cancel{L} \times \frac{0.12 \cancel{\text{mol } Ba(C_2H_3O_2)_2}}{\cancel{L}} \times \frac{2 \text{ mol } C_2H_3O_2^-}{\cancel{\text{mol } Ba(C_2H_3O_2)_2}} = 0.090 \text{ mol } C_2H_3O_2^-$$

$pK_a = 4.74$ $\text{pH} = 4.74 + \log \dfrac{0.090 \cancel{\text{mol}/0.640 \text{ L}}}{0.058 \cancel{\text{mol}/0.640 \text{ L}}} = 4.74 + 0.19 = \underline{4.93}$

14-77 The addition of the NaOH reacts with part of the acetic acid to produce sodium acetate. The net ionic equation for the reaction is

$HC_2H_3O_2 + OH^- \rightarrow C_2H_3O_2^- + H_2O$

0.20 mol of OH^- reacts with 0.20 mol of $HC_2H_3O_2$ to produce 0.20 mol of $C_2H_3O_2^-$ Thus 1.00 - 0.20 = 0.80 mol of $HC_2H_3O_2$ remains and 0.20 mol of $C_2H_3O_2^-$ is formed from the addition of 0.20 mol of OH^-.

$pK_a = 4.74$ $\text{pH} = 4.74 + \log \dfrac{0.20 \cancel{\text{mol/V}}}{0.80 \cancel{\text{mol/V}}} = 4.74 - 0.60 = \underline{4.14}$

14-78 $Mg(OH)_2(s) \rightleftharpoons Mg^{2+}(aq) + 2OH^-(aq)$ $K_{sp} = [Mg^{2+}][OH^-]^2$

pH = 10.50 pOH = 14.00 - 10.50 = 3.50 $[OH^-]$ = -inv log $[OH^-]$

$[OH^-] = 3.2 \times 10^{-4}$ $[Mg^{2+}] = [OH^-]/2 = 1.6 \times 10^{-4}$

$K_{sp} = [1.6 \times 10^{-4}][3.2 \times 10^{-4}]^2 = \underline{1.6 \times 10^{-11}}$

14-80 $BaSO_4(s) \rightleftharpoons Ba^{2+}(aq) + SO_4^{2-}(aq)$ $K_{sp} = [Ba^{2+}][SO_4^{2-}] = 1.1 \times 10^{-10}$

Let x = molar solubility, then $[Ba^{2+}] = [SO_4^{2-}] = x$

$[x][x] = 1.1 \times 10^{-10}$ $x = 1.0 \times 10^{-5}$ mol/L

$$1.0 \times 10^{-5} \text{ mol } BaSO_4 \times \frac{233.4 \text{ g } BaSO_4}{\text{mol } BaSO_4} = \underline{0.023 \text{ g } BaSO_4/L}$$

15

Nuclear Chemistry

Review Section A *Natural Occurring Radioactivity*

OUTLINE

15-1 Radioactivity

1. Spontaneous decay - alpha particles, beta particles, gamma rays, positron emission, and electron capture

OBJECTIVES

Define radioactivity and the identity of the five types of radiation discussed.

2. The nuclear equation

Explain what is illustrated by a nuclear equation and show how a specified isotope changes when it undergoes one of the five types of radiation discussed.

15-2 Rates of Decay of Radioactive Isotopes

1. The rate of decay and half-life

Describe the concept of half-life as it relates to a radioactive decay series.

15-3 Effects of Radiation

1. The effects of ionizing radiation

Describe how the three natural types of radiation affect nearby matter.

15-4 Detecting Radiation

1. Film badges, Geiger counters, and scintillation counters

Illustrate how various devices detect and measure radiation.

SUMMARY OF SECTIONS 15-1 THROUGH 15-4

Questions: *Who discovered natural radiation? How was radiation characterized? How are nuclear changes represented? How does radiation affect matter and how is it detected and measured?*

The nucleus of an atom is not necessarily stable and unchanging. For over a century it has been known that some nuclei are **radioactive isotopes**. These isotopes are unstable and disintegrate to form the nuclei of other elements and high energy particles or high energy forms of light. Three types of natural **radiation** are illustrated by the nuclear equations below. A **nuclear equation** indicates the original nucleus or nuclei on the left and the product nuclei and particles on the right.

Alpha (α) particles

$${}^{238}_{92}U \longrightarrow {}^{234}_{90}Th + {}^{4}_{2}He$$

[He nucleus is called an alpha (α) particle. It lowers the atomic number of the original isotope by two and the mass number by four.]

Beta (β) particles

$${}^{14}_{6}C \longrightarrow {}^{14}_{7}N + {}^{0}_{-1}e$$

[An electron from the nucleus is called a beta (β) particle. It raises the atomic number of the original isotope by one.]

Gamma (γ) rays

$${}^{60}_{27}Co \longrightarrow {}^{60}_{27}Co + \gamma$$

[High energy light from the nucleus is called a gamma ray (γ). It does not affect the atomic number or mass number of the original isotope.]

In addition to these there are two other forms of radiation that occur either rarely in nature (electron capture) or only in a few artificially produced isotopes.

Positron emission

$$^{26}_{14}Si \longrightarrow ^{26}_{13}Al + ^{0}_{+1}e \leftarrow$$

[A positron is a positively charged electron. It lowers the atomic number of the original isotope by one.]

Electron capture

$$^{179}_{73}Ta + ^{0}_{-1}e \longrightarrow ^{179}_{72}Hf \leftarrow$$

[When an isotope captures an orbital electron, the atomic number of the original isotope increases by one as in positron emission.]

Each radioactive isotope has a definite and unchanging rate of decay. The rate of decay of an isotope is indicated by its **half-life ($t_{1/2}$)**, which is the time required for one-half of a given sample to decay. Half-lives vary from fractions of a second to billions of years. For example, $^{238}_{92}U$ is a radioactive isotope that is still present because its half-life is about the same as the age of the Earth. This isotope is the beginning nucleus of a **radioactive decay series** that ends with a stable isotope of lead.

Radiation affects living tissue because of its ability to form ions as the particles or rays penetrate matter. Gamma radiation is considered the most dangerous type of radiation mainly because of its ability to penetrate through large quantities of matter. This destructive ability of gamma radiation can be put to constructive use when it is focused on malignant (cancer) cells thus causing their destruction. Alpha and beta emitters are not very dangerous when outside the body but are potentially very serious when ingested. In close contact with tissue, these isotopes can cause cancer.

Radiation is detected in basically three ways. The first is **film badges** which take advantage of the ability of radiation to penetrate and expose film. Film badges can detect radioactive dosage for those who work in the nuclear industry. More precise measurements are made with **Geiger counters** which work by detecting the ionization caused by radiation, and **scintillation counters**, which work by the use of phosphors that emit light when they interact with radiation.

NEW TERMS

Alpha particle	Half-life
Beta particle	Nuclear equation
Electron capture	Positron emission
Film badge	Radiation
Gamma ray	Radioactive decay series
Geiger counter	Scintillation counter

SELF-TEST

A-1 Multiple Choice

___ 1. Which of the following is a beta particle?

(a) $^{-1}_{0}e$ (b) $^{0}_{1}e$ (c) $^{4}_{2}He$ (d) $^{0}_{-1}e$ (e) $^{1}_{0}H$

___ 2. If $^{234}_{92}U$ emits an alpha particle, the product is which of the following?

(a) $^{230}_{90}U$ (b) $^{234}_{91}Np$ (c) $^{230}_{90}Th$ (d) $^{232}_{90}Th$ (e) $^{238}_{94}Pu$

___ 3. If the half-life of a particular isotope is 2.0 days, how much will remain of an 8.0-g sample in 8.0 days?

(a) none (b) 4.0 g (c) 0.5 g (d) 1.9 g (e) 2.0 g

___ 4. Sodium-20 emits a positron. The isotope remaining is

(a) magnesium-20 (b) neon-21
(c) magnesium-19 (d) neon-20

___ 5. Gallium-68 is an isotope formed as a result of electron capture. The original isotope was

(a) germanium-68 (b) gallium-69
(c) arsenic-68 (d) germanium-69

___ 6. A radioactive isotope decays by converting a neutron into a proton. This isotope emitted which of the following?

(a) alpha particle (b) beta particle
(c) gamma ray (d) positron

A-2 Matching

___ alpha particle

___ electron capture

___ beta particle

___ gamma ray

___ positron emission

___ half-life

(a) a high-energy form of light

(b) half of the time it takes a sample of a radioactive isotope to decay

(c) an electron absorbed from outside the nucleus

(d) a proton emitted from a nucleus

(e) a helium nucleus

(f) the time it takes half of a sample of a radioactive isotope to decay

(g) a proton with a negative charge

(h) a process whereby a nucleus gives off a particle to form a nucleus with the next lower atomic number

(i) an electron emitted from a nucleus

A-3 Problems

1. Complete the following nuclear equations. (You will need a periodic table.)

(a) $^{135}_{53}I \longrightarrow ^{135}_{54}Xe + \_\_\_\_$

(b) $^{232}_{90}Th \longrightarrow ^{228}_{88}Ra + \_\_\_\_$

(c) $^{20}_{8}O \longrightarrow \_\_\_\_ + ^{0}_{-1}e$

(d) $^{80}_{37}Rb + ^{0}_{-1}e \longrightarrow \_\_\_\_$

(e) $^{50}_{25}Mn \longrightarrow \_\_\_\_ + ^{0}_{+1}e$

2. In order for a long-lived isotope to still exist in any appreciable amount, it must have a half-life around the age of the Earth. If the Earth is about 5×10^9 years old (five billion) and a certain isotope has a half-life of 1×10^9 (one billion) years, about what percent of the original amount is still present?

Review Section B *Induced Nuclear Changes and Its Uses*

OUTLINE	OBJECTIVES
15-5 Nuclear Reactions	
1. Nuclear reactions and accelerators	
2. Synthetic heavy elements	*Describe how nuclear reactions are initiated and how heavy elements are artificially synthesized.*
15-6 Uses of Radioactivity	
1. Carbon dating, neutron activation analysis, uses of gamma radiation, and PET scans	*Explain how radioactivity is used to date carbon artifacts, kill bacteria and cancer cells, and detect tumors.*
15-7 Nuclear Fission	
1. Fission and nuclear chain reactions	*Describe nuclear fission and how a chain reaction occurs.*
2. Power from nuclear reactors	*Explain how fission is controlled in a nuclear reactor.*

15-8 Nuclear Fusion

1. Fusion: a future source of energy

Differentiate between fission and fusion and discuss the problems associated with controlled nuclear fusion.

SUMMARY OF SECTIONS 15-5 THROUGH 15-8

Questions: *How are elements changed into other elements by artificial means? What are the uses of these discoveries? How is an atom "split?" What is the origin of the power of the Sun?*

Besides naturally occurring nuclear reactions, artificial nuclear reactions have been studied for over 60 years. The following nuclear equation illustrates how the **transmutation** of one element to another occurs from the reaction of an alpha particle and an isotope of aluminum.

$$^{27}_{13}\mathrm{Al} + ^{4}_{2}\mathrm{He} \longrightarrow ^{30}_{15}\mathrm{P} + ^{1}_{0}\mathrm{n}$$

Notice in the previous nuclear equation that the total of the mass numbers on the left (27 + 4) equals the total of the mass numbers on the right (30 + 1). The total charge on the left (13 + 2) also equals the total charge on the right (15 + 0). Remember that the charge on the nucleus determines the identity of the element and vice versa (e.g., 15 = P). With the invention of the particle accelerator, which provides extra energy to the colliding nuclei, many new heavy elements have been artificially synthesized. The particle accelerator is needed to supply the positively charged colliding nuclei with enough energy to overcome the strong force of repulsion between them. The following nuclear equation illustrates how a new element was made from naturally occurring elements by use of a particle accelerator.

$$^{238}_{92}\mathrm{U} + ^{12}_{6}\mathrm{C} \longrightarrow ^{244}_{98}\mathrm{Cf} + 6^{1}_{0}\mathrm{n}$$

There are many important uses of radioactivity. One familiar application is **carbon dating** whereby the proportion of radioactive carbon present in a sample indicates its age. A nuclear reaction known as **neutron activation analysis** produces radioactive isotopes from stable ones. This process has many medical applications. Gamma radiation preserves food and destroys cancer cells. Positron emissions are used in an important medical test known as a **PET scan**.

In the 1930's a new type of nuclear reaction was discovered whereby a very heavy nucleus ($^{235}_{92}\mathrm{U}$) was split into two similar-sized parts by reaction with a neutron.

$$^{235}_{92}\mathrm{U} + ^{1}_{0}\mathrm{n} \longrightarrow ^{139}_{56}\mathrm{Ba} + ^{94}_{36}\mathrm{Kr} + 3^{1}_{0}\mathrm{n} + \text{energy}$$

A nuclear reaction where a large nucleus is split into two smaller nuclei is known as **fission**. In a nuclear fission reaction, a chain reaction occurs because more neutrons are produced than consumed. Thus the reaction is self-sustaining. Also, since product nuclei weigh less than the original nuclei, the reaction produces a large amount of energy by conversion of mass ($E = mc^2$). This reaction forms the basis of the atomic bomb but is controlled in nuclear plants for the generation of electrical power.

Another type of nuclear reaction was shown to be feasible in the late 1940s. In this case two light nuclei undergo **fusion** to form a heavier nucleus. Since a mass loss also occurs, significant energy is released.

$$^{3}_{1}\mathrm{H} + ^{2}_{1}\mathrm{H} \longrightarrow ^{4}_{2}\mathrm{He} + ^{1}_{0}\mathrm{n} + \text{energy}$$

This reaction is the basis of the hydrogen bomb but has not yet been controlled for use in the generation of electrical power, although much research is currently underway. The fusion reaction, however, is the original source of almost all of our energy since it is the reaction that continuously generates energy in the Sun.

NEW TERMS

Carbon dating	Neutron activation analysis
Fission	PET scan
Fusion	Transmutation

SELF-TEST

B-1 Multiple Choice

___ 1. Which of the following isotopes is used in carbon dating?

(a) ^{12}C (b) ^{11}C c) ^{14}C (d) ^{13}C

___ 2. When stable isotopes are neutron activated they

(a) become alpha particle emitters (b) become beta particle emitters
(c) become positron emitters (d) undergo fission

___ 3. Which of the following are detected outside the body in a PET scan?

(a) positrons (b) electrons (c) gamma rays (d) alpha particles

___ 4. Which of the following reactions represents a fission reaction?

(a) $^{1}_{1}H + ^{2}_{1}H \longrightarrow ^{3}_{2}He$

(b) $^{239}_{94}Pu + ^{1}_{0}n \longrightarrow ^{92}_{38}Sr + ^{144}_{56}Ba + 4^{1}_{0}n$

(c) $^{239}_{94}Pu \longrightarrow ^{235}_{92}U + ^{4}_{2}He$

(d) $^{238}_{92}U + ^{1}_{0}n \longrightarrow ^{239}_{93}Np + ^{0}_{-1}e$

(e) $^{14}_{7}N + ^{4}_{2}He \longrightarrow ^{17}_{8}O + ^{1}_{1}H$

___ 5. Which of the equations in problem 4 above represents a fusion reaction?

___ 6. Which of the following is a problem in the use of fusion power?

(a) It produces large amounts of radioactive products.
(b) It uses uranium, which is rare.
(c) It can be difficult to control.
(d) It requires a very high temperature to get started.

B-2 Problems

(a) $^{96}_{42}Mo + ^{2}_{1}H \longrightarrow$ ____ $+ ^{1}_{0}n$

(b) $^{27}_{13}Al + ^{2}_{1}H \longrightarrow ^{25}_{12}Mg +$ ____

(c) $^{108}_{48}Cd + ^{1}_{0}n \longrightarrow$ ____ $+ \gamma$

(d) $^{235}_{92}U + ^{1}_{0}n \longrightarrow ^{90}_{38}Sr + ^{144}_{54}Xe +$ ____

(e) $^{249}_{98}Cf + ^{18}_{8}O \longrightarrow$ ____ $+ 4^{1}_{0}n$

Review Section C

CHAPTER SUMMARY SELF-TEST

C-1 Matching

___ a proton

___ undergoes fission

___ an alpha particle

___ an isotope of sodium

___ detects formation of ions

___ a positron

___ an electron

___ an effect of radiation

(a) $^{238}_{92}U$

(b) Scintillation counter

(c) $^{24}_{11}X$

(d) beta particle

(e) $^{0}_{+1}e$

(f) $^{1}_{1}H^{+}$

(g) formation of ions

(h) a neutral particle with a mass of about 1 amu

(i) high energy electromagnetic radiation

(j) $^{235}_{92}U$

(k) $^{23}_{12}X$

(l) $^{4}_{2}He^{2+}$

(m) Geiger counter

Answers to Self-Tests

A-1 Multiple Choice

1. **d** $^{0}_{-1}e$

2. **c** $^{230}_{90}Th$ The mass number decreases by four, the atomic number by two. The new element with atomic number 90 is Th.

3. **c** 0.5 g

4. **d** neon-20

5. **a** germanium-68

6. **b** A beta particle increases the atomic number by one when a neutron is converted into a proton.

A-2 Matching

e An alpha particle is a helium nucleus.

c Electron capture is a process whereby an electron is absorbed from outside the nucleus.

i A beta particle is an electron emitted from a nucleus.

a A gamma ray is a high-energy form of light.

h Positron emission is a process whereby a nucleus gives off a particle to form a nucleus with the next lower atomic number.

f Half-life is the time it takes half of a sample of radioactive isotope to decay.

A-3 Problems

1. (a) $^{0}_{-1}e$ (b) $^{4}_{2}He$ (c) $^{20}_{9}F$ (d) $^{60}_{36}Kr$ (e) $^{50}_{24}Cr$

2.

	Time elapsed	Fraction of sample remaining	Percent of sample remaining
(beginning)	0	1	100.0%
	1×10^9 yr.	1/2	50.0%
	2×10^9 yr.	1/4	25.0%
	3×10^9 yr.	1/8	12.5%
	4×10^9 yr	1/16	6.2%
(present)	5×10^9 yr.	1/32	3.1%

B-1 Multiple Choice

1. **c** ^{14}C

2. **b** Isotopes become neutron rich and then decay by beta emission.

3. **c** Positrons are emitted but combine with electrons and form two gamma rays, which exit the body and are detected.

4. **b** $^{239}_{94}Pu + ^{1}_{0}n \longrightarrow ^{92}_{38}Sr + ^{144}_{56}Ba + 4^{1}_{0}n$

5. **a** $^{1}_{1}H + ^{2}_{1}H \longrightarrow ^{3}_{2}He$

6. **d** It requires a very high temperature to get started.

B-2 Problems

(a) $^{97}_{43}Tc$ (b) $^{4}_{2}He$ (c) $^{109}_{48}Cd$ (d) $2\,^{1}_{0}n$ (e) $^{263}_{106}Sg$

C-1 Matching

f A proton is the same as $^{1}_{1}H^{+}$.

j $^{235}_{92}U$ undergoes fission.

l An alpha particle can be represented as $^{4}_{2}He^{2+}$.

c An isotope of sodium is $^{24}_{11}X$.

m A Geiger counter detects the formation of ions.

e $^{0}_{+1}e$ is a positron.

d A beta particle is an electron.

g An effect of radiation is the formation of ions.

16

Organic Chemistry

Review Section A *Hydrocarbons*

OUTLINE	OBJECTIVES
16-1 Bonding in Organic Compounds	
1. The C-H bond in organic compounds	
2. Isomers in organic compounds	*Write the Lewis structures of selected organic compounds including isomers when possible.*

3. The geometry of CH_4, C_2H_6, C_3H_8
4. Condensed structures — *Write condensed and fully condensed formulas for specified hydrocarbons.*

16-2 Alkanes

1. A homologous series: alkanes — *Give the formulas of simple alkanes from the general formula of the homologous series.*
2. Common and IUPAC names of alkanes — *Determine the IUPAC name of simple and branched alkanes.*
3. Alkanes and petroleum — *Describe how petroleum is refined and the uses for the various components that are obtained.*

16-3 Alkenes

1. A homologous series: alkenes — *Write the formulas of simple alkenes from the general formula.*
2. Polymers from alkenes — *Describe how alkenes are used to make polymers and the names and uses of representative polymers of this type.*

16-4 Alkynes

1. A homologous series: alkynes — *Write the formulas of simple alkynes from the general formula.*

16-5 Aromatic Compounds

1. Aromatics: cyclic hydrocarbons
2. Derivatives of benzene — *Illustrate the difference in structure and chemical reactivity between benzene and an open-chain alkene.*

SUMMARY OF SECTIONS 16-1 THROUGH 16-5

Questions: Why does carbon form so many compounds? What distinguishes the several types of hydrocarbons? What is petroleum?

Organic chemistry is a subdivision of chemistry concerned with the millions of compounds containing carbon. It is an enormous topic. Thus we can only scratch the surface in this one chapter and then only if the various types of compounds can be conveniently classified into groups. To be able to classify organic compounds, we must first appreciate some of the subtleties of their Lewis structures. (This topic should be reviewed in Chapter 6 if necessary.) For example, we find that there are two organic compounds (called **isomers**) of the formula C_2H_6O. The location of the oxygen atom in the two compounds has large effects on the properties of the two compounds.

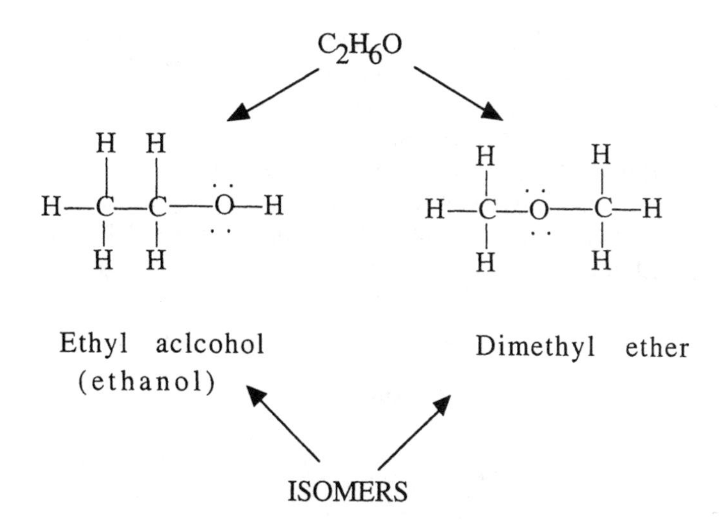

The first general class of organic compounds discussed is the **hydrocarbons**, which contain only the elements carbon and hydrogen. There are four types of hydrocarbons: **alkanes**, **alkenes**, **alkynes**, and **aromatics**. Alkanes contain only single covalent bonds and are thus known as **saturated** hydrocarbons. Open-chain (no ring) alkanes form a **homologous series** with the general formula, C_nH_{2n+2}.

Alkanes (and other organic compounds) have common names and IUPAC names. There are some generalizations in the common names, but IUPAC names are the more systematic. In this method the longest carbon chain is the root name, with branch hydrocarbons, called **substituents,** named according to their **alkyl group** designation. The names of alkyls (an alkane minus one hydrogen) are listed in the text. An example for an isomer of $C_{11}H_{24}$ is as follows:

methyl group

CH_3

numbers designate the location of the substituent

1 2 3 4 5 6 7 8

$CH_3 - CH_2 - CH - CH - CH_2 - CH_2 - CH_2 - CH_3$

$CH_2 - CH_3$

ethyl group

IUPAC name: 4-ethyl-3-methyloctane

Alkanes are the principal components of natural gas and petroleum. Since the lighter the alkane, the lower the boiling point (or the higher the volatility), petroleum is separated into components in a refinery. Each fraction has its special use, as described in the text. Alkanes recovered from petroleum can be changed in processes known as **alkylation**, **reforming**, and **cracking**. Most other organic chemicals used to make the vast number of commercial products of the petrochemical industry originate from the alkanes in petroleum.

Unlike alkanes, other hydrocarbon series contain a double bond (alkene) or a triple bond (alkynes). These are **unsaturated** hydrocarbons. Examples of each are as follows:

Hydrocarbon	General Formula	Example	Name
alkane	C_nH_{2n+2}	$C_3H_8(CH_3 - CH_2 - CH_3)$	propane
alkene	C_nH_{2n}	$C_3H_6(CH_3 - CH = CH_2)$	propene
alkyne	C_nH_{2n-2}	$C_3H_4(CH_3 - C \quad CH)$	propyne

A very important commercial use of alkenes is in a process where the unsaturated alkenes (**monomers**) are joined together in long saturated chains called **polymers**. Polymers are better known to us as the various types of plastics. Alkynes are important commercially in the preparation of various alkenes which in turn are used to make plastics.

Another important class of hydrocarbons is known as an aromatic. In these compounds carbon atoms (usually six) form a ring that can be written with alternating single and double bonds. The basic compound is benzene (C_6H_6). There are also many **derivatives** of benzene where alkyl groups or **hetero atoms** are substituted for one or more hydrogens on the ring.

Benzene  (carbon and hydrogens not shown)

NEW TERMS

Alkane	Hydrocarbon
Alkene	Isomer
Alkylation	Monomer
Alkyl group	Organic chemistry
Alkyne	Polymer
Aromatic compound	Reforming
Cracking	Saturated hydrocarbon
Derivative	Substituent
Hetero atom	Unsaturated hydrocarbon
Homologous series	

SELF-TEST

A-1 Multiple Choice

____ 1. Which of the following is a open-chain alkane?

(a) C_4H_8
(b) C_3H_8O
(c) C_2H_2
(d) C_5H_{12}
(e) CH_2O

____ 2. Which of the following hydrocarbons does not have isomers?

(a) C_8H_{18}
(b) C_2H_4
(c) C_4H_{10}
(d) C_8H_{16}
(e) C_6H_{10}

____ 3. Which of the following is the formula of an alkyl group?

(a) CH_4
(b) C_2H_6
(c) C_3H_7
(d) C_2H_4
(e) C_2H_2

_____ 4. Which of the following may contain a triple bond?

(a) C_6H_{10}
(b) C_6H_{12}
(c) C_6H_{14}
(d) C_2H_4
(e) C_4H_8

_____ 5. Which of the following is the root name of the hydrocarbon shown below?

```
    C—C
    |
C—C—C—C—C—C
  |       |
  C—C     C
```

(a) octane
(b) heptane
(c) pentane
(d) hexane
(e) decane

_____ 6. Which of the following is the name of the C_2H_5 group?

(a) ethane
(b) ethane
(c) ethyl
(d) acetylene
(e) propyl

_____ 7. Which of the following hydrocarbons is a principal component of gasoline?

(a) CH_4
(b) C_2H_6
(c) C_8H_{18}
(d) C_8H_{16}
(e) C_8H_{14}

_____ 8. The hydrocarbon $C_{12}H_{26}$ can be made into gasoline by which of the following processes?

(a) alkylation
(b) cracking
(c) reforming
(d) refining
(e) polymerization

_____ 9. Which of the following is a true statement about benzene?

(a) It is a saturated hydrocarbon.
(b) It is not as reactive as straight-chain alkenes.
(c) It is a major component of petroleum.
(d) It is a gas at room temperature.

_____ 10. Which of the following is used as a monomer in the production of certain polymers?

(a) alkanes
(b) alkenes
(c) alkynes
(d) aromatics

A-2 Matching

____ normal isomer of an alkene

____ iso isomer of an alkane

____ alkyl group

____ alkene

____ benzene

____ alkyne

____ derivative of benzene

(a) C_6H_4 (d) C_6H_6 (g) C_6H_{10}

(b) $C_6H_5(CH_3)$ (e) C_6H_{13} (h) C_6H_{11}

(c) C_6H_{12} (f) $C(CH_3)_4$ (i) $CH_3(CH_2)_3CH_3$

A-3 Problems

Fill in the missing hydrogens and give the IUPAC name for the following:

Name

1. C—C—C—C ____________________

2.
```
              C
              |
    C—C—C—C—C
          |
          C
          |
        C—C—C
```
____

3. C—C═C—C ____________________

4. C≡C—C—C ____________________

5. C—C (attached to a benzene ring) ____________________

Review Section B

OUTLINE

16-6 Organic Functional Groups

1. The center of reactivity

OBJECTIVES

Distinguish between a compound containing a functional group and an alkane.

16-7 Alcohols

1. The hydroxyl group: alcohols

16-8 Ethers

1. Hydrocarbon oxides: ethers

16-9 Aldehydes and Ketones

1. The carbonyl group: aldehydes and ketones

16-10 Amines

1. Organic compounds as bases: amines

16-11 Carboxylic Acids, Esters, and Amides

1. The carboxyl group: carboxylic acids
2. Derivatives of carboxylic acids: esters and amides

Give the unique structural feature, an example, and a common use of each of the eight types of organic compounds with hetero atom functional groups.

SUMMARY OF SECTIONS 16-6 THROUGH 16-11

Questions: *What distinguishes one type of organic compound from another? What are the origins and uses of some familiar organic compounds?*

Alkanes are comparatively unreactive hydrocarbons. Because of the presence of double and triple bonds in unsaturated hydrocarbons, however, alkenes and alkynes are more reactive. The multiple bonds in these cases are the centers of reactivity and are thus called **functional groups**. When atoms other than carbon or hydrogen (hetero atoms) are introduced into a hydrocarbon, they also serve as functional groups. Most of the more important functional groups contain oxygen and/or nitrogen. Some common groups are **hydroxyl group** (OH), **carbonyl group** (CO), and a **carboxyl group** (COO). The presence of a functional group determines, to a large extent, the type of chemical reactions of the entire molecule. Thus chemical activity such as the analgesic action of aspirin or the smell of oil of wintergreen is an effect of the functional group or groups attached to the hydrocarbon. As mentioned, the Lewis structures of molecules such as ethyl alcohol and dimethyl ether may not look very different, but whether there is an H or another C attached on an oxygen makes a profound difference in the compound's chemistry. Although we have not discussed the chemical reactions of the functional groups (to a significant extent), the types of compounds in each group have everyday familiarity. The eight classes of compounds, the characteristic functional group, and an example of each are as follows:

Class	Functional Group	Example	Name
Alcohol	$—\ddot{\underset{..}{O}}—H$	CH_3OH	methanol or methyl alcohol*
Ether	$—\ddot{\underset{..}{O}}—$	$C_2H_5—O—C_2H_5$	diethyl ether
Amine	$—\underset{..}{\overset{\vert}{N}}—$	$(CH_3)_2NH$	dimethyl amine
Aldehyde	$—\overset{:O:}{\overset{\Vert}{C}}—H$	$CH_3—\overset{O}{\overset{\Vert}{C}}—H$	ethanal or acetaldeyde*
Ketone	$—\overset{:O:}{\overset{\Vert}{C}}—$	$CH_3—\overset{O}{\overset{\Vert}{C}}—CH_3$	propanone or acetone*
Carboxylic Acid	$—\overset{:O:}{\overset{\Vert}{C}}—OH$	$CH_3—\overset{O}{\overset{\Vert}{C}}—OH$	ethanoic acid or acetic acid*
Ester	$—\overset{:O:}{\overset{\Vert}{C}}—\ddot{\underset{..}{O}}—$	$CH_3—\overset{O}{\overset{\Vert}{C}}—O—C_2H_5$	ethyl ethanoate or ethyl acetate*
Amide	$—\overset{:O:}{\overset{\Vert}{C}}—\underset{..}{\overset{\vert}{N}}—$	$CH_3—\overset{O}{\overset{\Vert}{C}}—NH_2$	ethanamide or acetamide*

* The common name

NEW TERMS

Alcohol
Aldehyde
Amide
Amine
Carbonyl group
Carboxylic acid
Ester
Ether
Functional group
Hydroxyl group
Ketone

SELF-TEST

B-1 Functional Groups

Write the class of compound for each of the following.

(a) 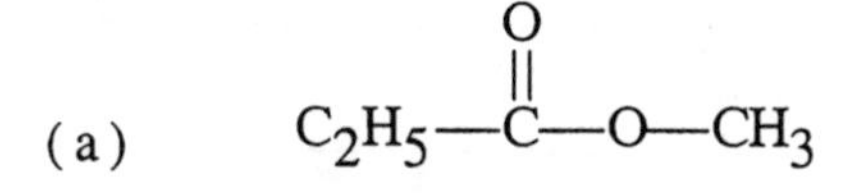

____

(b) $(CH_3)_3N$ ______________________

(c) C_4H_7OH ______________________

(d) $C_4H_7-\overset{O}{\overset{\|}{C}}-CH_3$ ______________________

(e) C_5H_{10} ______________________

(f) ______________________

(g) HCHO ______________________

(h) HO— —$\overset{O}{\overset{\|}{C}}$—$NH_2$ ______________________

(i) $C_{10}H_{21}$-COOH ______________________

(j) —O— ______________________

(k) C_5H_8 ______________________

B-2 Uses of Organic Compounds

For each of the following uses, indicate the class or classes of organic compounds that are appropriate:

antifreeze ____________________

anesthetic ____________________

analgesic ____________________

disinfectant ____________________

solvent ____________________

flavoring ____________________

polymers ____________________

Review Section C

CHAPTER SUMMARY SELF-TEST

C-1 Matching

____________ the major component of natural gas

____________ a monomer

____________ a derivative of a carboxylic acid

____________ an aromatic group or substituent

____________ used in alcoholic beverages

____________ has base characteristics in water

____________ name ends in "one"

____________ has acid characteristics in water

____________ a saturated hydrocarbon found in petroleum

____________ two compounds that would combine to form an ester

____________ used to make monomers for eventual polymerization

____________ an ether with the same formula as propyl alcohol

____________ two compounds that would combine to form an amide

(a) CH_3OH	(e) CH_3COCH_3	(i) CH_3CONH_2	(m) C_2H_2
(b) C_2H_5COOH	(f) CH_4	(j) CH_3NH_2	(n) $CH_3CH_2OCH_3$
(c) C_2H_4	(g) C_6H_5	(k) CH_3CHO	(o) C_2H_5Cl
(d) C_2H_5OH	(h) C_7H_{16}	(l) CH_3OCH_3	

C-2 Problems

Fill in the missing hydrogens and give the name for the following: (In all of the following, carbon forms four bonds, nitrogen three, and oxygen two.)

Name

1. C—C—C=O ____________________

2. C—C—C(=O)—O—C—C ____________________

3. C—C≡C—C—C ____________________

4. C—C—C—N—C ____________________

5. C—C(=O)—C—C ____________________

6. C—C—C (with a benzene ring attached to the middle carbon) ____________________

7. C—C(—C)—C(—C—C)—C ____________________

8. C—C—C—C—O ____________________

9. C—C—C(=O)—N ____________________

10. C—C—C(=O)—O ____________________

11. C—O—C—C—C ____________________

12. C—C=C—C—C ____________________

Answers to Self-Tests

A-1 Multiple Choice

1. **d** C_5H_{12} ($n = 5$, $2n + 2 = 12$)

2. **b** C_2H_4 The only possible structure is

$$\begin{matrix} H & & & & H \\ & \diagdown & & \diagup & \\ & & C{=}C & & \\ & \diagup & & \diagdown & \\ H & & & & H \end{matrix}$$

3. **c** C_3H_7 An alkyl group has the general formula C_nH_{2n+1}.

4. **a** C_6H_{10} Alkynes have triple bonds. The general formula is C_nH_{2n-2}.

5. **a** octane. The longest possible chain has eight carbon atoms.

6. **c** ethyl

7. **c** C_8H_{18} (The other C_8 hydrocarbons are not alkanes.)

8. **b** cracking. Cracking breaks large molecules into small ones.

9. **b** Benzene is not as reactive as straight chain alkenes.

10. **b** alkenes

A-2 Matching

i The normal (n) isomer of an alkane is $CH_3(CH_2)_3CH_3$.

f The iso isomer of an alkane is $C(CH_3)_4$.

e An alkyl group is C_6H_{13}. ($C_6H_{14} - H = C_6H_{13}$)

c An alkene is C_6H_{12}. (C_nH_{2n})

d The formula for benzene is C_6H_6.

g An alkyne is C_6H_{10}. (C_nH_{2n-2})

b A derivative of benzene is $C_6H_5(CH_3)$. (methylbenzene or toluene)

A-3 Problems

		Name
1.	$CH_3—CH_2—CH_2—CH_3$	butane
2.	$CH_3—CH_2—CH(CH_2—CH(CH_3)—CH_3)—CH(CH_3)—CH_3$	2-methyl, 4-isopropylhexane
3.	$CH_3—CH{=}CH—CH_3$	2-butene
4.	$CH{\equiv}C—CH_2—CH_3$	1-butyne
5.	$C_6H_5—CH_2—CH_3$	ethylbenzene

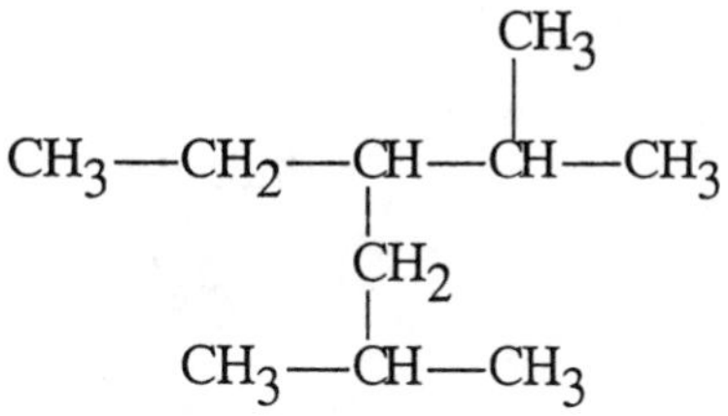

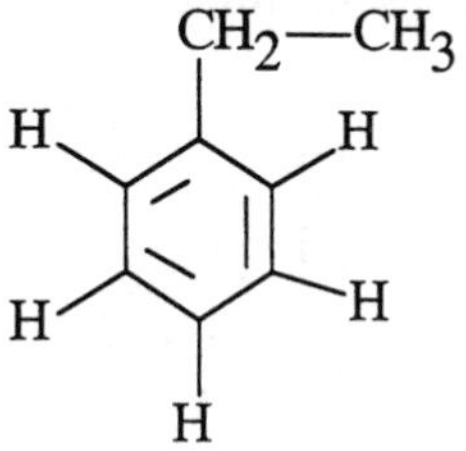

B-1 Functional Groups

(a)	$C_2H_5—C(=O)—O—CH_3$	Ester
(b)	$(CH_3)_3N$	Amine
(c)	C_4H_7OH	Alcohol
(d)	$C_4H_7—C(=O)—CH_3$	Ketone
(e)	C_5H_{10}	Alkene
(f)	(naphthalene ring structure)	Aromatic
(g)	HCHO	Aldehyde

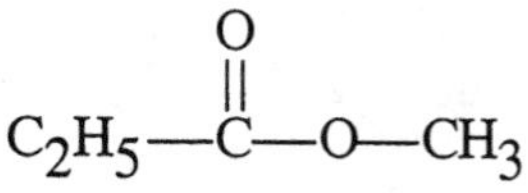

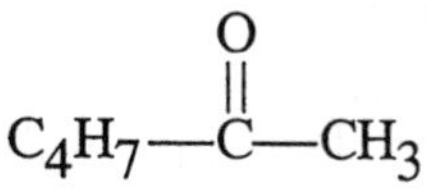

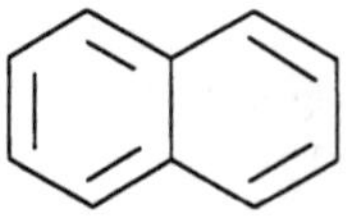

(h) 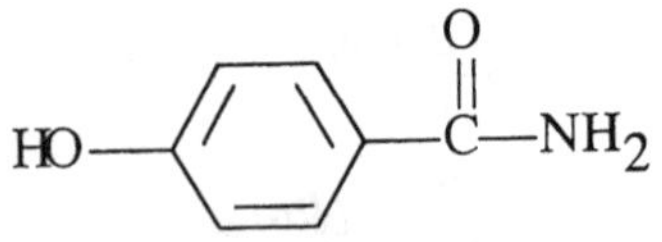 HO— —$\overset{\mathrm{O}}{\overset{\|}{\mathrm{C}}}$—$NH_2$ Alcohol and Amide

(i) $C_{10}H_{21}COOH$ Carboxylic acid

(j) 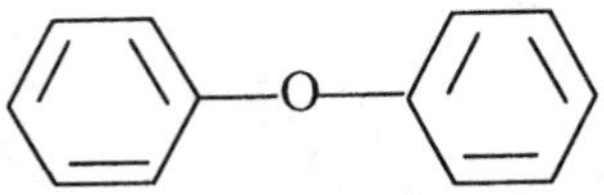 —O— Ether

(k) C_5H_8 Alkyne

B-2 Uses of Organic Compounds

Antifreeze - alcohols

Anesthetic - ether

Analgesic - amines, carboxylic acids, esters, amides

disinfectant - alcohols, aldehydes

solvent - aldehydes, ketones, alcohols (also ether)

flavoring - aldehydes, ketones, esters

polymers - aldehydes, esters, amides, alkenes

C-1 Matching

f The major component of natural gas is CH_4 (methane).

c C_2H_4 (ethylene) is the monomer of polyethylene.

i CH_3CONH_2 is an amide, which is a derivative of a carboxylic acid.

g An aromatic substituent is C_6H_5. (C_6H_6 - H = C_6H_5)

d C_2H_5OH (ethanol or ethyl alcohol) is found in alcoholic beverages.

j CH_3NH_2 is an amine which, like ammonia, has basic characteristics in water.

e CH_3COCH_3 is propanone. It is a ketone.

b C_2H_5COOH is a carboxylic acid which, like acetic acid, has acidic characteristics in water.

h C_7H_{16} is a saturated hydrocarbon found in petroleum.

a & b An alcohol (CH_3OH) and a carboxylic acid (C_2H_5COOH) combine to form an ester.

m C_2H_2 (acetylene) is used to make an alkene which is a monomer.

n $CH_3CH_2OCH_3$ is an ether with the same formula as propyl alcohol (C_3H_8O).

b & j A carboxylic acid (C_2H_5COOH) and an amine (CH_3NH_2) combine to form an amide.

C-2 Problems

		Name
1.	$CH_3—CH_2—CH{=}O$	propanal or propionaldehyde
2.	$CH_3—CH_2—C({=}O)—O—CH_2—CH_3$	ethyl propionate
3.	$CH_3—C{\equiv}C—CH_2—CH_3$	2-pentyne
4.	$CH_3—CH_2—CH_2—NH—CH_3$	methylpropylamine
5.	$CH_3—C({=}O)—CH_2—CH_3$	ethylmethyl ketone or butanone
6.	$CH_3—CH(—CH_3)—C_6H_5$ (benzene ring with five H)	isopropylbenzene
7.	$CH_3—CH(CH_3)—CH(CH_2—CH_3)—CH_3$	2,3 dimethylpentane
8.	$CH_3—CH_2—CH_2—CH_2—O—H$	n-butyl alcohol

9. $CH_3-CH_2-C(=O)-NH_2$ propionamide

10. $CH_3-CH_2-C(=O)-O-H$ propionic acid

11. $CH_3-O-CH_2-CH_2-CH_3$ methylpropyl ether

12. $CH_3-CH=CH-CH_2-CH_3$ 2-pentene

MAR 1 5 2004